ARTHROPOD NATURAL ENEMIES IN ARABLE LAND

I

Proceedings of the First EU Workshop on Enhancement, Dispersal and Population Dynamics of Beneficial Insects in Integrated Agrosystems: "Estimating population densities and dispersal rates of beneficial predators and parasitoids in agroecosystems", held at the University of Aarhus (Denmark) 21-23 October 1993

ARTHROPOD NATURAL ENEMIES IN ARABLE LAND

I

Density, Spatial Heterogeneity and Dispersal

Edited by Søren Toft and Werner Riedel

ACTA JUTLANDICA LXX:2
Natural Science Series 9

AARHUS UNIVERSITY PRESS

Printed in England by Cambridge University Press
ISBN 87 7288 492 4
ISSN 0065 1354 (Acta Jutlandica)
ISSN 0105 6824 (Natural Science Series)

AARHUS UNIVERSITY PRESS
University of Aarhus
DK-8000 Aarhus C
Fax (+ 45) 8619 8433

73 Lime Walk
Headington, Oxford OX3 7AD
Fax (+ 44) 1865 750 079

Box 511
Oakville, Conn. 06779
Fax (+ 1) 203 945 9468

Foreword

This book is the first of three volumes that will form the proceedings of a series of workshops, initiated and organized by C.J.H. Booij and funded as a Concerted Action by the European Community/European Union. All three workshops deal with the biology of arthropod predators and parasitoids and their role in limiting pest insects in European agriculture, as well as the methodological problems associated with studies of these questions. The common title of the three workshops is:

ENHANCEMENT, DISPERSAL AND POPULATION DYNAMICS OF BENEFICIAL PREDATORS AND PARASITOIDS IN INTEGRATED AGROECOSYSTEMS.

The first workshop took place at University of Aarhus, Denmark, during 21-23 October 1993. It was announced as an EC-workshop under the following headline:

Estimating population densities and dispersal rates of beneficial predators and parasitoids in agroecosystems

The following two meetings, announced as EU-workshops, are being held in December 1994 in Wageningen, The Netherlands, and in November 1995 at University of Bristol, U.K. Each has its own subject area, defined as:

1994: *Estimating survival and reproduction of beneficial predators and parasitoids in relation to food availability and quality of the habitat*, and

1995: *Analysing and modelling of population dynamics of beneficial predators and parasitoids in agroecosystems.*

In his proposal for these workshops, Kees Booij presented the following arguments:

"Integrated pest control is based on an optimal utilization of the natural control agents within the agroecosystem in combination with selective use of pesticides when necessary. Though the role of beneficial insects in the suppression of insect pests and factors which affect their abundance is well documented, very little is known about the population dynamics of beneficial species. Such information, however, is essential if in the future we like to manipulate insect predators and parasites under field conditions, or at least to predict the effects that changing agricultural systems have on their populations and effectiveness.

Recent developments in ecological theory and increasing evidence from field studies indicate, that the population dynamics of predators and parasites in agricultural systems should be studied not only in individual fields, but also at the farm and the landscape levels. The main reason for this is that many of the species involved are adapted to the dynamic nature of agroecosystems and, as a consequence move frequently not only between different crops and fields, but also between fields and neighbouring (semi-)natural habitats. This frequent movement of insect populations makes it extremely difficult to predict the population size of any insect species within a given field.

Realizing this is a key problem in the dynamics of beneficial insects in agriculture, several research groups in Europe have started to study dispersal of predators and other useful insects in relation to scale and infrastructure of the agricultural landscape. In several European countries, considerable efforts are being made to study how field margins and farming systems affect the density of beneficial insects. These studies all face a range of similar (methodological) problems, such as how to sample the insects adequately, how to measure dispersal and how to develop suitable models for predicting the ways insect populations change under a range of conditions.

In order to answer some of these questions, and to develop standard methods so that the results obtained throughout Europe can be compared directly between countries, there is a considerable need to set up a network, through which European science in this field can be advanced considerably by the various participants pooling their resources and information.The advantage of such a network is that much of the information for the development of adequate methods is rarely sufficient on its own to merit publication. Hence, this information can only be obtained from direct dialogue between scientists. The work described in this proposal is intended to establish a network and through a series of workshops, to discuss methodological problems and how they can be solved by the scientists in cooperative trials."

These proceedings document the first steps taken toward reaching that goal.

During the detailed planning of the Århus workshop, it was felt that besides the problems of estimating densities and dispersal rates, spatial heterogeneity was a tightly connected phenomenon and much neglected in terms of research. It was therefore decided to organize the workshop in three sessions instead of two.

The workshop consisted of two parts: First, original papers and posters were presented, relating to the three subjects, density, spatial heterogeneity and dispersal, respectively. This was followed by group discussions, where the participants joined the session closest to their original contribution. Each session was headed by a session leader. The purpose of these discussion sessions was to sum up the current status of research on the subject and to point out gaps in our present knowledge and directions for future research. Finally, each session leader presented the results of his group discussion in plenum. Two of the sessions resulted in separate articles, included in these proceedings at the end of the respective sections.

The original contributions submitted for these proceedings were planned and executed independently of the present context. Several of them cover questions relating to more than one of the session headlines, which follows naturally from the fact that these subjects are intimately related biologically. The practical solution for the organizing of the book has been to include them in the session in which they were presented at the workshop. However, in order to have related papers grouped together, a few articles have been placed otherwise.

Several people have contributed to this book. We are grateful to all the workshop participants who contributed (anonymously) to the mutual reviewing of the manuscripts. We are also indebted to several people who did not participate but anyway have read and commented on one or several manuscripts: Jørgen A. Axelsen, Mike Downes, Mikael Münster-Swendsen, and Matthias Schaefer. Very special thanks are due to Nils Skyberg for his help with all computer problems during the editing and layout procedure. Charlotte Sørensen contributed to the smooth running of the workshop with help in all practical matters.

Århus, February 1995

Søren Toft   Werner Riedel

Contents

SPATIAL HETEROGENEITY

DISCUSSION PAPER

DISPERSAL

DENSITY

Estimating population densities of spiders in cereals

K.D. Sunderland & C.J. Topping[1]

Horticulture Research International, Littlehampton, West Sussex, BN17 6LP, England
[1] Present address: Scottish Agricultural College, 581 King Street, Aberdeen, AB9 1UD, Scotland

Abstract
A method for estimating "absolute" population density of spiders was used in winter wheat in 4 years. This method, consisting of suction-sampling plus surface-searching unit areas of crop, showed that suction sampler (D-vac) efficiency varied with site/year, season and degree of weed cover. Various spider taxa were recovered with a different efficiency depending on the vertical stratum they occupied. Results of the "absolute" density method were compared with other published methods in terms of their ranges in sampling precision and coefficients of variation. It was concluded that the "absolute" method is an efficient and reliable technique suitable for use in intensive programmes for comparing spider populations between sites, seasons and years.

Key words: Spiders, Araneae, population density, absolute density, D-vac, suction sampler, sampling efficiency, sampling precision, coefficient of variation, winter wheat, *Lepthyphantes tenuis*, Linyphiidae

Introduction

There is a growing global trend to find ways of controlling pests without recourse to excessive applications of pesticides and to develop pest control techniques that are compatible with sustainable agriculture. One approach to achieving these ends is to make better use of natural enemies in crops. This need has stimulated studies into the population ecology of polyphagous predators, of which spiders are an important component. Reliable estimates of population density are the cornerstone of investigations into the natural ecology of spiders in agroecosystems, of the side-effects of pesticides on spider populations and of modifications to farming systems designed to promote natural pest control. The aim of this paper is to evaluate a method for

Arthropod natural enemies in arable land · I *Density, spatial heterogeneity and dispersal*
S. Toft & W. Riedel (eds.). *Acta Jutlandica* vol. 70:2 1995, pp. 13-22.
 ISBN 87 7288 492

estimating the "absolute" density of spiders in agroecosystems and to compare its advantages and disadvantages with other methods.

Materials and methods

Sites

Spider density was measured in 1990 in a 17 ha field of winter wheat (cv. Pastiche) in southeast England. In 1991 and 1993 a 3 ha field of winter wheat (cv. Riband in 1991 and cv. Haven in 1993) was used, which was 24 km from the 1990 study field. The 1990 field had a chalk with flints soil, whilst the 1991/1993 field had a silty brick earth soil. Both fields were treated with fungicide, herbicide and fertiliser, following normal farming practice; insecticides were not applied to those parts of the fields used for sampling spiders. Plant density appeared to be similar throughout the study, but numbers of tillers per m^2 were not measured.

Sampling spider density

The "absolute" density sampling method used (Sunderland & Topping 1993, Topping & Sunderland 1994) was a modification of that described by Sunderland et al. 1987. Each sample unit consisted of a spider collection within a defined area using a vacuum sampler (D-vac) immediately followed by careful searching of the ground just sampled by the D-vac and collection of spiders with a pooter. A 1 m long steel tube was fitted to the D-vac sampling head (diameter 33 cm) to prevent spiders being crushed during sampling. D-vac sampling times per unit were negligible for $0.1 m^2$ and a couple of minutes for $0.5 m^2$. Destructive hand-searching per $0.5 m^2$ took approximately 10 man minutes (Topping & Sunderland 1994); this is an average figure, the time required varied with season, vegetation density and spider abundance. Searching time per $0.1 m^2$ unit was proportionally shorter. Spiders were collected from under weeds and stones, around the base of cereal plants and from the soil to a depth of c. 3 cm. After an initial collection period the ground was left undisturbed for a few minutes to allow any remaining spiders to reveal themselves by movement. Cereal plants, together with their superficial root systems, were removed, whenever this was necessary for efficient searching and collection of spiders. In 1990 and 1991 samples were taken at approximately weekly intervals and each consisted of fifteen $0.5 m^2$ sample units in a stratified random sampling programme. The data used in this paper relate to the periods 29 March to 8 October 1990 and 23 April to 24 September 1991. Sampling was more restricted in 1993 and confined to 30 x 0.1 m^2 on 4 June, 45 x 0.1 m^2 on 20 July and 10 x 0.5 m^2 on 4 August. D-vac sample units were either live-sorted immediately or kept for up to 3 days at 9°C before being live-sorted. Losses of spiders during cold-storage were assessed by comparing groups of ten D-vac units taken on the same day; one set was sorted immediately and another after 3 days cold-storage. The efficiency of live-sorting was assessed, for 31 randomly selected sample units, by microscopic examination of the material remaining after live-sorting.

Results

There was no significant reduction in spider density during 3 days cold storage (immediate sorting mean 25.4, 95% Confidence Limits (CL, 1.96 x SE) ± 4.7, cf. 3 days storage mean 24.8, 95% CL ± 4.5) and the efficiency of live-sorting was 98.1% (95% CL ± 6.7), indicating that, overall, cold-storing and live-sorting recovered c. 95% of the spiders sampled. Of the 43 spiders overlooked during live-sorting 18% were adult or sub-adult, 42% were juveniles and 40% were hatchlings.

The procedure of ground-searching an area that had just been sampled by D-vac provided an opportunity to assess the efficiency of the D-vac (defined here as number of spiders in D-vac as a percentage of number of spiders in D-vac plus ground search) under a variety of conditions. Table 1 shows D-vac efficiencies for the main sampling programmes in 1990 and 1991. Juveniles were sampled more efficiently than adults, and the foliage-inhabiting linyphiid sub-family Linyphiinae more efficiently than the ground-living Erigoninae. The dominant species in 1991, *Lepthyphantes tenuis* (Blackwall), a member of the

Table 1. Efficiency of the D-vac (number in Dvac as % of total) in winter wheat in 1990 and 1991.

	1990			1991		
Category	N[1]	Mean[2]	95% CL[3]	N[1]	Mean[2]	95%CL[3]
Adult[4]	18	54.1	45.9-62.1	16	79.2	74.1-83.9
Juvenile[4]	18	81.1	72.8-88.7	16	94.3	92.7-95.8
Erigoninae[4]	18	33.1	19.3-46.9	16	71.6	62.6-81.6
Linyphiinae[4]	18	76.1	69.9-82.2	16	92.4	90.5-94.3
L. tenuis	17	81.5	73.5-88.3	13	93.5	88.0-97.4
Bare				10	97.6	95.1-99.3
Weedy				10	85.4	81.4-89.0

[1] N = number of sample dates; 15 x 0.5 m^2 sampled on each date
[2] back-transformed from angular transformation
[3] 95% confidence limits (1.96 x SE), back-transformed from angular transformation
[4] from Topping & Sunderland 1994

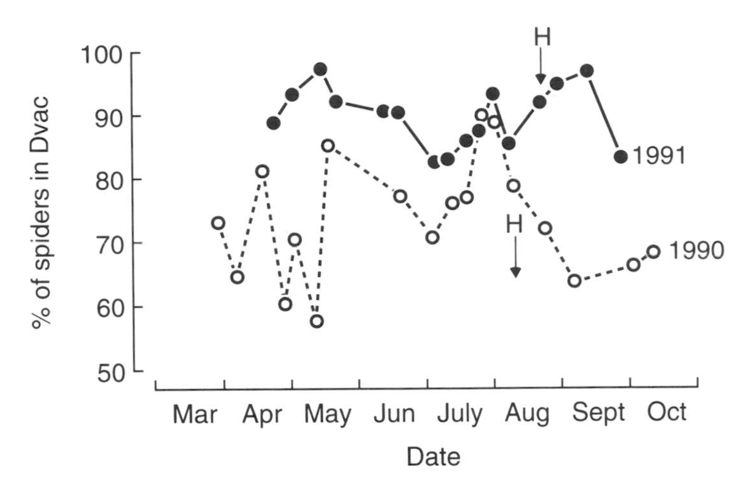

Fig. 1. D-vac efficiency (number of spiders in D-vac as percentage of total spiders) in winter wheat in 1990 (open circles) and 1991 (closed circles), based on 15 x 0.5 m^2 per sample date. H = harvest.

sub-family Linyphiinae, was also sampled more efficiently than the Erigoninae. Efficiency was reduced in weedy patches (> 75% weed cover, Topping & Sunderland 1994), between 9 July and 9 September 1991, as compared to bare ground. Fig. 1 shows the variation in efficiency during 1990 and 1991. Efficiency was lower in 1990 than 1991 (either due to the site or the year), and the seasonal pattern was dissimilar in the two years. Spider species composition was similar in the two years (Sunderland & Topping 1993). Surprisingly, there did not appear to be a change in efficiency following harvest of the cereal crop. In 1993 efficiency varied between 81% and 93%, June to August (Table 2) and was in the same order as 1990 and 1991, for this period. Limited data (from

Table 2. Efficiency of the D-vac (number in Dvac as % of total) in winter wheat in 1985 and 1993.

Year	Category	Site	Sample size	Date	% in Dvac
1985[1]	Total spiders	Field 1	10 x 0.1m^2	30 May	91.4
1985[1]	Total spiders	Field 1	10 x 0.1m^2	28 June	77.8
1985[1]	Total spiders	Field 2	10 x 0.1m^2	30 May	90.0
1985[1]	Total spiders	Field 2	10 x 0.1m^2	28 June	65.0
1993	Total spiders	Field 3[2]	30 x 0.1m^2	4 June	88.4
1993	*L. tenuis*	Field 3[2]	30 x 0.1m^2	4 June	82.1
1993	Total spiders	Field 3[2]	45 x 0.1m^2	20 July	80.8
1993	Total spiders	Field 3[2]	10 x 0.5m^2	4 Aug	92.6

[1] data from Sunderland et al. (1987)

[2] the same field as was sampled in 1991 (see Table 1)

Sunderland et al. 1987) for two winter wheat fields sampled by the "absolute" density method in 1985 showed a reduction in efficiency from May to June (Table 2).

Sampling precision (SE/mean) for total spiders in March to October 1990 was 0.1 - 0.2. In 1991 it was 0.1 - 0.4 until May, 0.1 until the end of July,

Table 3. Coefficients of variation for a selection of spider density sampling programmes.

Programme	Sample size	Area	Category	Number of dates or plots	Coefficient of variation Median	Range
1 (1990)	15 x 0.5m^2	7.5m^2	Total	18	0.61	0.38-0.84
1 (1991)	15 x 0.5m^2	7.5m^2	Total	16	0.55	0.29-1.01
1 (1991)	15 x 0.5m^2	7.5m^2	*L. tenuis*	16	0.77	0.35-1.50
2 (1993)	10 x 0.1m^2	1.0m^2	Total	3		0.48-0.57
2 (1993)	10 x 0.1m^2	1.0m^2	*L. tenuis*	3		0.79-0.85
2 (1993)	15 x 0.1m^2	1.5m^2	Total	3		0.40-0.73
2 (1993)	10 x 0.5m^2	5.0m^2	Total	1		0.30
3 (Field 1)	10 x 0.1m^2	1.0m^2	Total	2		0.73-0.74
3 (Field 2)	10 x 0.1m^2	1.0m^2	Total	2		0.65-0.67
4 (maize)	30 x 156cm^2	0.5m^2	Total	7	3.87	1.46-5.76
4 (ryegrass)	30 x 156cm^2	0.5m^2	Total	7	3.93	1.50-5.81
5	30 x 156cm^2	0.5m^2	*O. fuscus* ♂	13	3.25	1.87-5.50
5	30 x 156cm^2	0.5m^2	*O. fuscus* ♀	13	1.73	1.04-5.63
6	30 x 156cm^2	0.5m^2	*E. atra* ♂	11	3.28	1.47-5.50
6	30 x 156cm^2	0.5m^2	*E. atra* ♀	13	2.81	1.22-4.93
7(wheat)	50 x 0.16m^2	8.0m^2	micryphantids	1		0.52
7(wheat)	40 x 0.04m^2	1.6m^2	micryphantids	1		0.66
7(hay 1)	80 x 0.04m^2	3.2m^2	micryphantids	1		1.01
7(hay 1)	40 x 0.04m^2	1.6m^2	micryphantids	1		1.20
7(hay 2)	40 x 0.04m^2	1.6m^2	micryphantids	1		1.41
7(hay 2)	20 x 0.16m^2	3.2m^2	micryphantids	1		0.74
8	15 x 625cm^2	0.9m^2	*E.arctica* ad.	29	1.71	0.61-4.00
9	* 0.1m^2	*	*T. terricola*	14	0.67	0.46-1.07

Index to sampling programmes:

1 - Sunderland & Topping (1993), winter wheat, "absolute" density
2 - Sunderland (unpublished), winter wheat, "absolute" density
3 - Sunderland et al. (1987), winter wheat, "absolute" density
4 - Alderweireldt (1987), quadrats handsorted then extracted
5 - DeKeer & Maelfait (1987), pasture, sampling as 4
6 - DeKeer & Maelfait (1988), pasture, sampling as 4
7 - Nyffeler & Benz (1988), spiders in webs counted in quadrats *in situ*
8 - Van Wingerden (1977), dune grassland, handsort then extracted
9 - Workman (1978), grass heath, bowl extractor, from data grouped according to % tussock cover (* N = 2-140)

Index to species

L. tenuis - Lepthyphantes tenuis (Blackwall), *O.fuscus - Oedothorax fuscus* (Blackwall)
E. atra - Erigone atra (Blackwall), *E. arctica - Erigone arctica* (White)
T. terricola - Trochosa terricola Thorell, micryphantids - mainly genus *Erigone, Oedothorax*

then 0.1 - 0.2 until the end of September. This was in spite of an increase in SEs in July and August 1991 due to aggregation of spiders in weedy areas of the crop (Topping & Sunderland 1994). Sampling precision for *L. tenuis* in 1991 was 0.3 - 1.0 during a period of low density until May, 0.1 - 0.2 until mid-August and 0.3 thereafter. Sampling precision for the samples taken in 1993 was 0.1 - 0.2 for total spiders and 0.2 - 0.3 for *L. tenuis*. The range in sampling precision for all "absolute" density measurements was therefore 0.1 - 1.0.

The coefficient of variation (CV = S/Mean) is a dimensionless measure of sampling variability that allows comparison between sites and years. Table 3 shows CVs for 1990-1993 from this study together with data from other comparable studies. CVs for total spiders were in the range 0.29 - 1.01, and CVs for a single species, *L. tenuis,* were 0.35 - 1.50. CVs calculated from the data of Sunderland et al. (1987) were 0.65 - 0.74, so the range for all "absolute" density samples currently available is 0.29 - 1.50.

Discussion

The greater relative efficiency of the D-vac in sampling juvenile than adult spiders probably relates more to the stratum occupied by the spiders than to their size. The D-vac used in this study was capable of picking up moderately large carabid beetles with masses many times greater than adult linyphiid spiders. Juvenile spiders tend to be more exposed in the foliage compared with adults, many of which hide under weeds and stones during the daytime. This is consistent with the finding that efficiency is higher for *L. tenuis* and other Linyphiinae, which tend to build webs above ground (Sunderland, Fraser & Dixon 1986), than for the Erigoninae, which are more likely to build webs on the ground. It is also consistent with the finding that efficiency was reduced in weedy patches in 1991; these patches were known to harbour a higher spider density than the surroundings (Topping & Sunderland 1994) and many of these spiders would have been unavailable to the D-vac under a thick mat of weeds. Thick vegetation would also reduce airflow rates through the D-vac thereby reducing its efficiency. The reasons for the considerable variation in efficiency between years/sites (Tables 1 & 2) are not known, but such variation can probably be attributed to differences in soil structure (spiders can hide under loose stones or small clods of earth), weediness and the density and growth form of the crop. Such variation amply justifies the use of ground searching in combination with suction sampling; the combination of techniques will be much more robust than either alone at providing standardised estimates of density, with little bias, in a range of cropping situations. The method will not, however, provide truly "absolute" estimates of density in all circumstances. Large, fast-moving spiders, such as Lycosidae, are likely to flee from the sampling quadrat at the first disturbance. Mark-release-recapture (e.g. Hackman 1957) or catch per unit effort (e.g. Greenstone 1979) methods might be suitable for estimating density of these spiders under certain conditions, but the sexes should be treated separately if pitfalls are used for recapture (Samu & Sárospataki 1995). Alternatively, areas of habitat could be isolated, without disturbance, by activating isolation equipment from a distance (e.g. Turnbull & Nicholls 1966), but this method is limited to small scale studies on suitable habitats. Soils which are prone to cracking under dry conditions could harbour spiders that remain unsampled. The proportion of spider populations that reside inside cracks during the daytime is unassessed (Sunderland & Chambers 1983, Toft, Vangsgaard & Goldschmidt 1995). Flooding techniques (e.g. Basedow et al. 1988) might be appropriate here, if the cracks are not too deep. Sampling cannot be carried out under wet conditions and there could be difficulties on very sandy soils because of the large quantity of soil picked up by the suction sampler.

Sampling precision for the "absolute" density method in 1990-93 varied from 0.1 to 1.0. This is similar to the range recorded for other sampling programmes (details of the programmes are in Table 3) i.e van Wingerden (1977); 0.2 - 1.0, DeKeer & Maelfait (1987,1988); 0.2 - 1.0, Alderweireldt (1987); 0.3 - 1.0. Better sampling precision was achieved by Nyffeler & Benz (1988); 0.1 - 0.2 and Workman (1978); up to 0.1. The range of coefficients of variation (CV) for the "absolute" method was 0.29 - 1.50, which compares favourably with other sampling programmes (Table 3). It is noticeable, for example, that CVs for the studies of Alderweireldt (1987) and DeKeer & Maelfait (1987, 1988) tend to be higher, ranging from 1.04 to 5.81 (Table 3), with high CVs for total spiders and individual species and in three different crops. This is probably due either to their sample unit (156 cm^2) or the area taken per sample (0.5 m^2), both of which are the smallest out of the selection of sampling programmes presented in Table 3. It is possible that 156 cm^2 is small in relation to the scale of graininess of spider aggregation and that many sample units of this size would contain no spiders, whilst a few (that happened to hit an aggregation) would contain relatively large numbers. A full study would be needed to determine the most efficient and profitable (in terms of acceptable CV) combination of number and size of sample units per sample for determining spider density in agroecosystems. The CV can be used to estimate the number of sample units per sample (n_p) needed for a given level of sampling precision (p); $n_p = (CV/p)^2$ (Mukerji & Harcourt 1970). Vickerman (1985) reported that, for juvenile linyphiids at a density of 100 m^{-2} in cereals, 6 D-vac sucks would be needed for a p of 0.2. Using the method of handsorting quadrats followed by extraction in a Tullgren-Berlese extractor, Alderweireldt (1987) would have needed 53 - 829 quadrats to achieve p of 0.2 over the range of density encountered for total spiders in maize. Using the "absolute" density method for total spiders in winter wheat, 2 - 25 sample units would have been required for the same level of precision. This figure would increase to 56 sample units for *L. tenuis* at low density.

It can be concluded that the "absolute" density method described here is a reliable and efficient technique, suitable for intensive sampling programmes. It is, however, less suitable for extensive sampling because it is labour-intensive (c. 3 man hours for ten 0.5 m^2 sample units, excluding sorting (Topping & Sunderland 1994)). Pitfall trapping cannot be used as a substitute to give an index of abundance because the relationship between pitfall catch and density is unreliable (Topping & Sunderland 1992, Dinter & Poehling 1992, Dinter 1995, Sunderland et al. 1995). Since there is no perfectly reliable method applicable to extensive sampling the vacuum net could be used, but variation in its efficiency (as described here) should be taken into account when interpreting results.

Acknowledgements

We are grateful to Sarah Ellis, Steve Long and Anette Weiss, who helped with the sampling and sorting in 1993. KDS was funded by the Ministry of Agriculture Fisheries & Food and CJT by the Natural Environment Research Council.

References

Alderweireldt, M. 1987. Density fluctuations of spiders on maize and Italian ryegrass fields. *Meded. Fac. Landbouwwet. Rijksuniv. Gent* 52: 273-282.

Basedow, T., Klinger, K., Froese, A. & Yanes, G. 1988. Aufschwemmung mit Wasser zur Schnellbestimmung der Abundanz epigaischer Raubarthropoden auf Ackern. *Pedobiologia* 32: 317-322.

De Keer, R. & Maelfait, J.P. 1987. Life history of *Oedothorax fuscus* (Blackwall, 1834) (Araneae, Linyphiidae) in a heavily grazed pasture. *Rev. Ecol. Biol. Sol* 24: 171-185.

De Keer, R. & Maelfait, J.P. 1988. Observations on the life cycle of *Erigone atra* (Araneae, Erigoninae) in a heavily grazed pasture. *Pedobiologia* 32: 201-212.

Dinter, A. 1995. Estimation of epigeic spider population densities using an intensive D-vac sampling technique and comparison with pitfall trap catches in winter wheat. In: Toft, S. & Riedel, W. (eds.) *Arthropod natural enemies in arable land · I*, pp. 23-32. Aarhus.

Dinter, A. & Poehling, H.M. 1992. Spider populations in winter wheat fields and the side-effects of insecticides. *Asp. Appl. Biol.* 31: 77-85.

Greenstone, M.H. 1979. A line transect density index for wolf spiders (*Pardosa* spp.), and a note on the applicability of catch per unit effort methods to entomological studies. *Ecol. Entomol.* 4: 23-29.

Hackman, W. 1957. Studies on the ecology of the wolf spider *Trochosa ruricola* Deg. *Societatis Scientiarum Fennica Commentationes Biologicae* 16: 1-34.

Mukerji, M.K. & Harcourt, D.G. 1970. Design of a sampling plan for studies on the population dynamics of the cabbage maggot *Hylemya brassicae* (Diptera: Anthomyiidae). *Can. Ent.* 102: 1513-1518.

Nyffeler, M. & Benz, G. 1988. Prey and predatory importance of micryphantid spiders in winter wheat fields and hay meadows. *J. Appl. Ent.* 105: 190-197.

Samu, F. & Sárospataki, M. 1995. Estimation of population sizes and "home ranges" of polyphagous predators in alfalfa using mark-recapture: an exploratory study. In: Toft, S. & Riedel, W. (eds.) *Arthropod natural enemies in arable land · I*, pp. 47-55. Aarhus.

Sunderland, K.D. & Chambers R.J. 1983. Invertebrate polyphagous predators as pest control agents: some criteria and methods. In: R. Cavalloro (ed.), *Aphid Antagonists.* A.A. Balkema, Rotterdam, pp. 100-108.

Sunderland, K.D., Fraser, A.M. & Dixon, A.F.G. 1986. Distribution of linyphiid spiders in relation to capture of prey in cereal fields. *Pedobiologia* 29: 367-375.

Sunderland, K.D., Hawkes, C., Stevenson, J.H., Mcbride, T., Smart, L.E., Sopp, P.I., Powell, W., Chambers, R.J. & Carter, O.C.R. 1987. Accurate estimation of invertebrate density in cereals. *Bull. SROP/WPRS* 1987/X/1, 71-81.

Sunderland, K.D. & Topping, C.J. 1993. The spatial dynamics of linyphiid spiders in winter wheat. *Mem. Queensl. Mus.* 33: 639-644.

Sunderland, K.D., De Snoo, G.R., Dinter, A., Hance, T., Helenius, J., Jepson, P., Kromp, B., Lys, J.-A., Samu, F., Sotherton, N.W., Toft, S. & Ulber, B. 1995. Density estimation for beneficial predators in agroecosystems. In: Toft, S. & Riedel, W. (eds.) *Arthropod natural enemies in arable land · I*, pp. 133-162. Aarhus.

Toft, S., Vangsgaard, C. & Goldschmidt, H. 1995. Distance methods used to estimate densities of web spiders in cereal fields. In: Toft, S. & Riedel, W. (eds.) *Arthropod natural enemies in arable land · I*, pp. 33-45. Aarhus.

Topping, C.J. & Sunderland, K.D. 1992. Limitations to the use of pitfall traps in ecological studies exemplified by a study of spiders in a field of winter wheat. *J. Appl. Ecol.* 29: 485-491.

Topping, C.J. & Sunderland, K.D. 1994. Methods for quantifying spider density and migration in cereal crops. *Bull. Br. arachnol. Soc.* 9: 209-213.

Vickerman, G.P. 1985. Sampling plans for beneficial arthropods in cereals. *Asp. Appl. Biol.* 10: 191-198.

Turnbull, A.L. & Nicholls, C.F. 1966. A "quick trap" for area sampling of arthropods in grassland communities. *J. Econ. Ent.* 59: 1100-1104.

Wingerden, W.K.R.E. van. 1977. *Population dynamics of Erigone arctica (White) (Araneae, Linyphiidae).* Thesis (Free University, Amsterdam). 147p.

Workman, C. 1978. Life cycle and population dynamics of *Trochosa terricola* Thorell (Araneae: Lycosidae) in a Norfolk grass heath. *Ecol. Ent.* 3: 329-340.

Estimation of epigeic spider population densities using an intensive D-vac sampling technique and comparison with pitfall trap catches in winter wheat

Axel Dinter

Institute for Plant Diseases and Plant Protection, University of Hannover, Herrenhäuserstraße 2, D-30419 Hannover, Germany

Abstract
An intensive D-vac sampling technique was used to estimate the population densities of spiders (Araneae) in winter wheat from 1989 to 1991. Comparison of intensive D-vac sampling with hand-search and heat-extraction of soil samples, plus recapture rates of dye-marked spiders released in enclosed field areas for a day, suggested that the D-vac technique provides a good estimation of spider population density. In contrast, pitfall trapping, conducted at the same time as D-vac sampling, overestimated males and species of Erigoninae and Lycosidae compared to their estimated absolute density. Statistical analyses indicated that pitfall catches are not an adequate measure for the detection of density dynamics of spider populations or that of single species in winter wheat.

Key words: Araneae, D-vac sampling, suction apparatus, sampling efficiency, pitfall traps, density, abundance, winter wheat

Introduction

Ecological studies on population dynamics of spiders or other arthropods are more frequently based on data from pitfall catches rather than on absolute density estimations. This is probably because pitfall traps are easy to handle and yield large numbers of individuals and species. In contrast, methods for the estimation of absolute densities are very laborious, frequently depending on technical requirements and therefore often restricted to longer sampling intervals. In addition, most methods for the estimation of absolute densities provide fewer individuals than pitfall traps. Despite the supposed advantages of pitfall trapping, the interpretation of the results is problematical and there has been much criticism of this method (Adis 1979, Topping & Sunderland 1992).

Arthropod natural enemies in arable land · I *Density, spatial heterogeneity and dispersal*
S. Toft & W. Riedel (eds.). *Acta Jutlandica* vol. 70:2 1995, pp. 23-32.
 ISBN 87 7288 492

The aim of this study was to quantify the efficiency of an intensive D-vac technique for the estimation of absolute spider densities in winter wheat and to compare this technique with other methods. Furthermore, the degree to which pitfall catches of spiders are related to estimates of their absolute density was tested.

Materials and Methods

From 1989 to 1991 the population dynamics of spiders were investigated in winter wheat fields near Göttingen (Lower Saxony, Germany) from April/May until July/August. Two to three plots, located in the same field, were studied each year (1989: 0.48 ha, n = 3; 1990: 0.72 ha, n = 2; 1991: 0.84 ha, n = 2). Estimations of spider abundance were achieved by sampling field areas, enclosed by a biocoenometer of 0.25 m², using a D-vac suction machine (Dietrick 1961) with a small nozzle (Ø 20 cm, originally Ø 33 cm) to provide a high suction effect. After the biocoenometer was put on an undisturbed area, the plants inside the frame were shaken and cut off 10 cm above ground level. Then the area was vacuumed for one minute and the material collected was removed from the gauze bag. Finally, the same area was vacuumed again for another two minutes. There should be a high suction effect in the second period of sampling even if plant material and soil diminished the air speed of the D-vac during the first period. The pooled samples were kept cool and hand-sorted for spiders in the laboratory. D-vac sampling was done between 9 a.m. and 6 p.m. and D-vac samples were taken in a distance of at least 4 m from each other to minimize the move away of spiders by disturbance of the previous D-

Table 1. Numbers of spiders collected by two successive intensive D-vac sampling procedures on the same enclosed field areas and the effects of varying the time between the first and second D-vac sampling periods. The experiment was repeated on five dates.

Date of 1st D-vac sampling	13 July 1989	17 July 1989	12 June 1990	18 May 1990	26 April 1990
Number of D-vac samples (0.25 m²)	8	8	4	4	4
Time between 1st and 2nd D-vac sampling	5 min	5 min	1 day	3 days	4 days
1st D-vac sampling					
males	10	26	8	3	6
females	19	27	28	14	7
juveniles	80	215	275	88	14
2nd D-vac sampling					
males	-	-	-	-	3
females	1	1	1	-	1
juveniles	2	4	54	10	1

vac sampling procedure and the approach of the investigator to the next sampling area. In 1989 eight D-vac samples per plot and in 1990/91 twelve D-vac samples per plot were taken seven to ten times during the study. At the same time spiders were caught by means of 10 pitfall traps per plot (plastic cups, Ø 8.3 cm, ethylene glycol as preservative, 'Pricol' as detergent) emptied at intervals of 3 to 8 days.

To test whether estimates of spider densities would increase as a result of repeated D-vac sampling of the same area, a second sample was taken again just after the first or one to four days later for comparison. The D-vac technique was compared with visual hand-search in the field (each 10 samples of 0.25 m^2, 45 min hand-searching for spiders/0.25 m^2 field area, 27 July 1991) and heat-extraction of soil samples (34 soil samples of 0.03 m^2 and 12 D-vac samples, 31 May 1991). In addition, the efficiency of the intensive D-vac technique was investigated by the recapture of dye-marked spiders which had been released in gauze-enclosed field areas one day before (four males and four females of *Erigone atra* (Blackwall) and one female of Oedothorax apicatus (Blackwall), 12 replicates of 0.25 m^2, 22 and 24 July 1991). Regression analyses were done using the SAS package (release 6.04).

Results

Measurement of the efficiency of the intensive D-vac technique, and comparison with other methods

Most of the spiders collected by D-vac sampling were still alive when the samples were sorted in the laboratory. Repeated D-vac sampling of the same area gave only a slight increase in the number of spiders collected (Table 1).

Table 2. Comparison of spider density (individuals / 0.25m^2 ± s.d.) estimated by intensive D-vac sampling and hand-search in winter wheat, 27 July 1991 ($n = 10$). Different letters (horizontally) indicate significant differences between intensive D-vac sampling and hand-search: $p < 0.05$, U-test.

	D-vac	Hand-search
Total	30.7 ± 8.7 a	22.2 ± 7.4 b
Juveniles	22.6 ± 7.8 a	15.2 ± 4.8 b
Males	4.2 ± 1.8 a	3.6 ± 2.8 a
Females	3.8 ± 3.0 a	3.4 ± 1.5 a
Erigone atra	2.7 ± 1.5 a	2.0 ± 1.1 a
Bathyphantes gracilis	1.4 ± 1.5 a	1.7 ± 2.7 a
Lepthyphantes tenuis	1.7 ± 1.1 a	1.2 ± 1.2 a
Oedothorax apicatus	1.0 ± 1.1 a	0.8 ± 1.3 a

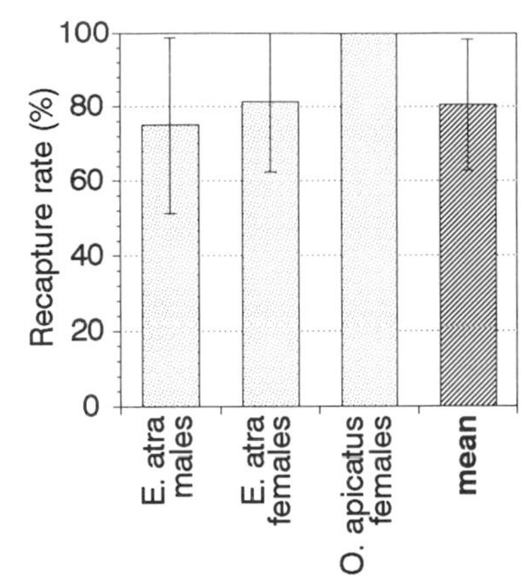

Fig. 1. Recapture rates (% ± s.d.) of dye-marked spiders (four males and four females of *E. atra* and one female of *O. apicatus* per replicate) which were released in enclosed winter wheat areas (0.25 m², n = 12) one day before intensive D-vac sampling.

The comparison of the intensive D-vac technique and visual hand-search revealed an identical spectrum of dominant species. Numbers of juvenile spiders per sample were significantly higher by intensive D-vac sampling than by hand-search but the numbers of male or female spiders were not different (U-test, $p < 0.05$) (Table 2). Heat-extraction of soil samples and intensive D-vac sampling also detected the same dominant species but the estimate of juvenile density was several times higher by heat-extraction than by intensive D-vac sampling (Table 3). In the recapture experiments a mean efficiency of the intensive D-vac technique of 80% for adults of *E. atra* and females of *O. apicatus* was found. The released females of *O. apicatus* were completely recaptured and adults of *E. atra* were caught at a rate of 75% for males and 81% for females (Fig. 1).

Table 3. Comparison of spider density (individuals / m²) estimated by intensive D-vac sampling (12 x 0.25 m²) and heat-extraction of soil samples (34 x 0.03 m²) in winter wheat, 31 May 1991.

	D-vac	Heat-extraction
Total	50.0	147.9
Juveniles	35.0	134.8
Males	5.3	3.8
Females	9.7	9.4
Erigone atra	5.7	3.8
Porrhomma microphthalmum	3.7	1.9
Meioneta rurestris	3.3	1.9
Lepthyphantes tenuis	1.0	2.8

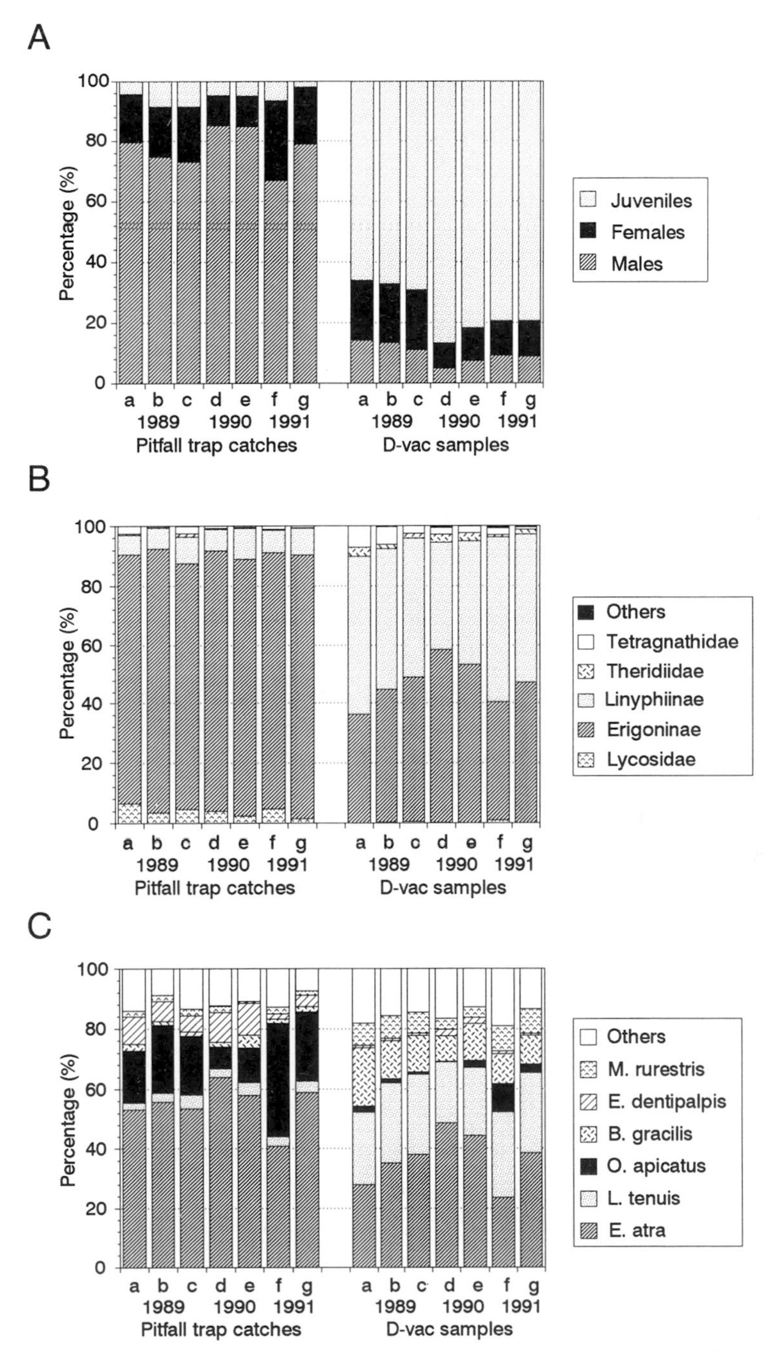

Fig. 2. Comparison of catches of intensive D-vac sampling and pitfall traps with regard to composition of life stages (A), families (B) and species (C) in seven winter wheat plots from 1989 to 1991.

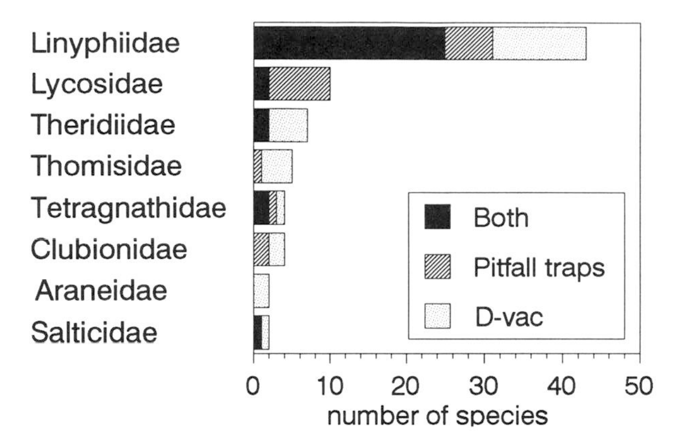

Fig. 3. Number of species collected by intensive D-vac sampling and pitfall traps in seven winter wheat plots from 1989 to 1991. Total identified species = 78.

Comparison of D-vac samples and pitfall trap catches

Juvenile spiders were the dominant group (66% - 87%) in the D-vac samples throughout the study. The percentages of male and female spiders varied between 5% and 20%. Pitfall trap catches were characterized by males (up to 85% of all spiders) while females and spiderlings were caught in small numbers (Fig. 2A). Species of Erigoninae (*Erigone atra*, *E. dentipalpis* (Wider) and *Oedothorax apicatus*) and of Lycosidae mainly influenced the pitfall catches. In contrast, Linyphiinae (*Lepthyphantes tenuis* (Blackwall), *Bathyphantes gracilis* (Blackwall) and *Meioneta rurestris* (C.L.Koch)) were more numerous in the D-vac samples than the pitfalls (Fig. 2B+C). During the study, 32 identified species were common to both sampling methods (Fig. 3). D-vac sampling detected 27 and pitfall trapping 19 additional species.

Population dynamics and the relationship of pitfall catch to abundance

From April/May the density of spiders increased in all years. In some plots this trend continued until the end of the investigation period (harvest) while in other plots, for example in 1990, density decreased in July and August. The pattern of pitfall catch was characterized by peaks in May and July and a period of reduced catch in June (Fig. 4).

Regression analysis between pitfall catch and density was calculated for individual years and for all three years combined. Fig. 5 shows the linear regression lines for *E. atra* males and illustrates the variation of this relationship between years. For *E. atra* males and other spider groups significant linear rela-

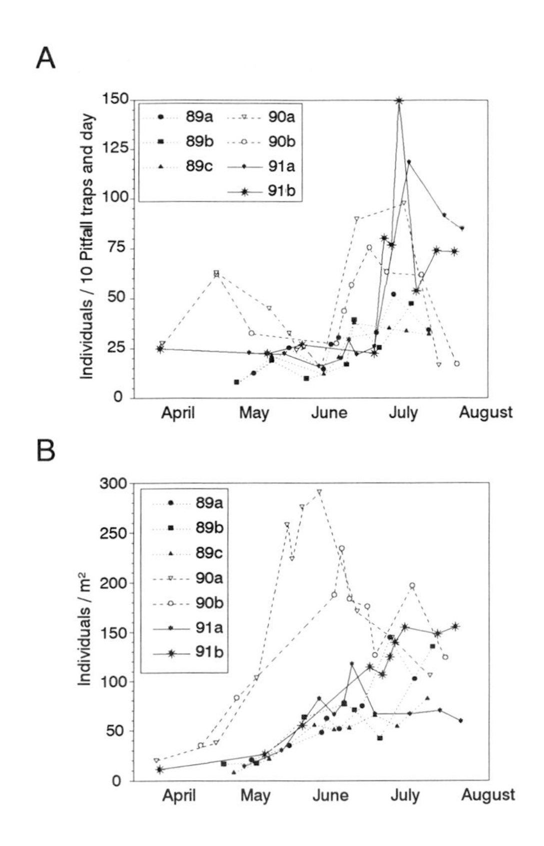

Fig. 4. Pitfall catch (A) and absolute density (B) for total spiders in 7 winter wheat plots from 1989 to 1991.

tionships were found but generally the fitted regressions were characterized by low coefficients of determination (Table 4). Sex-specific different coefficients of determination were calculated for *E. atra* and the relationship between pitfall catch and density was found to be weaker for females than for males.

Discussion

The intensive D-vac technique proved to be an appropriate tool for a quite realistic estimation of the density of dominant spider species in winter wheat. The difference in the numbers of juveniles in D-vac samples in comparison to those in the soil samples extracted by heat may have resulted from the emergence of spiderlings from cocoons in the soil samples during the first days of non-lethal warming. In fact, cocoons were found in nearly all D-vac samples

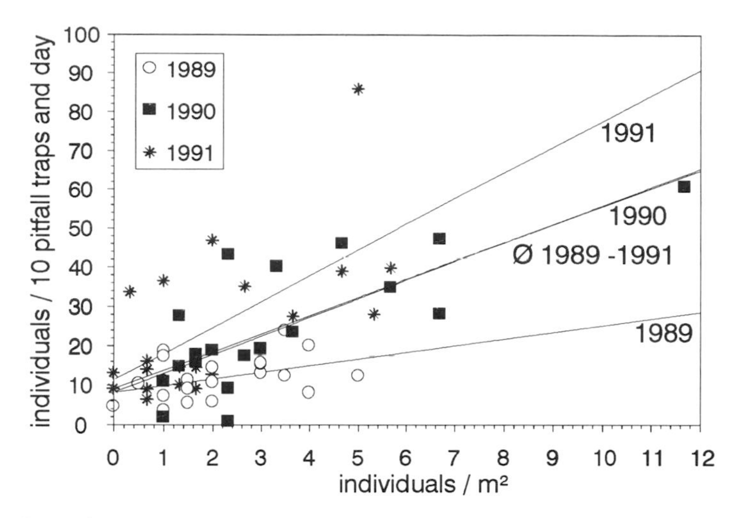

Fig. 5. Relationship between density (estimated by intensive D-vac sampling) and pitfall catch for *E. atra* males for each of the years 1989 to 1991, and for the years combined (Ø 1989 - 1991).

and several spiderlings emerged from these cocoons if they were kept in the laboratory for some days (unpubl. data). The approximately equal densities of adult spiders detected by both methods represents an important result. Duffey (1980) measured the efficiency of the D-vac by heat-extracting turfs which had already been D-vac sampled. He found low D-vac efficiencies (7% to 49%) for spiders in grassland using a nozzle with a diameter of 33 cm and a suction period of 10 seconds. But he did not present separate data for juvenile and adult spiders nor did he discuss the problem of an impairment of his results by the emergence of spiderlings from cocoons. Nevertheless, the intensive D-vac sampling technique seems to be a more effective method for the estimation of spider densities, especially epigeic ones, than the much used "original" D-vac technique. Lycosids were also vacuumed by the intensive D-vac sampling technique, but if they are present only in small densities large numbers of samples are necessary for an exact estimation of their density. Furthermore it is possible that single lycosids, which are able to run very fast, succeed to escape the approaching investigator. This can result in an underestimation of their real densities. Therefore, for the estimation of lycosid densities other methods should be used as reviewed by Sunderland et al. (1995).

In comparison with other methods, D-vac sampling allows many samples to be taken in a short time. D-vac samples can be frozen and sorted later, if it is not possible to sort them immediately. Disadvantages of the intensive D-vac technique are the long time spent on hand-sorting in the laboratory (10 - 60 min per sample) and the restriction to favourable (dry) weather conditions.

Table 4. Relationships between density (x) and pitfall catch (y) of various spider groups from 1989 to 1991.

Spider group	Year	Regression line	r^2	p	df
Total spiders	1989	y = 9.35 +0.28x	0.58	0.000	21
	1990	y = 55.34 -0.06x	0.04	0.420	17
	1991	y = 21.41 +0.37x	0.20	0.050	18
	1989 - 1991	y = 30.02 +0.11x	0.07	0.045	60
Erigoninae	1989	y = 6.71 +1.59x	0.29	0.008	21
	1990	y = 5.57 +2.40x	0.48	0.000	17
	1991	y = 12.73 +4.10x	0.33	0.009	18
	1989 - 1991	y = 10.16 +2.39x	0.26	0.000	60
Linyphiinae	1989	y = 0.76 +0.11x	0.41	0.001	21
	1990	y = 2.11 +0.19x	0.14	0.109	17
	1991	y = 0.71 +0.41x	0.43	0.002	18
	1989 - 1991	y = 1.33 +0.21x	0.18	0.000	60
E. atra males	1989	y = 8.16 +1.70x	0.18	0.041	21
	1990	y = 9.13 +4.64x	0.58	0.000	17
	1991	y = 11.36 +6.62x	0.40	0.003	18
	1989 - 1991	y = 8.25 +4.75x	0.38	0.000	60
E. atra females	1989	y = 0.77 +0.06x	0.04	0.368	21
	1990	y = 1.05 +0.09x	0.09	0.226	17
	1991	y = 0.79 +0.15x	0.11	0.153	18
	1989 - 1991	y = 0.76 +0.12x	0.13	0.004	60
L. tenuis adults	1989	y = 0.48 +0.08x	0.24	0.029	18
	1990	y = 0.37 +0.25x	0.40	0.003	17
	1991	y = 0.41 +0.37x	0.32	0.022	14
	1989 - 1991	y = 0.46 +0.22x	0.23	0.000	53

Substantial changes in vegetation density or structure are also factors that can affect D-vac efficiencies (Duffey 1980). However, in this study, these limitations can be considered as unimportant because the efficiency of the intensive D-vac technique was still high even when the vegetation reached maximum growth in July.

In contrast to the intensive D-vac samples, pitfall catches misrepresented the real population structure of spiders and overestimated males as well as species of Erigoninae and Lycosidae. Catches of spiderlings, the majority of the spider population in this study, were almost completely lacking in pitfalls. These results suggest that the interpretation of pitfall trap catches for an estimation of changes in the real abundance of spiders seems to be quite restricted. For example, results of own field experiments on insecticide side-effects on *E. atra* in winter wheat investigated by intensive D-vac sampling and pitfall trapping were not identical for the two methods. Sprayings of the pyrethroid fenvalerate caused less of a decline in the density (estimated by intensive D-vac sampling) of *E. atra* than the numbers caught in pitfall traps (Dinter & Poehling 1992a, 1992b). The present study also showed that the abundance of spiders could not be exactly correlated with pitfall catches. It may be supposed that several additional factors, such as temperature, humidity, spatial resistance, reproductive behavior or species interactions (predation, competition) influence the activity pattern of spider species and the numbers caught in pitfall traps. Therefore, the possibilities for an estimation or explicit modelling of the absolute density dynamics of spiders using pitfall data seem to be limited. Investigations of population dynamics require the quantitative analysis of absolute densities. Intensive D-vac sampling fulfills this requirement more closely than pitfall trapping.

Acknowledgements

The author likes to thank H.-M. Poehling and K.D. Sunderland for critical reading the manuscript and correcting the English. The study was supported by the German Research Council, DFG grant Po 207/6-2.

References

Adis, J. 1979. Problems of interpreting arthropod sampling with pitfall traps. *Zool. Anz. Jena* 202: 177-184.

Dietrick, E.J. 1961. An improved backpack motor fan for suction sampling of insect populations. *J. Econ. Entomol.* 54: 394-395.

Dinter, A. & Poehling, H.-M. 1992a. Spider populations in winter wheat fields and the side-effects of insecticides. *Asp. Appl. Biol.* 31: 77-85.

Dinter, A. & Poehling, H.-M. 1992b. Freiland- und Laboruntersuchungen zur Nebenwirkung von Insektiziden auf epigäische Spinnen im Winterweizen. *Mitt. Dtsch. Ges. Allg. Angew. Ent.* 8: 152-160.

Duffey, E. 1980. The efficiency of the Dietrick vacuum sampler (D-vac) for invertebrate population studies in different types of grassland. *Bull. Ecol.* 11: 421-431.

Sunderland, K.D., De Snoo, G.R., Dinter, A., Hance, T., Helenius, J., Jepson, P., Kromp, B., Lys, J.-A., Samu, F., Sotherton, N.W., Toft, S. & Ulber, B. 1995. Density estimation for beneficial predators in agroecosystems. In: Toft, S. & Riedel, W. (eds.) *Arthropod natural enemies in arable land · I*, pp. 133-162. Aarhus.

Topping, C.J. & Sunderland, K.D. 1992. Limitations of the use of pitfall traps in ecological studies exemplified by a study of spiders in a field of winter wheat. *J. Appl. Ecol.* 29: 485-491.

Distance methods used to estimate densities of web spiders in cereal fields

Søren Toft, Claus Vangsgaard & Henrik Goldschmidt

Department of Zoology, University of Aarhus, Building 135,
DK-8000 Århus C, Denmark.

Abstract
Distance methods (Closest Individual and Nearest Neighbour methods) have been used to estimate population densities of sheet-web spiders in cereal fields in Denmark. Density is estimated from distances from a random point in the field to the web(s) around it. In several investigations these methods have produced density figures higher than any previously published from cereal fields in Europe. These results have been partly confirmed by soil extractions and suction sampling, though the latter method consistently gave far lower densities. The uncertainty of the density estimate is improved if distances from the random point to several webs are measured, but little further improvement is obtained after the fifth closest web. Distance methods are fast and easy to perform in the field and there is no subsequent laboratory work required.

Key words: Sheet-web spiders, Linyphiidae, population density, cereal fields, distance methods, vacuum sampling, soil extractions.

Introduction

Distance methods, e.g. the Closest Individual Method, Nearest Neighbour Method etc. (Southwood 1978, Krebs 1989), have been developed for fast and easy estimation of the density of trees, then subsequently applied to animal species, including invertebrates. Standard distance methods described by Krebs (1989) measure the distance from a random point to the (first) nearest individual. However, Keuls et al. (1963, see also Southwood 1978) provides a formula for density estimation where the distances measured are to the nth nearest individual, which should improve reliability. By comparison, the nearest neighbour method measures the distance between a random individual and its nearest neighbour (Krebs 1989). All formulas that translate these distances into density estimates require random distribution of individuals. However, Diggle's compound estimator (Krebs 1989) is regarded as robust against a range of deviation from randomness. It requires that both point-to-nearest-individual and individual-to-nearest-neighbour distances are measured concurrently.

***Arthropod natural enemies in arable land* · *I** *Density, spatial heterogeneity and dispersal*
S. Toft & W. Riedel (eds.). *Acta Jutlandica* vol. 70:2 1995, pp. 33-45.
 ISBN 87 7288 492

Like plants, web spiders are sedentary creatures and their density can be estimated from that of their webs, though so far with unknown accuracy. The spider fauna of temperate agricultural systems is dominated by sheet-web spiders of the family Linyphiidae (Sunderland 1987, Toft 1989); this family may comprise more than 90% of the total fauna. Uninhabited spider webs quickly degenerate (1-2 days in the spider *Lepthyphantes tenuis* in winter wheat, F. Samu, pers. comm.) and at any point in time a small fraction of the population are floaters (Toft 1988, F. Samu, pers. comm.). Therefore the number of intact webs can be assumed to approximate the number of active web spiders in the habitat. Because of the prominence of sheet-web builders agricultural systems are better suited for use of a web count method than most other habitats; the absence of a litter layer and often low density of weeds, creating a bare soil surface between the crop plants, adds to this. Still, web counts should underestimate total spider density, since 1) even if only the typical web spider species are considered, adult males do not have their own webs but wander between the females' webs; 2) some web spider species have rudimentary webs or may forage without a web for part of their life cycle; 3) several web spider species may be partly subterranean, inhabiting soil cracks and crevices, especially in dry periods; 4) non-web building spiders, of which the wolf spiders (Lycosidae) and *Pachygnatha degeeri* (Tetragnathidae) may sometimes be of some importance in cereal systems, cannot be recorded with these methods.

Even with these shortcomings our practical experiences lead us to believe these methods are potentially useful, because large data sets can be gathered in a short time. It is necessary to know, however, how distance-based density estimates relate to other standard quantitative sampling methods.

One of us (CV) has applied a distance method through two years in a spring barley field. The other two (ST and HG) subsequently used such methods in a winter wheat field, from which soil samples were concurrently collected for extraction. They also completed a special sampling session with the purpose of comparing the distance methods with a version of vacuum suction sampling.

Methods

Procedure for distance measures

In our use of the method we have followed the procedure given by Southwood (1978). A stick was placed at a random point (the "sampling point") within the field plot. In order to make the webs more easily observable the area surrounding the stick was dusted with potato starch. This was done either by shaking a pair of socks filled with potato starch over the ground (Eberhard 1976) or by using a special dusting device (Carico 1977). Thorough search was made for webs close to the sampling point. Thin metal marker sticks were stuck into the soil at the center of each web, and distances from the sampling point to the five or ten closest of these marker sticks were measured to the nearest centimeter. In some cases the distance from the first web to its nearest neighbour was also measured.

At high spider densities the area close to the ground, especially in the spaces between the dead basal leaves along the rows of the crop, was criss-crossed by a mess of silken threads. This necessitated a criterion of a clearly visible web sheet, though reduced webs and recently initiated webs would then be systematically ignored. A possible source of overestimation was that large webs sometimes extended on both sides of a crop row with sheet fractions also in the small spaces within the rows, each of which might be counted as separate small webs. However, only a small proportion of webs was large enough or so positioned to allow such mistakes to occur.

Another source of underestimation is the consistent use of the tenth web as the cutoff point, even if the eleventh or still later webs were situated at the same distance.

Distance samples

During 1991 and 1992 web densities were estimated in a conventionally managed, but unsprayed spring barley field at Rønde near Århus, Denmark. The field (named Stegelykke) forms part of the research area of the National Environmental Research Institute at Kalø. Seven to ten estimates were obtained weekly from early spring to harvest time in August. In 1991 the distance to the ten nearest webs were measured, in 1992 to the five nearest webs.

In 1993 the distance methods were used in a conventionally managed winter wheat field (no insecticide treatment) belonging to Ødum Experimental Station, also near Århus, Denmark. Distances to the five or, in most cases, the ten nearest webs, as well as the nearest neighbour of the first web, were measured for each sample. Five to ten samples were taken each sampling date in a stratified random manner.

Quantitative soil samples

These were taken during 1993 in the same winter wheat field as the above-mentioned distance samples. Soil samples of 15*20=300 cm^2 were taken using an ash-pan being pushed horizontally through the soil at a depth just below the roots of the crop (c. 5 cm). During transport care was taken to disturb the soil as little as possible. In the laboratory the samples were extracted in a large-scale version of the Tullgren funnel. On each occasion seven to fifteen stratified random samples were taken, as above.

Vacuum sampling

Between 19 and 21 August 1993, 26 estimates were made in a winter wheat field with the distance methods as described above. This field is part of a system research area at the Ødum Experimental Station. The particular field belonged to the integrated management system. It was subdivided into 24 small plots subjected to different combinations of organic material input level (high/low), ploughing depths (10/20 cm), and fertilizer level (-/+ inorganic N added to a basic level of organic manure), with each combination replicated three times. (Two plots were sampled on both days, because it was feared that heavy rains influenced the samples taken on the first day; the second samples turned out to be nearly identical to the first, however).

Prior to the web distance procedure, an area of 0.12 m^2 with the sampling point at the center, was enclosed by means of a 30 cm high PVC ring, dug c. 5 cm into the ground. Immediately following the web measurements the enclosed area was vacuum cleaned using a Vax 2000 vacuum cleaner (nozzle Ø 38mm) driven by a generator. This equipment was modified to reduce the damage to soft-bodied animals like spiders in two ways: First, the animals were collected wet (in water) as a glass jar was inserted along the cleaner's nozzle, using the pooter design. Second, the part of the nozzle between the tip and the glass jar was lined inside with fine cloth to avoid smashing the animals on the nose spiralling. These modifications were designed by E. Gravesen.

We used the following sample procedure: First, all wheat tillers inside the enclosed area were shaken and then cut away at 5-10 cm height and the whole area vacuum cleaned several times for about five minutes, taking care to get into crevices and in between plants, etc. This was followed by an undisturbed 5 min period during which only animals observed moving were sucked up. Finally, the whole area was vacuum cleaned again. We expected this procedure to be as thorough as the D-vac sampling followed by ground search recommended by Sunderland & Topping (1995).

Analysis of distance data

Before using distance data for density estimation one must determine the spatial pattern of the data. This was done using 1) Eberhardt's index for point-to-nearest-web distance measures ($I_E=(s/av)^2+1$, where s is the standard deviation and av the mean), and 2) Hopkins' index $h=\Sigma(x_i^2)/\Sigma(r_i^2)$, where both closest individual distances x_i and nearest neighbour distances r_i were available. These tests are described in detail in Krebs (1989), including criteria for determining the dispersion pattern from the index values.

Krebs (1989) also provides the following formulae by which distances can be converted into absolute densities:

For point-to-nearest-web distances: $N_1=n/\pi\Sigma(x_i^2)$;
for web-to-nearest-neighbour distances: $N_2=n/\pi\Sigma(r_i^2)$.

Provided the webs are randomly distributed both formulae give unbiased estimates of the population density. For cases deviating from random, Krebs (1989) suggests computing Diggle's compound estimator N_3, which is the geometric mean of N_1 and N_2 given above and whose standard error can also be estimated (Krebs 1989).

For distance measures taken from the random point to the qth closest web we have used the formula provided by Keuls et al. (1963): $N_4=(q-1)/\pi r_q^2$. For the results reported here q was either five or ten. The effect of increasing q on the variance of the results is analysed at the end.

Results and discussion

Dispersion pattern of webs and spiders

For the seasonal spring barley data 1992 both the Eberhardt index and the Hopkins index accepted random distributions of webs on 15 out of 17 dates. For the seasonal winter wheat samples 1993 the same was true for seven out of nine dates. Also the large August 1993 sample in winter wheat conformed to the random distribution. We have therefore accepted the Poisson assumption to be generally fulfilled and are able to apply the above formula as unbiased density estimates. Still, the N_1 and N_2 estimates often deviate considerably from each other, for which reason we use Diggle's estimator N_3 as a compromise.

For the soil extraction data the variance-to-mean ratio (s^2/avg) has been calculated as an index of dispersion. On all dates this figure is larger than two (range 1.9-5.9), indicating an aggregated distribution of spiders. The seemingly contradictory results from the two types of data will be discussed below.

Seasonal estimates in spring barley

The N_4 values (Keuls' formula) revealed a consistent seasonal pattern of density variation for years and crops. In spring barley (Fig. 1) the early densities both years lie between 1 and 10 per m^2. Procedures connected with sowing of the crop reduced numbers to less than 1 per m^2. Through the growing season densities increase exponentially, reaching a maximum above 300 m^{-2} in 1991 (Vangsgaard, unpublished) and above 600 m^{-2} in 1992 (Fig. 1).

Seasonal estimates in winter wheat

In the winter wheat densities (N_4) are about twice as high as in spring barley during June (Fig. 1, 100-150 m^{-2} compared to c. 50-100 m^{-2} in spring barley) probably because the field remains relatively undisturbed during spring. The generally lower growth rate and the summer fluctuations in the 1993 winter wheat data compared to 1992 spring barley should probably both be explained by the variable weather conditions of 1993, characterized by long spells of cold, rainy weather; 1992 was an extremely stable summer with more sunshine hours and higher mean temperatures than any other year of the century.

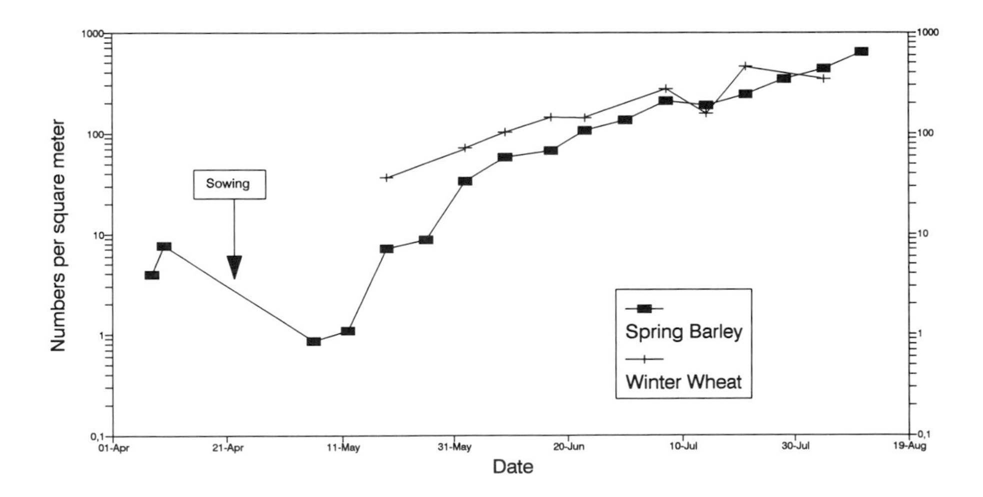

Fig. 1. Seasonal variation in the population of sheet-web spiders in a spring barley field 1992 and in a winter wheat field 1993, Denmark, as revealed by the distance method (Keuls' formula, q = 5 for spring barley, q = 10 for winter wheat).

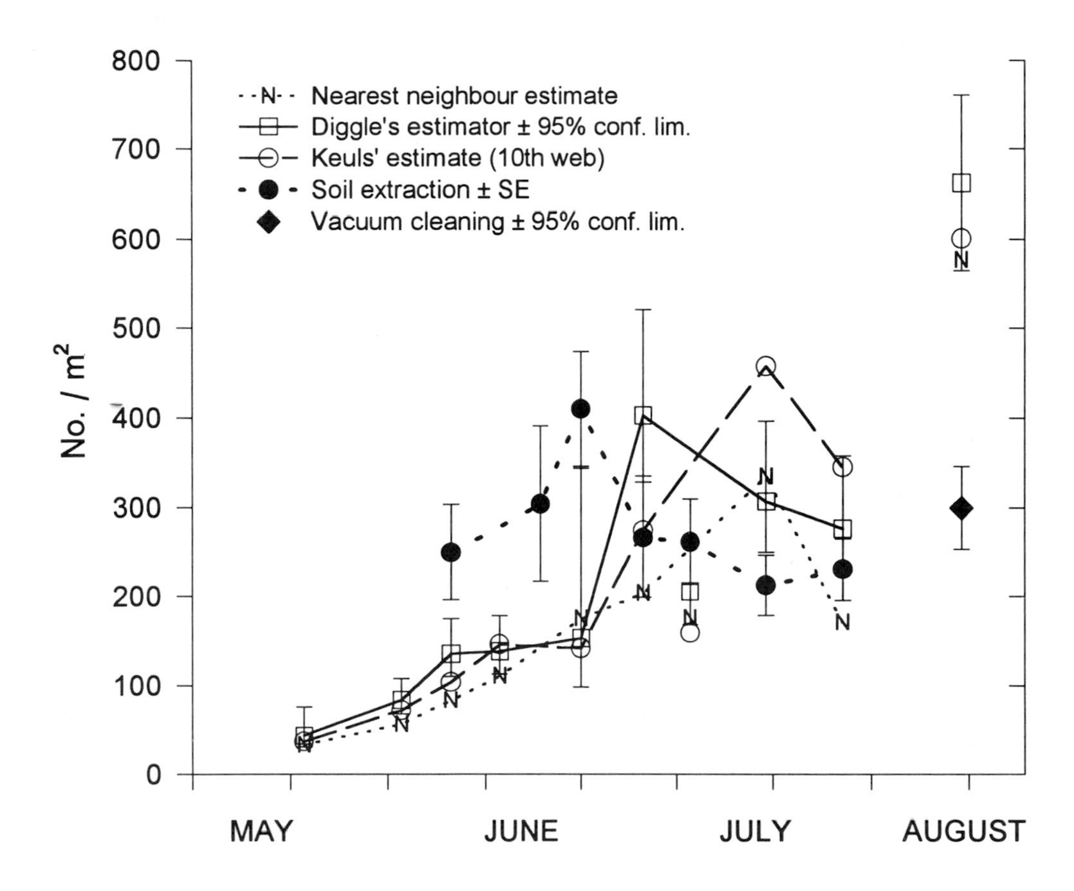

Fig. 2. Seasonal variation in the population of web spiders in a conventional winter wheat field 1993 as revealed by various sampling and distance methods. The three distance estimates are based on exactly the same observation series. (The distance estimates from 12 July were unduly low because of bad weather and therefore not included when the curves were drawn). The data presented to the right in the figure (late August) are from a different winter wheat field, see text.

Fig. 2 illustrates some consistency in the general seasonal pattern in winter wheat, with an early spring increase leading to a peak, and a subsequent summer decline. However, there is considerable divergence with respect to the timing of the peak and the magnitude of fluctuations, when the various ways of computing the densities are compared, even though the calculations are based on parts of the same data series. Comparing the distance methods with soil extraction (Fig. 2) the latter reveals even higher numbers during the spring months (300-400 m^{-2}), while the opposite is true later in the summer. Maxima of 400-450 m^{-2} were recorded depending on the method of calculation; however the field was harvested before the "true" maximum was expected to occur (hatching of the eggs of the summer generation of spiders took place in mid-August, see below). The low values on 12 July were the result of heavy rains during the preceeding period. The fact that numbers rise again indicates that many spiders must have been without webs for a period. As expected distance procedures are sensitive to heavy rains that destroy the webs; we have no indication, however, that mild showers or winds have any disturbing effect.

The high numbers of spiders in the soil extractions in June compared to the distance estimates requires an explanation. In both years 1992 and 1993 the spring was warm and dry, creating a hard, compact soil with deep crevices. We tend to believe (see also Sunderland & Topping 1995) that many spiders retreated into these crevices, at least by day, not only to avoid the heat and dry conditions, but also because their main prey (especially the Collembola) may have done the same. The data indicate that this may be a serious source for underestimation by the web count methods.

If this is true, it may also explain the discrepancy in the dispersion pattern of spiders and webs. As webs are only recorded on the surface, these measurements are not influenced by spiders aggregating in the crevices.

It is pertinent to discuss whether the seasonal pattern of density variation obtained with the distance methods is biologically meaningful. It is known that the species forming the core of the linyphiid community of cereal fields have breeding periods in spring (May-June) (DeKeer & Maelfait 1988, Toft 1989), and the following generation has a new breeding period in July-August. From this phenology pattern, a density peak is expected when egg sacs of the spring generation have just hatched, i.e. in June, and another larger peak in late August when egg sacs of the summer generation are hatching. The latter was recorded with the distance method producing a figure of 776 m^{-2}; even this may not be the true peak as all egg sacs had not yet hatched at the time these samples were taken (see below). Unfortunately no soil extractions were made at that time. The expected spring peak is perhaps observed in the soil extractions, whereas the distance method peaks must be too late to result from recently hatched young. One reason for not seeing the expected spring peak may be that some spiders move into crevices at this time, as already suggested; another possibility is that webs of the earliest instars are too small to be recorded with the same efficiency as those of later instars. Some of the increase in spider numbers may thus reflect the growth process, spiders' webs gradually growing into observable size. Only more detailed studies, applying web counts and soil extractions to the same population through a full season, can solve these questions.

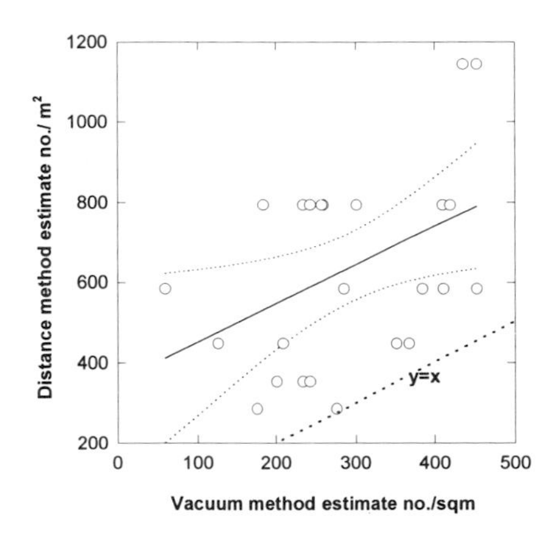

Fig. 3. Correspondence between vacuum cleaner sampling and a distance method estimate (Keuls' formula for 10th web) obtained simultaneously at the the same place. Solid line: regression; dotted lines: 95% confidence limits.

Distance methods compared with vacuum suction

Diggle's estimator N_3 produced an average density of 776 spider webs m^{-2} for the 24 experimental plots, a value only slightly higher than obtained with Keuls' formula N_4 (av. 662 m^{-2}) and with the nearest neighbour estimate N_2 (av. 576 m^{-2}) (Fig. 2, right side). Vacuum cleaning, on the other hand, gave an average of only 300 m^{-2}.

In spite of this there is a significant positive correlation (Pearson $r=0.435$, $p=0.026$) between the density indicated by the vacuum method and the Keuls' estimate N_4 (Fig. 3). All 26 distance estimates lie above the corresponding vacuum values. Though the regression line runs parallel to the equality line, it is considerably displaced. A more reasonable relationship would be a line through the origin; even so the distance method produced estimates more than 300 individuals m^{-2} higher than those derived from suction sampling. This difference probably reflects the fact that the sampling was done at a time when egg sacs of the summer generation of *Erigone atra* were hatching *en masse*, resulting in an abundance of extremely small sheet-webs at the soil surface, the producers of which may have escaped the suction sampler. Another explanation may be that the dispersion pattern of the spiders changes with spatial scale, which makes it problematic to extrapolate from a small to a larger area. At the densities existing at the time these samples were taken, the area enclosing the ten closest webs was much smaller than the area sampled by the vacuum cleaner. In future uses of distance methods, small scale heterogeneity needs to be considered more carefully.

In the winter wheat field sampled, two neighbouring plots always had high and low fertilizer levels and otherwise identical treatments. Both methods gave significantly higher spider densities at the low fertilization level (paired t-tests; $p=.001$ (vacuum sampling), $p=.018$ (distance method, Keuls' estimate)). This may be related to the denser weed growth in the low fertilizer plots, or to the weaker-developed crop (Goldschmidt & Toft, unpublished).

Extractions are often criticized for the possibility that egg sacs in the sample may hatch during the extraction process due to the heating. In the present study we had a similar experience with vacuum sampling: Egg sacs deposited on the vegetation were torn apart by the suction force and the young released into the sample. Proof of this was that three samples contained a small number of individuals of the incomplete ("larval") instar, that cannot possibly have emerged from the egg sac by themselves. Juveniles of first complete instar likewise made to emerge by force would be impossible to distinguish from self emerged ones.

How many webs should be measured?

Fig. 4B shows the empirically estimated mean densities and 95% confidence limits obtained with the Keuls' formula applied to the August winter wheat data, when the second up to the tenth web is used for the estimate. There is a downward trend in the estimates as more webs are included, but this is relatively small. More important, there is a strong reduction in the uncertainty of the estimate, illustrated by the coefficient of variation, CV (Fig. 4A). Clearly CV is levelling off at the higher web numbers. It appears from this that there is little to be gained by extending the observations past the fifth web.

Conclusion

The density estimates obtained with the distance method are consistent in the sense that they correlate with a vacuum suction method used at the same place, and the method also reflects the effects of at least some agricultural management practices on web spider densities. It may produce very high density estimates, perhaps because it requires the researcher to concentrate his attention on a very small area compared to normal ground search procedures, and perhaps because the objects scrutinized don't hide when disturbed.

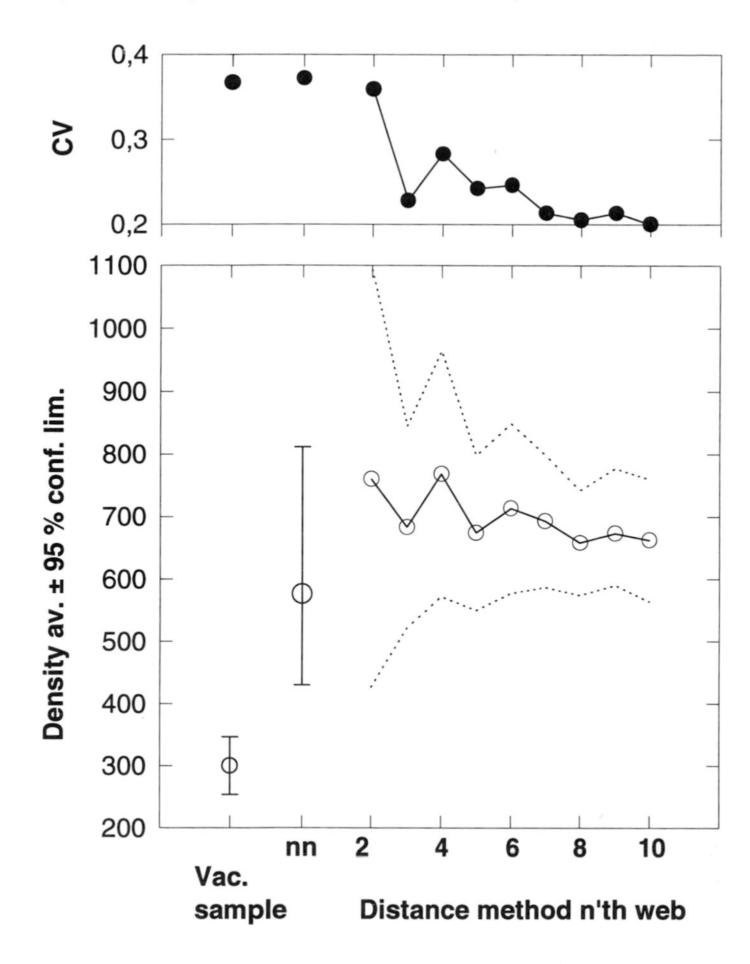

Fig. 4. Variation in results obtained with the distance method (Keuls' formula) as a function of the number of webs included in each sample. To the left: Vacuum cleaner samples and nearest neighbour estimates included for comparison. A. Coefficient of variation CV (SD/av.). B. Means ± 95% confidence limits, for distance samples indicated by hatched lines around the averages.

The maximum density estimates we have obtained by the distance method are higher than any published records from cereal fields that we know of (cf. review by Sunderland & Topping 1993). Comparison with two other methods that both involve collecting the animals and thus will always give underestimates, tends to show that they are not completely unrealistic. Our vacuum cleaner also revealed densities higher than the maximum values listed by Sunderland & Topping (1993, see also Dinter 1995) for cereal systems. The same was true for the soil extraction samples, even at a time when distance estimates were quite low. Considering also all the reasons why distance methods applied to spider webs are supposed to underestimate true densities (see Introduction), and the few potential causes for overestimation, the high numbers obtained in our applications of the method indicate that spider densities in Danish cereal fields are truly higher than those reported from other European countries.

The somewhat divergent results obtained by the various distance methods even when used simultaneously at the same spots may be troublesome. However, their confidence ranges overlap and with comparable sample sizes the estimates are not more variable than those obtained with standard direct methods (soil extraction, suction sampling). Unfortunately our results do not allow us to identify one of the distance methods as superior to the others; however, because of the reduced variance of the estimates when more webs around the sampling point are included, all other things equal the Keuls' procedure should be the most reliable.

The statistical assumption of random distribution may be problematic for general usage of the distance methods. At the very high densities it might be better to make visual web counts within small quadrats instead of measuring distances, in order to escape this requirement.

We have found that the distance method is easy to apply in the field and many samples can be obtained in a short time span. At reasonably high densities a sample including the ten nearest webs can be taken in less than 15 minutes. The result is obtained in the field with no subsequent sorting work in the laboratory. Another advantage is that the sampling procedure can easily and with low extra time costs be widened to incorporate a series of very important details about the animals: web area, height, collection of web owner (species and instar identification), etc.

Acknowledgements

Thanks are due to the Department of Flora and Fauna, Kalø, and to Karl Rasmussen and Gunnar Mikkelsen for providing access to the fields at Kalø and Ødum, respectively; further to Børge Lysgaard for working facilities and information. The studies at Ødum were financed by a grant from the Danish Environmental Research Programme to the Centre for Agricultural Biodiversity.

References

Eberhard, W.G. 1976. Photography of orb webs in the field. *Bull. Br. arachnol. Soc.* 3: 200-204.

Carico, J.E. 1977. A simple dusting device for coating orb webs for field photography. *Bull. Br. arachnol. Soc.* 4: 100.

DeKeer, R. & Maelfait, J.-P. 1988. Observations on the life cycle of *Erigone atra* (Araneae, Erigoninae) in a heavily grazed pasture. *Pedobiologia* 32: 201-212.

Dinter, A. 1994. Estimation of epigeic spider population densities using an intensive D-vac sampling technique and comparison with pitfall trap catches in winter wheat. In: Toft, S. & Riedel, W. (eds.) *Arthropod natural enemies in arable land · I*, pp. 23-32. Aarhus.

Keuls, M., Over, H.J. & de Wit, C.T. 1963. The distance method for estimating densities. *Statistica Neerlandica* 17: 71-91.

Krebs, C.J. 1989. *Ecological methodology.* Harper & Row, New York.

Southwood, T.R.E. 1978. Ecological Methods. Methuen & Co., London.

Sunderland, K.D. 1987. Spiders and cereal aphids in Europe. *Bull.SROP/WPRS* 1987/X/1: 82-102.

Sunderland, K.D. & Topping, C.J. 1993. The spatial dynamics of linyphiid spiders in winter wheat. *Memoirs of the Queensland Museum* 33: 639-644.

Sunderland, K.D. & Topping, C.J. 1994. Estimating population densities of spiders in cereals. In: Toft, S. & Riedel, W. (eds.) *Arthropod natural enemies in arable land · I*, pp. 13-22. Aarhus.

Toft, S. 1988. Interference by web take-over in sheet-web spiders. *XI. Europäisches Arachnologisches Colloquium, Berlin 1988* (Ed. Joachim Haupt): 48-59.

Toft, S. 1989. Aspects of the ground-living spider fauna of two barley fields in Denmark: species richness and phenological synchronization. *Ent. Meddr.* 57: 157-168.

Estimation of population sizes and "home ranges" of polyphagous predators in alfalfa using mark-recapture: an exploratory study

F. Samu[1] & *M. Sárospataki*[1,2]

[1]Plant Protection Institute of the Hungarian Academy of Sciences
P.O.Box 102 Budapest, H-1525 Hungary
[2]Agricultural University of Gödöllô, Department of Zoology and Ecology,
Gödöllö, Hungary

Abstract
The ground living predatory arthropod populations in an alfalfa field were investigated by marking and recapturing animals using live trapping pitfalls. Twelve pitfalls were set up in the field in a grid arrangement. The total capture and recapture data were used to calculate population sizes using the Petersen-Lincoln index, while the grid location and the time elapsed to recaptures were used to assess the "home range" of the animals. The study focused on the dominant predatory arthropods: *Harpalus rufipes* (Coleoptera, Carabidae), *Cicindela campestris* (Coleoptera, Cicindelidae), and *Pardosa agrestis* (Araneae, Lycosidae). There were few recaptures of beetles and of female wolf spiders. The resulting large error of the Petersen-Lincoln estimates precluded the drawing of final conclusions on the population sizes of all predators, except male *P. agrestis*. The size of the trapping grid permitted assessment of the "home range" only for male wolf spiders. The resulting estimate was a density of 4 spiders/m^2 and a maximum "home range" of about 300 m^2. Studies to estimate densities of such mobile animals seem only to be feasible if a sufficient number of recaptures can be achieved and the spatial scale of the trapping is comparable with the "home range" of the species studied.

Key words: Population estimation, mark-recapture, movement activity, predatory arthropods, wolf spiders

Introduction

The assessment of population size, "home range" and mobility of ground living arthropod predators is a difficult task for the field biologist, since something

Arthropod natural enemies in arable land · I *Density, spatial heterogeneity and dispersal*
S. Toft & W. Riedel (eds.). *Acta Jutlandica* vol. 70:2 1995, pp. 47-55.
 ISBN 87 7288 492

intangible is being measured. Collection methods useful in faunistic research often fail to give even a relative estimate of population sizes. Absolute density estimating methods, like enclosing an area or D-vac sampling can also fail in the case of highly mobile animals, which might escape on the approach of the investigator.

Pitfall trapping is a cost effective, widely used method of sampling ground living arthropod populations (Uetz & Unzicker 1976). The ecological interpretation of pitfall trap data has generated much controversy in the entomological literature (Lövei & Samu 1987, Halsall & Wratten 1988, Topping & Sunderland 1992). Conclusions have often been drawn on the relative abundance of the species caught. However, abundance is not the only factor which determines how many individuals from a local arthropod population will be caught. Catches from traps such as pitfalls, light or malaise traps, depend also on the mobility, "home range" and trap avoidance behaviour of the populations studied. Mobility has a simple effect on catches. The more mobile an animal, all other factors being constant, the higher the chance that it crosses the trap's perimeter and gets caught. Greater "home range", on the other hand, lowers the chances that the particular area of the trap opening will be encountered. Mobility and "home range" interact with each other. A species with a large "home range" and with high mobility can be caught with the same frequency as another species with small "home range" and small mobility. Trap avoidance behaviour can vary between different species. Some spiders build a web across the trap opening, and are underrepresented in the caught material (Topping 1993). Walking behaviour (fast or slow; to what extent the animals walk on the low vegetation or directly on the soil surface) acts also as a kind of trap avoidance and determines what proportion of the animals will fall in the trap. Trap avoidance can be regarded as a species-specific constant factor, at least within developmental stage and sex, while mobility can vary a lot due to weather, hunger and sexual activity even within a species or stage.

Marking studies can be useful to separate the combined information contained in pitfall trap data. Marking and recatching the animals (as in classical mark-recapture studies) can, by the dilution principle, give an idea of the population size of the species (Seber 1982). If many appropriately spaced traps and individual marking are used, the distance covered and time elapsed between two catches can be combined to give information on the "home range" of the animals (Otis et al. 1978).

Here we report on a small-scale study of this type in which ground living predatory arthropods of an alfalfa field were sampled by a grid of live trapping pitfalls, and then marked. The trapping data as such gave us the traditional information obtainable from a usual pitfall trapping session. From the recapture rate we attempted to estimate the population sizes of the focal species, while from the spatial location of recaptures within the grid, and from the time elapsed between captures, we tried to assess the "home range" of the animals.

Materials and methods

Live trapping pitfall traps of 36 cm^2 upper area were planted in a 3x4 grid, 3 m apart from each other. This catching unit was unfenced and situated in the middle of a 150 m wide alfalfa field. Traps were visited daily. Trapped animals were marked individually with a "patch mapping" method (Samu et al., unpublished) using enamel paint (initially acrylic paint in the case of spiders). Plant material was placed on the bottom of the traps to serve as a refuge for the animals, and thus decrease predation and cannibalism among trapped animals. Marked animals were set free 0.5 m away from the trap where they were caught. The date and place of release of recaptured animals were identified by the marking pattern. For the safety of the identification, besides applying a unique marking to every individual, a new colour was used for each day. Over 1000 predatory arthropods were marked during the three weeks of study (30 July - 21 August 1991). The most numerous species were the carabid beetle *Harpalus rufipes* (De Geer), the tiger beetle *Cicindela campestris* L., and the wolf spider *Pardosa agrestis* Westring. Since virtually all recaptures occurred in these species, further analysis focused on them. Male and female wolf spiders (sexes are readily identifiable in the field) had different trappability and recapture rates, and so were treated separately. Weather data was obtained from the daily weather records of the Hungarian Meteorological Institute.

Population sizes of the studied species were calculated from the Petersen-Lincoln index modified for multiple releases and recaptures (Begon 1979, Demeter & Kovács 1991):

$$N = \frac{\sum M_i n_i}{\left(\sum m_i\right) + 1}$$

where N is the population size, M_i is the total number of marked animals at time i, n_i is the number of animals captured at time i, and m_i is the number of marked animals in that capture.

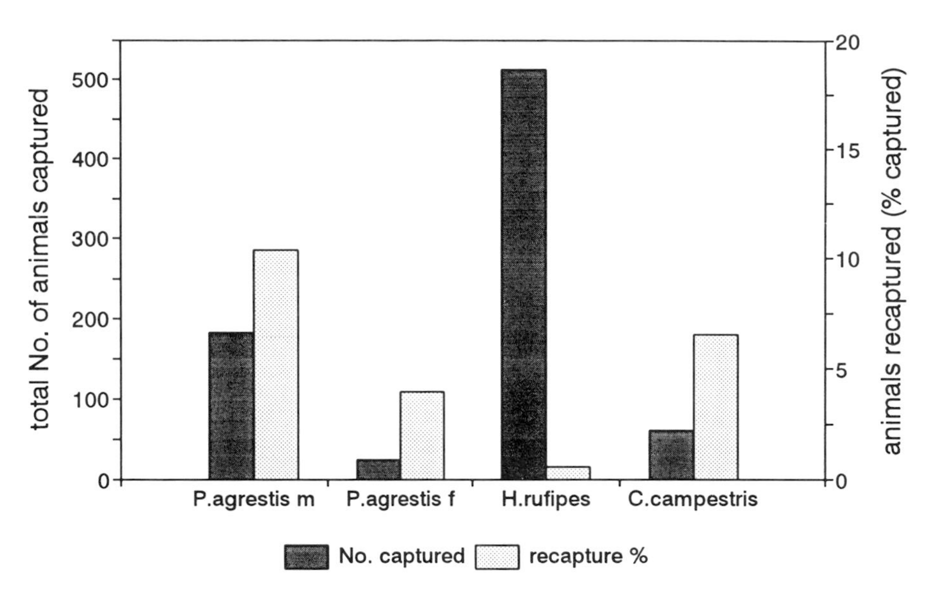

Fig. 1. Total number of individuals caught in the 12 pitfalls and recapture rate (percentage of total catch) of the three focal species.

Results

A total of 781 individuals of the three species investigated were marked and released during the 3 weeks duration of the study. *Harpalus rufipes* made up 65% of the captured animals. Daily captures of this beetle were 32 ± 15.4 (mean ± S.D.). However, this species also had the smallest overall recapture rate, only 0.59% (Fig. 1). *Pardosa agrestis* was represented in the catches with a fairly high number, but there was a great difference between the daily catches of males and females (11.4 ± 9.03 vs. 1.6 ± 2.57). Overall recapture rate was around 10% for the males, and around 4% for the females (Fig. 1). *Cicindela campestris* was caught in smaller numbers (3.8 ± 3.52), but it was relatively frequently recaptured (recapture rate: 6%).

Variability in daily catches was very high. The time series of catches of *P. agrestis* (both sexes) and *C. campestris* changed together (Fig. 2). *H. rufipes* showed peak catches shifted by a day compared to the other two species. Over such a short period of time population changes can be ruled out; thus the fluctuations in daily catches must have reflected changes in the mobility level of the animals. Mobility of the animals on the other hand could be influenced by weather factors. Cold fronts indeed seemed to explain much of the variation observed (Fig. 2). *H. rufipes* showed peak activity at the arrival of the fronts

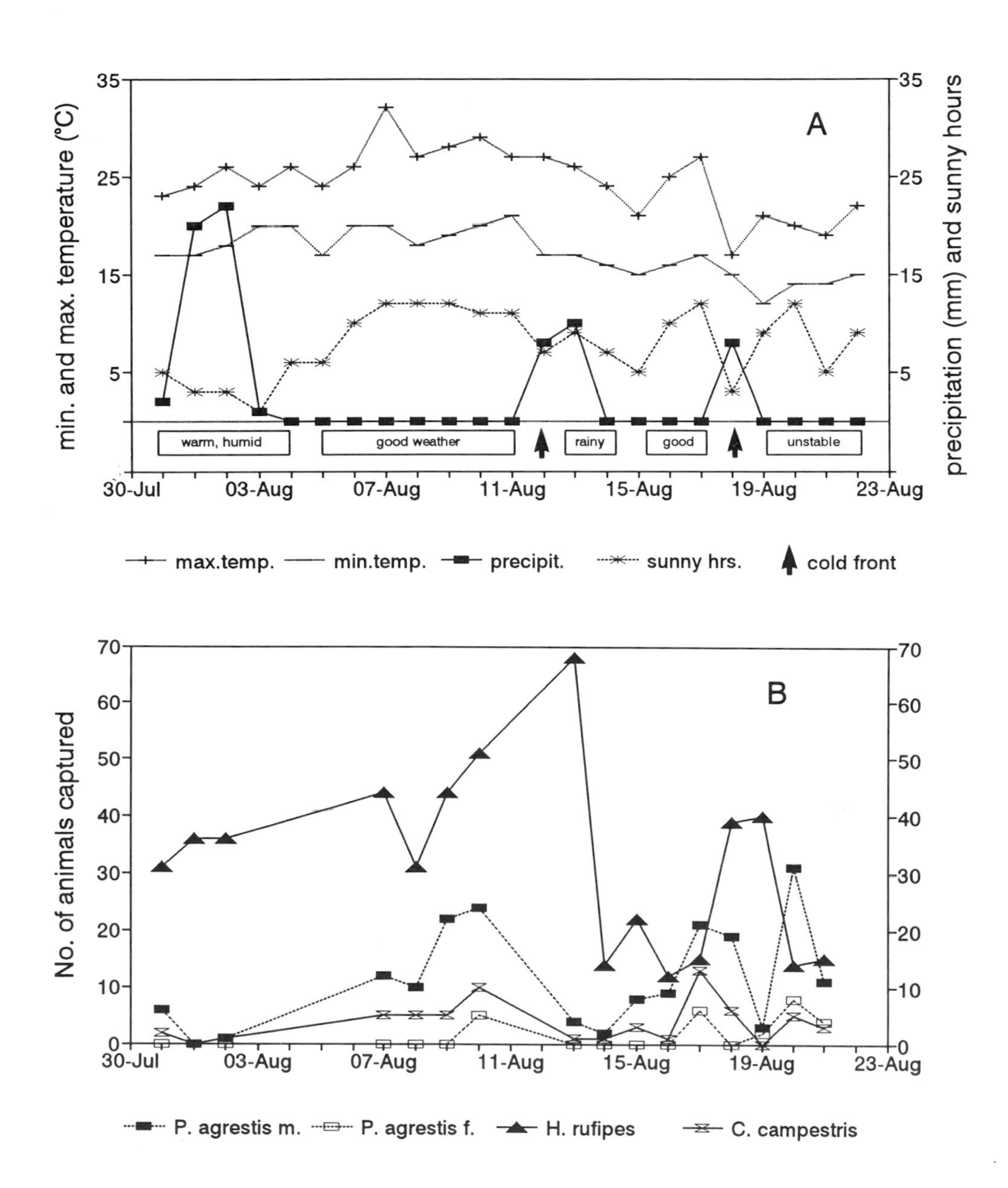

Fig. 2. A. Variation in weather during the study period. B. Daily No. of animals caught in the pitfalls. Data, even in the case of the two interruptions of the continuous observation, refer to the catch of the preceding 24 hours.

and then declined, while the wolf spider and the tiger beetle species had an activity minimum from the beginning of the fronts. Catches in all groups were high during the good weather periods.

The distance at which marked animals were recaptured, measured from the release trap, can provide an estimate of the distances these animals can travel between captures (Table 1). Of the four tiger beetles recaptured, all were recaught on the day following release, three in the same trap, one in a neighbouring trap. *H. rufipes* was recaptured only on three occasions, twice on the next day, and once 11 days after the marking. In the latter case the exact release site was not determinable. In *P. agrestis* 65% of the males were recaught in the same trap. Contrary to the tiger beetles, these spiders were not always recaught on the next day, but on the average 2.3 days after marking. The remaining recaptures occurred either in the closest or the second closest neighbouring trap (maximum distance 6.7 m). Mean time elapsed between capture and recapture, considering all recaptures, was also 2.3 days. There was no correlation between the time to recapture and the distance covered during this time (d.f.=15, r=0.24, N.S.).

Estimation of the population size gave the highest value in the case of *H. rufipes*, where the highest pitfall catches were associated with the lowest recapture rates. The *Pardosa* population was also shown to be substantial, but estimations for males and females were very different. The smallest population size was calculated for the tiger beetles (Table 1).

Discussion

Mark-recapture studies can give information about the size of a population. Precision of the estimation by the Petersen-Lincoln index, or indeed any other model, is largely dependent on the number of animals recaptured. In the present study low recapture rates are reflected by the large standard deviations associated with the population estimates. Therefore care should be taken in the interpretation of these values, with the possible exception of male wolf spiders. These figures give rather the order of magnitude of the population sizes. Apparently a larger scale sampling is needed to study the relatively more mobile carabid populations than the more sedentary wolf spiders.

Daily variation in the mobility of animals might substantially change catches of the pitfalls from day to day. Such very short-term activity changes are often caused by changing weather conditions. Weather fronts have been shown to change flight activity of carabids significantly (Honek 1988, Kádár & Szentkirályi 1991). In mark-recapture situations weather unavoidably changes

Table 1. Mean and standard deviation of the distance covered and time elapsed between marking and recapture, and the estimate of mean population size by the Petersen-Lincoln index.

	P. agrestis ♂♂ mean ± SD	*P. agrestis* ♀♀ mean ± SD	*H. rufipes* mean ± SD	*C. campestris* mean ± SD
Distance covered (m)	2.14 ± 2.94	6.00 ± -	3.00 ± -	0.75 ± 1.30
Days to recapture	2.32 ± 1.45	4.00 ± -	4.33 ± 1.41	1.00 ± 0.00
Population size*	701 ± 165.6	116 ± 153.5	23144 ±15846	284 ±162.5

* Estimated by the Petersen-Lincoln index weighted by the number of recaptures.

the capture rate during the study. If sufficient data is available, mark-recapture models less sensitive to such changes (e.g. the Jolly-Seber model; Seber, 1982) would be preferred.

To know not only the population size, but the density of a population as well, the area which the animals can cover, i.e. their "home range" has to be known. The grid arrangement of the traps and the individual marking aimed at this goal, but success varied with species, as was the case with population size.

In *Harpalus rufipes* the large pitfall catches and small recapture rate indicated a very large population. One recapture after 11 days suggested that the population was not in migration. This species is regarded as very mobile and capable of flight (Kádár & Lövei 1992). The location of the recaptures in the grid could not be considered in calculating "home range", because they were too few, and in the light of its high mobility it is possible that the sampling grid did not completely encompass the "home range". The lack of information on "home range" precludes any conclusion on the density of the carabid population.

Adult males of the wolf spider *Pardosa agrestis* were trapped more frequently than females. Since there is no indication that sex ratio would be (at least at the initial phase of adulthood) different from 50% (Edgar 1972; pers. observation), different catches must be attributed to the different activity and possibly different trap avoidance behaviour of the sexes (Hallander 1967, Topping & Sunderland 1992). *Pardosa* species are considered to employ a "sit-and-wait" predatory strategy. The pattern of movement consists of long motionless waiting, then short bursts of movements (Ford 1978). Even if this pattern is different for males in the mating period, Hallander's (1967) studies on two other *Pardosa* species demonstrated that average daily distances covered even in the mating period did not exceed 2.5 m on the average and 12 m at the maximum. The present data demonstrate that most recaptures occurred in the same or in the neighbouring trap. This suggests that the 9 m by 12 m sampling grid represented approximately the same dimensions as the "home range" of the spiders. Thus a rough estimate of 10 m activity radius, and a "home range" of

300 m^2 is perhaps not unrealistic. This allows a rough estimation of the density of *P. agrestis*, namely c. 4 spiders (of both sexes) per square meter. Such figures are very approximate, but even if they were not, we need to know more about the spatial and temporal variation in density. However, such approximate figures can be helpful in giving the order of magnitude on which other studies can be based.

Less information could be extracted from the capture and recapture data of the tiger beetle *Cicindela campestris*. Despite being rather mobile, like *H. rufipes*, a modest recapture rate allowed a calculation of population size. However, all recaptures occurred on the day after the release of the marked animals, and these were either in the same or in the neighbouring trap. Such a recapture pattern unfortunately does not exclude the possibility that the population was actually migrating. In this case all recaptures could have been the result of the beetles falling back in the same or neighbouring trap in the vicinity of which they were released. This, besides making the population size estimation less reliable, also prevents the next stages of the deduction, the calculation of "home range" and density of the tiger beetles.

Good-sighted, mobile, but small animals will pose problems to biologists in the future, too. Some sophisticated methods are now available for tracking a few individuals (Baars, 1979; Mascanzoni & Wallin, 1986), but these are still only applicable to relatively large, hard-bodied invertebrates. Pitfall trapping together with marking is an accessible, but labour-intensive method, and as the present example has shown, finding the right scale is essential for obtaining sound density data of natural populations.

Acknowledgements

We are grateful to Dr. K.D. Sunderland and F. Kádár for the helpful comments on the manuscript, and to Dr. V. Rácz for the weather data. The work was supported by the OTKA grant No. F5042.

References

Baars, M. A. 1979. Patterns of movement of radioactive carabid beetles. *Oecologia (Berl.)* 44: 125-140.

Begon, M. 1979. *Investigating animal abundance*. Edward Arnold, London.

Demeter, A. & Kovács, G. 1991. *Estimation of the size of animal populations* (in Hungarian). Academic Press, Budapest.

Edgar, W. D. 1972. The life cycle of the wolf spider *Pardosa lugubris* in Holland. *J. Zool. Lond.* 168: 1-7.

Ford, M. J. 1978. Locomotory activity and the predation strategy of the wolf spider *Pardosa amentata* (Clerck) (Lycosidae). *Anim. Behav.* 26: 31-35.

Hallander, H. 1967. Range and movements of the wolf spiders *Pardosa chelata* O. F. Muller and P. pullata Clerck. *Oikos* 18: 360-364.

Halsall, N. B. & Wratten, S. D. 1988. The efficiency of pitfall trapping for polyphagous predatory Carabidae. *Ecol. Entomol.* 13: 293-299.

Honek, A. 1988. The effect of crop density and microclimate on pitfall trap catches of Carabidae, Staphylinidae and Lycosidae in cereal fields. *Pedobiologia* 32: 233-242.

Kádár, F. & Lövei, G. L. 1992. Light trapping of carabids (Coleoptera: Carabidae) in an apple orchard in Hungary. *Acta Phytopathol. Entomol. Hung.* 27: 343-348.

Kádár, F. & Szentkirályi, F. 1991. Influences of weather fronts on the flight activity of ground beetles (Coleoptera, Carabidae). *Proc. 4th ECE/XIII. SIEEC, Gödöllö*, pp. 500-503.

Lövei, G. L. & Samu, F. 1987. Estimation of the number of carabid species occurring on a wet meadow. *Acta Phytopathol. Entomol. Hung.* 22: 399-402.

Mascanzoni, D. & Wallin, H. 1986. The harmonic radar: a new method of tracing insects in the field. *Ecol. Entomol.* 11: 387- 390.

Otis, D.L., Burnham, K.P., White, G.C. & Anderson, D.R. 1978. Statistical interference from capture data on closed animal populations. *Wildlife Monographs* 62: 1-135.

Seber, G. A. F. 1982. *The estimation of animal abundance.* Griffin, London.

Topping, C. J. 1993. Behavioural responses of three linyphiid spiders to pitfall traps. *Entomol. Exp. Appl.* 68: 287-293.

Topping, C. J. & Sunderland, K. D. 1992. Limitations to the use of pitfall traps in ecological studies. *J. Appl. Ecol.* 29: 485- 491.

Uetz, G. W. & Unzicker, J. D. 1976. Pitfall trapping in ecological studies of wandering spiders. *J. Arachnol.* 3: 101-111.

A rapid method for handling and marking carabids in the field

C.F.G. Thomas

IACR-Long Ashton Research Station, Department of Agricultural Sciences, University of Bristol, Long Ashton, Bristol BS18 9AF, England

Abstract
The ability to mark large numbers of carabids in the field for recapture for either dispersal or density studies can be a critical factor affecting the success of an experiment. After trying a variety of methods we have developed a technique that enables large numbers of beetles to be quickly and permanently marked in the field without the need to accumulate populations in the laboratory and the consequent problems of feeding and reacclimatisation prior to release.

Marking beetles

Marking methods have been thoroughly reviewed (Southwood 1978). Paints and inks have the advantage that a range of colours can be used to code for number or release site but the disadvantage is that the marks seldom persist on waxy elytra for more than four to eight weeks (e.g. Lys & Nentwig 1991), particularly for those species that burrow in the soil during the day. Although paint adheres better to areas of the elytra which have been scraped (Murdoch 1963), for studies of dispersal carried out over longer periods, permanent marking methods are more suitable. We have found that cauterisers stress the animals (induce vomiting) and believe that this stress may influence their activity after release (e.g. Greenslade 1964). Cauterisers are also difficult to handle in the field where wind cools the tip; batteries require frequent recharging or replacement.

A model makers drill (MB140, Minicraft Ltd UK) was recommended (Sandrine Petit pers. comm.) which has proved to be ideal for inscribing digital numbers on the elytra of *Carabus* spp. in the laboratory when used with a 1mm drill bit under a dissecting microscope (Naomi Chinn pers. comm.). In the field, the drill can be powered by battery; we routinely use a sealed, lead acid rechargeable (gel type) 12V,2.8Ah battery (Yuasa Battery (UK) Ltd). We have tried several drill bits and have found that with a little practice, marks can be

Arthropod natural enemies in arable land · I *Density, spatial heterogeneity and dispersal*
S. Toft & W. Riedel (eds.). *Acta Jutlandica* vol. 70:2 1995, pp. 57-59.
 ISBN 87 7288 492

made precisely with an abrasive (sanding) disc supplied with the drill. This can be used to abrade small areas of the elytra and pronotum in predefined areas corresponding to coded values (e.g. after Sheppard et al. 1969). The abraded marks can be highlighted with paints if desired for rapid recognition in the field. The paint adheres better to the elytra after abrasion. We use extra-fine metallic silver paint markers (edding 780, Edding (UK) Ltd, St Albans, UK). We have not evaluated the effects of paint marking on losses due to predation.

Handling beetles

In order to avoid inadvertent amputation of fingers during marking with the abrasive disc, and to enable large numbers of beetles to be handled quickly, we have devised a simple restraining technique. A piece of wooden dowelling (diameter 23 mm — a broom handle) is cut to a convenient length of approximately 600 mm. Elastic bands (51 mm x 1.5 mm) are wound twice around the dowelling and arranged along the length at suitable intervals (say every 20 mm - 40 mm) according to the size of the beetle that is being handled. Fifteen to twenty beetles can be mounted on each dowelling rod by inserting the head under one turn of the elastic band which then restrains the animal by light pressure between the pronotum and elytra. The beetles may then be quickly marked and sexed. This technique can also be used to allow paint marks to dry undisturbed. The beetles seem to be unperturbed by this gentle and effective method of restraint.

Several prepared dowels can be used in the field. With an assistant one can be loaded with captures from dry pitfall traps while those on another are being marked. In this manner, several hundred beetles can be marked in a day. Recapture rates of over 50% from 298 releases of *Pterostichus melanarius* (Ill.) have been achieved over a three week period in an unenclosed experiment (Thomas, unpublished) in a cereal field covering an area of approximately 2400 m^2.

Precautions

Safety goggles are strongly recommended when using the abrasive disc drill attachment, particularly when marking species with pubescent elytra (e.g. *Harpalus rufipes* (Degeer)) as some hairs are shed and blown into the face.

The abrasive discs are quite brittle and can break if handled roughly. It is advisable to take a supply of spares into the field.

References

Greenslade, P.J.M. 1964. The distribution, dispersal and size of a population of *Nebria brevicollis* (F.), with comparative studies on three other Carabidae. *J. Anim. Ecol.* 33: 311-333.

Lys, J-A. & Nentwig, W. 1991. Surface activity of carabid beetles inhabiting cereal fields. Seasonal phenology and the influence of farming operations on five abundant species. *Pedobiologia* 35: 129-138.

Murdoch, W.W. 1963. A method for marking Carabidae (Col.) *Ent. Mon. Mag.* 99: 22-24.

Sheppard, P.M., Macdonald, W.W., Tonn, R.J. & Grab, B. 1969. The dynamics of an adult population of *Aedes aegypti* in relation to dengue haemorrhagic fever in Bankok. *J. Anim. Ecol.* 38: 661-702.

Southwood, T.R.E. 1978. *Ecological Methods* (2nd Edition). Chapman & Hall, London.

Validation of the use of pitfall traps to study carabid populations in cereal field headlands

Amanda Hawthorne

School of Environmental Sciences, University of East Anglia, Norwich, NR4 7TJ, England
Present address: Waltham Centre for Pet Nutrition, Waltham on the Wolds, Melton Mowbray, Leicestershire, LE14 4RT, England

Abstract
Carabid populations from three cereal headland treatments were compared using pitfall traps and absolute density estimates obtained by mark-release-recapture and trapping out of enclosures.

The capture efficiency of individual species in pitfall traps was evaluated in the laboratory. Efficiency varied from 25% to 60% and a significant difference was found between species.

Cereal field headland treatments differed greatly in vegetation density, species composition and structural complexity. The activity of selected species was therefore compared in each of the three treatments, but did not differ significantly.

Pitfall trap data were only found to reflect density estimates by mark-release-recapture for *Pterostichus melanarius*. This suggested that pitfall traps reflected changes in activity not abundance.

Population density estimates made from population enclosures within barriered plots over the whole field season were only weakly correlated with pitfall trap catches. However such seasonal pitfall trap data gave an indication of headland treatment preference by carabids.

The validity of using pitfall traps to study carabid populations in cereal field headlands is discussed.

Key words: Carabidae, pitfall traps, population density

Introduction

Pitfall traps have been the principle method by which carabid populations have been studied, despite the fact that catches are a function of the population size, activity and trapability of a species. Factors affecting any of these are liable to

Arthropod natural enemies in arable land* · *I *Density, spatial heterogeneity and dispersal*
S. Toft & W. Riedel (eds.). *Acta Jutlandica* vol. 70:2 1995, pp. 61-75.
 ISBN 87 7288 492

influence pitfall catches (Southwood 1978, Topping & Sunderland 1992). Therefore in order to interpret pitfall trap data it is important to evaluate all such factors.

To minimise the influence of these limitations, some factors, such as trap type, can be standardised within a study. But to relate this information to published studies, the capture efficiency of traps for individual species, under standardised conditions should be evaluated (Halsall & Wratten 1988). Indeed if pitfall catches are to be used to estimate relative abundance of species, it is important that all species are caught with similar efficiency by a trap (Topping & Sunderland 1992).

The main factors which cannot be controlled are differences between habitats, e.g. vegetation density and species composition and structural composition which can vary in density and structure, both between habitats and within a season. This has been shown to affect activity of beetles and their susceptibility to being caught in pitfall traps (Greenslade 1965, Honek 1988).

By studying the activity of species in habitats with different vegetation structure, and comparing pitfall trap catches with estimates of population density in these habitats, the validity of using pitfall traps for inter-habitat comparisons can be evaluated.

In this paper I present part of a comparative study of carabid populations from three cereal headland treatments. Both Conservation Headlands (CH) and Uncropped Wildlife Strips (UH) have been included in the management prescriptions for Breckland Environmentally Sensitive Area, in the eastern counties of England. Both headlands were situated in the outer 6 m of cereal fields adjacent and parallel to the field boundary. The Conservation Headlands are cropped as usual and receive only selected pesticide applications (Sotherton, Boatman & Rands 1989), whilst the Uncropped Wildlife Strips remain uncropped, receive limited pesticide applications and must not be limed, fertilised nor irrigated, thus allowing resident weed populations to flourish (MAFF 1993). Carabid beetles in these two treatments were compared with those in Sprayed Headlands (SH) which received the full complement of agrochemical applications.

Study Sites

Two replicates of each of the three headland treatments, 120 m x 6 m wide, were arranged in a randomised block design along one side of an 18.61 ha winter wheat field. The farm was situated on the southern edge of Breckland ESA at Ixworth Thorpe, Suffolk (Grid reference TL 931 739).

Carabid beetles were surveyed from 1990 to 1992, using pitfall traps.

These consisted of a plastic cup, 7 cm high and 6 cm diameter at the brim, filled to approximately 3 cm with one part ethylene glycol, two parts water and a drop of detergent. They were set in plastic sleeves in the ground to aid removal and so that the rim of the cup was level with the ground and covered with metal squares, to act as deterrents to birds and prevent dilution by rainwater, 15 x 15 cm at a height of approximately 3 cm.

Methods

Laboratory study of the capture efficiency of study pitfall traps for individual species

The aim of this experiment was to establish how efficiently pitfall traps caught several different species under controlled laboratory conditions.

Laboratory stocks of carabids were captured from the edge of a linseed field in standard, dry covered pitfall traps, or by hand from the surface of the soil. They were maintained in clear plastic containers (0.3 m x 0.2 m x 0.2 m) filled with moist soil, and kept at 18°C under 12h L / 12h D light regime and fed on complete dry dogfood.

The experimental arena was a soil-filled plastic tank (0.3 m x 0.2 m x 0.2 m high), five standard pitfall traps were placed in the tank to ensure maximum contact with traps (Fig. 1). Beetles, starved for 24 hours, were released individually into the arena and their behaviour upon first contacting the edge of a trap recorded as either "capture" or "avoidance". Fifty replicate specimens were used for *Bembidion lampros* (Herbst) and *Agonum dorsale* (Pontoppidan) but only 12 and 15 for *Harpalus affinis* (Schrank) and *Pterostichus melanarius* (Illiger) respectively. The experiment was conducted in daylight for diurnal species and under red light conditions in a dark room for nocturnally-active

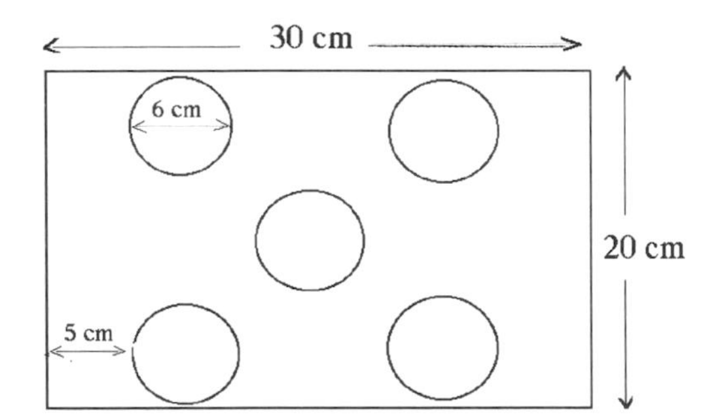

Fig. 1. Diagram of the arrangement of pitfall traps within a laboratory experimental arena, to determine the efficiency with which they capture carabid beetles.

species, which had previously had their circadian rhythm shifted by 12 hours over a two week period. All experiments were conducted within three weeks of capture.

The efficiency of a pitfall trap depends not only upon its ability to catch an individual, but also on its ability to retain it. This is particularly important when dry traps are used to estimate population density by mark-release-recapture of species such as *B. lampros*, which are dimorphic with respect to their ability to fly.

To investigate retention times similar arenas were used but with dampened circles of filter paper in the base of each pitfall trap to prevent beetles from dehydrating. Five individuals of small species or three of medium to large species were placed in the base of each trap. Individuals in different traps were marked with a different colour of typewriter correction fluid. This was applied to the elytra so that the original location of each individual was immediately obvious. Only one species was studied at a time, numbers remaining in each pot were recorded every 24 hours for three days. The experiment was repeated with the sides of traps dampened with a fine mist sprayer, to simulate dew formation.

Investigation into the activity of study species in the headland treatments

Activity was estimated in each headland treatment and in bare ground plots, which had been created by removing surface vegetation from four 6 m x 6 m randomly arranged blocks in one of the Uncropped Wildlife Strip treatments. Paraquat had been applied at field strength (0.5 g/litre) using a knapsack sprayer, six weeks prior to the start of the experiment, then any further emergent vegetation was removed by hand as seedlings.

Three replicate polythene barrels (0.37 m diameter x 0.43 m high, buried into the ground to 0.13 m) were randomly placed 3 m into each of the headland treatments and the bare ground plots on 11 July 1992. A single covered pitfall trap was established in the centre of each barrel and partially filled with ethyl glycol solution (Fig. 2). Barrels had been trapped out for two weeks prior to the introduction of a known, and marked population, to standardise conditions in all barrels. On 11 July 45 *H. affinis* were released into each barrel and on 21 July, 20 *Amara similata* (Gyllenhal), 20 *Amara aenea* (De Geer) and 17 *Amara plebeja* (Gyllenhal) were released into two randomly selected barrels from each treatment. The catch was collected daily from 12 July until 18 July and then every two days until 28 July. Vegetation cover was recorded for each barrel on 16 July.

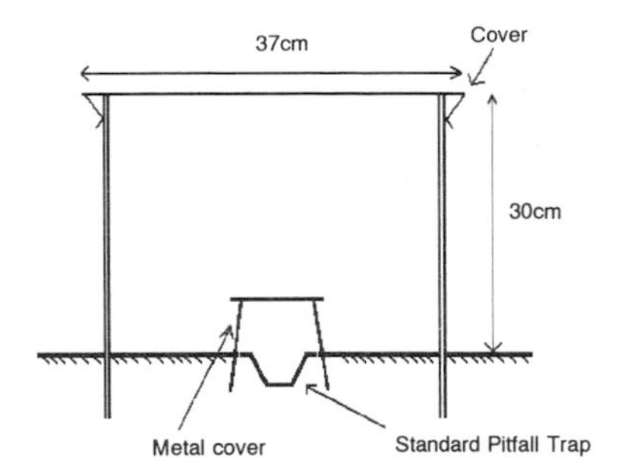

Fig. 2. Diagram of the experimental polythene barrel used to determine the activity of carabids in each headland treatment in the field.

A large number of individuals was found to be necessary in such a small arena to ensure recapture of individuals over a short time period and to limit any change in activity of individuals due to maturation. However this had the limitation of creating artificially high densities which may also have had a bearing on behaviour, due to constant interspecific interaction.

Comparison of pitfall catches with absolute density estimates

Pitfall trap catches were compared with absolute density measurements estimated on four occasions in 1991 by mark-release-recapture and over three months by trapping out population enclosures in 1992. Data were not used for species considered to fly, since population densities of these species would have been underestimated by mark-release-recapture because they could escape from traps.

Mark-release-recapture

Three mark-release-recapture grids of 5 m x 5 m were randomly established in each of the first replicates of each treatment, between 1 m and 5 m from the edge of the crop. Standard covered pitfall traps were placed at 0.5 m intersections within the grid. A mark-release-recapture programme was conducted at approximately four weekly intervals from 13 May until 2 August 1991. Between programmes traps remained in situ but were filled with soil. Whilst actively trapping, a piece of crumpled paper was placed in the bottom of each cup as a shelter for captured animals. Traps were checked at two day intervals, and captured individuals marked with a small date coded spot of coloured typewriter correction fluid on the elytra. Marked beetles were released between the capturing trap and its nearest neighbour, facing the field boundary. Beetles were marked for three consecutive dates and the population density estimated using the Jolly method (Jolly 1965). Bailey's correction factor was used (Bailey 1951) when the number of recaptures was low (less than 10). Population density estimates were then compared with numbers captured in pitfall traps over the same period.

Population enclosures

In 1992 4 m^2 enclosures were established from 2 m to 4 m from the field boundary at three locations in each headland treatment. These consisted of 2 m lengths of plywood, 0.4 cm thick and 25 cm high, buried to leave approximately 15 cm above ground. Corners were nailed to prevent the entrance or exit of carabids. From 6 May beetles were trapped out in nine ethyl-glycol filled pitfall traps. These were established with one in the centre of each quadrant, one at the centre of and adjacent to each barrier and one in the middle of the enclosure. All traps were emptied weekly until 27 July.

Results

Laboratory study of the capture efficiency of study traps for individual species

Observations on the types of behaviour elicited by an individual encountering a pitfall trap revealed two fundamentally different types of movement. Individuals engaged in fast unidirectional movements were most likely to fall into a trap. Those engaged in slower, more random movement, usually stopped and "investigated" the trap. This was achieved by suspending itself over the edge of the trap, whilst maintaining contact with the top of the trap with its hind tarsi. In addition the smaller species such as *B. lampros* were able to walk down inside the trap. Both types of "investigation" behaviour could result in capture or in escape, however species differed in their ability to escape. Table 1 indicates the percentage of successful encounters with a trap made by four common species. *H. affinis* was particularly adept at avoiding capture, whilst *A. dorsale* was more frequently captured whilst investigating a trap. A signifi-

Table 1. Results for the laboratory experiment to evaluate capture efficiency for first contact with a pitfall trap for individual species.

Species	No. individuals tested	No. captured	Percentage captured
Bembidion lampros	50	32	64
Agonum dorsale	50	26	52
Pterosticus melanarius	15	6	40
Harpalus affinis	12	3	25

Table 2. Accumulated mean catch after seven days following the release of beetles for each treatment compared with expected catch, calculated as the average from each habitat. Actual and expected values were compared using the Kolmogorov-Smirnov Test, significance levels are displayed.

	Harpalus affinis	*Amara similata*	*Amara aenea*	*Amara plebeja*	Total
No. released per barrel	45	20	20	17	
Uncropped Wildlife Strip	25.5	8.5	10.5	5.0	49.5
Conservation Headland	30.67	10.5	16.5	12.0	69.67
Sprayed Headland	33.67	16.0	12.0	5.0	66.67
Bare Ground	38.33	15.0	18.0	13.0	84.33
Total	128.14	50.0	57.0	35.0	270.17
Expected	32.035	12.5	14.25	8.75	
p-value	<0.05	<0.05	>0.05	<0.05	

cant difference ($\chi^2_{(3)}=7.227$, $P < 0.05$) was found in the efficiency of capture between species using a heterogeneity χ^2 test.

Only one individual of those placed within traps was observed to escape, that of *H. affinis* from a dry trap. No individuals were found to escape from traps with dampened sides, although *B. lampros* was observed to climb part way. Population estimates by mark-release-recapture of these species would not appear to be greatly affected by their ability to escape from a trap.

Investigation of activity of individuals in the headland treatments

Analysis was completed on the accumulated catch of each species in each barrel, seven days after beetles were released (Table 2). The Uncropped Wildlife Strips had the lowest cumulative catches of *H. affinis*, *A. similata* and *A. plebeja*, whilst catches were highest in the bare ground plots for *H. affinis*, *A. aenea* and *A. plebeja*. For each of these three species over half the number of beetles released were captured within 24 hours in the bare ground plot, whilst it took at least four days in the remaining treatments to capture the same quantity. A significant difference ($p < 0.05$) in the cumulative catch after seven days was recorded between the four treatments for *H. affinis*, *A. similata* and *A. plebeja* using a Kolmogorov-Smirnov test. However this test did not distinguish which treatments were different from each other and data sets were too small for further analysis of individual species, therefore data were pooled for all species. χ^2 tests were then used to compare pairs of treatments. The only significant difference found was that recapture rate was higher in the bare ground plots (BG) than in the Uncropped Wildlife Strip $\underline{BG>CH>SH}>UH$ ($\chi^2 = 5.86$ $P < 0.05$).

Table 3. Mean monthly values (with standard error) of the percentage of bare ground and total percentage cover of all dicotyledonous species in three headland treatments in 1991. Cover was estimated from 10 replicate quadrats (0.25m x 0.25m each), placed randomly in each treatment.

	Conservation Headland	Uncropped Headland	Sprayed Headland
% cover of bare ground			
April	20.1 ± 4.15	85.5 ± 1.7	28.5 ± 2.70
May	5.5 ± 2.41	39.5 ± 2.63	17.3 ± 3.34
June	12.0 ± 5.01	21.5 ± 2.70	28.0 ± 5.59
July	10.5 ± 2.93	5.5 ± 2.41	16.0 ± 2.21
% cover dicotyledonous vegetation			
April	34.8 ± 3.51	15.4 ± 2.66	18.3 ± 2.74
May	63.6 ±11.51	68.0 ± 8.73	5.9 ± 1.92
June	61.8 ± 6.99	74.1 ± 5.00	2.7 ± 1.04
July	13.9 ± 3.81	105.6 ± 9.23	2.6 ± 0.82

The percentage cover of vegetation in all three headland treatments was at a maximum in July, with the most complex habitat of all found in the Uncropped Wildlife Strip at this time. However in Spring the Uncropped Wildlife Strip was the least complex, consisting largely of bare ground, in the absence of either crop or weeds (Table 3). Despite these differences there were no significant difference in activity of carabids between the three headland treatments, nor between the Conservation Headland, Sprayed Headland and the bare ground in July. It was therefore unlikely that there would be a significant difference in activity of carabids between the three treatments at any time during the field season, thus allowing direct comparisons to be made.

Comparison of pitfall catches with absolute density estimates

Mark-release-recapture

Only a single species *P. melanarius* was captured in sufficiently high numbers to make a population density estimate on a number of occasions using mark-release-recapture. Pitfall trap catches were significantly ($r_{(8)} = 0.804$, $p < 0.05$) correlated with population density for *P. melanarius* (Fig. 3). No conclusions can be drawn for other individual species and no relationship was found when these species were pooled (Fig. 3). This is possibly because if there had been a relationship between the two methods for individual species, it could have been different from that of *P. melanarius*.

Population enclosures

Density estimates could only be made for species in those habitats where accumulated catches had levelled off in the weeks prior to the end of the trap-

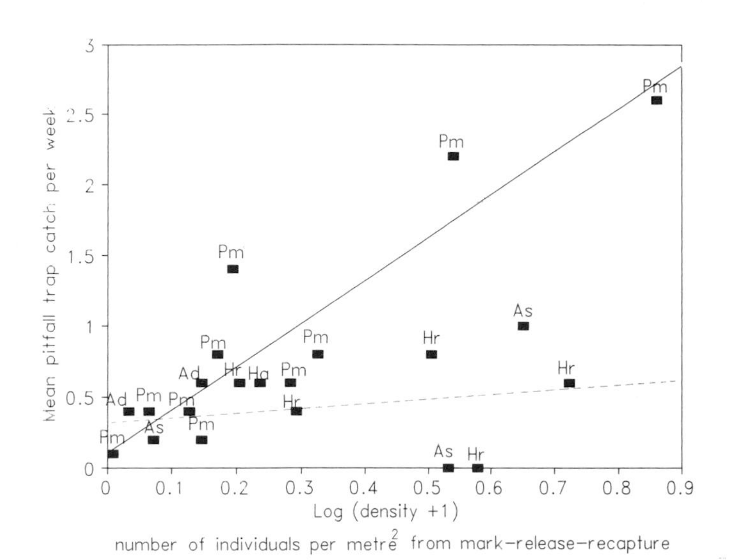

Fig. 3. The relationship between mean pitfall catch and population density estimated by mark-release-recapture for *P. melanarius* (Pm), *H. affinis* (Ha), *H. rufipes* (Hr), *A. dorsale* (Ad) and *A. similata* (As) for those weeks when a population density could be estimated in 1991. —— is the best line fitted for *P. melanarius* alone and ---- for the remainder of the species.

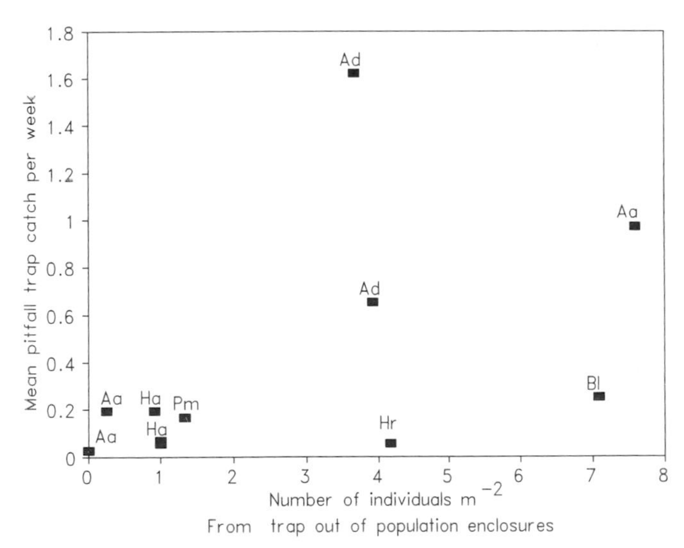

Fig. 4. The relationship between pitfall catch (as a mean of the weekly catch from 1 May to 27 July 1992) and population density estimated by trapping out 4 m^2 enclosed areas for *P. melanarius* (Pm), *H. affinis* (Ha), *H. rufipes* (Hr), *A. dorsale* (Ad), *B. lampros* (Bl) and *A. aenea* (Aa).

Table 4. Mean accumulated catch of study species in population enclosures between 6 May and 20 July 1992, compared with mean pitfall catch over the same period. *denotes a sample which was excluded from correlation analysis because individuals were still being captured from population enclosures at the end of the experiment. CH: Conservation Headland, UH: Uncropped Headland, SH: Spreyed Headland.

	Mean catch m^{-2}			Mean pitfall catch		
Treatments	CH	UH	SH	CH	UH	SH
Pterosticus melanarius	7.83*	7.75*	1.33	1.10	0.88	0.17
Harpalus rufipes	1.0	12.42*	4.17	0.05	0.33	0.06
Bembidion lampros	15.58*	18.08*	7.08	0.22	0.41	0.25
Agonum dorsale	3.67	3.92	5.83*	1.62	0.65	1.72
Harpalus affinis	1.0	28.5*	0.92	0.07	1.43	0.20
Amara aenea	0	7.58	0.25	0.03	0.97	0.19

ping period. Unfortunately this excluded data from those habitats which had the highest densities and the highest pitfall catches, because individuals continued to be captured from the population enclosures (Table 4). The "catch out" results were compared with those from pitfall catches, which were calculated as the mean catch week^{-1} for all replicates per week in a habitat, during the period between 6 May and 20 July (Fig. 4).

There were no significant correlations between these two sets of results, largely due to high pitfall catches of *A. dorsale* in the Conservation Headland. These could have resulted from higher activity of this species in this treatment than in other headland treatments. It was recognised that this method had its limitations because trapping out of population enclosures took many weeks. The original populations enclosed in May would not necessarily be comparable with populations outside barriered areas. However later in the season, data used in the analyses were for those species caught in relatively low densities and thus were trapped out quickly. Populations of these species within and outside the population enclosures should therefore not have differed greatly.

The distribution of carabid beetles between headland treatments

From pitfall catches over the whole of 1991 and of 1992 , *Harpalus rufipes* (De Geer) and *B. lampros* were significantly ($P < 0.0001$ and $P = 0.024$) and more frequently caught in the Uncropped Wildlife Strip. *P. melanarius* was more abundant in the Conservation Headland and Uncropped Wildlife Strip and *A. dorsale* was always more frequently caught in the Conservation and Sprayed Headland than the Uncropped Wildlife Strip.

High estimates of population density from mark-release-recapture reflected high pitfall catches for *B. lampros* and *H. affinis* and low densities for *A. dorsale* in the Uncropped Wildlife Strip.

Seasonal averages of pitfall traps and results from population enclosures and mark-release-recapture suggested that the Uncropped Wildlife Strip was the most important habitat for the majority of study species. It would therefore appear that where pitfall data were used over a number of months, they could reflect the relative abundance of species more accurately than individual weekly samples, possibly because species differ in activity over the season.

Discussion

Carabid beetles have been identified as potentially important predators of cereal aphids (Sunderland & Vickerman 1980). Their ability to influence populations of these pests depends upon their abundance and also on their activity in the vicinity of the pest, i.e. their "effective abundance" (Luff 1990).

In the Breckland ESA carabid populations in the edges of cereal fields may be increased through the use of Uncropped Wildlife Strips and Conservation Headlands. However these are very different habitats, compared to the Sprayed Headland and therefore in order to evaluate their importance any method used must be able to compare carabid populations effectively in each of these three headland treatments. Pitfall traps were selected as a simple means of sampling the carabid populations, but required a full evaluation of their effectiveness in order to interpret the results.

All beetles were not caught equally in pitfall traps. Species differ in the susceptibility with which they are captured, this can vary with trap material or the sex of the individual (Luff 1975). Efficiency in this study varied between 25% and 60%, which is comparable with results obtained by Luff (1975), but higher than those obtained by Halsall & Wratten (1988) when the maximum capture rate was 10% over a 12 hour period. The results presented here were for the response of an individual upon first contacting a trap, whereas those of Halsall & Wratten (1988) were based upon the number of encounters with a trap during a 12 hour period. Thus individuals which moved quickly and fell straight into the trap without investigation, would have been removed from the experiment, leaving individuals which investigate and escape, to contact the traps repeatedly. This may have given rise to low estimates of capture efficiency. Such differences in capture efficiency between species suggests that while catches of individual species can be compared between headland treatments, pitfall trapping is not a suitable method for comparing the relative abundance of different species within a habitat.

Differences between headland treatments can affect activity in a number of ways, but most can be attributed to differences in weed density (Honek 1988,

Powell et al. 1985). Weedy conditions support abundant and diverse invertebrates (Murdoch et al. 1972), which can be potential prey for carabids. In such habitats, for example the Uncropped Wildlife Strips in July, it is expected that there will be a large number of satiated individuals, which are less active than hungry beetles (Chiverton & Sotherton 1991). Activity can also be physically reduced due to the presence of plant stems causing individuals to reduce speed and turn more. Alternatively in the absence of vegetation, the soil surface can be 5 - 7°C warmer than in the presence of a crop (Honek 1988). Temperature is known to affect activity (Luff 1975, Baars 1979), therefore in habitats with high vegetation density, carabids will not be exposed to extremes of temperature and therefore activity might be expected to be lower than in habitats with less weed cover at ground level. Any or all of these differences between vegetated and bare ground can affect the number of carabids caught in pitfall traps. Therefore to test for such differences, density was held constant in experimental arenas and activity of species in different habitats compared. *H. affinis*, *A. similata, A. aenea* and *A. plebeja* activity were not found to differ significantly between headland treatments, although those in the Uncropped Wildlife Strips were the least active. This lack of difference in activity between headland treatments suggested that pitfall traps were a suitable means of comparing populations of *H. affinis*, *A. similata, A. aenea* and *A. plebeja* in these headland treatments.

The extent to which activity may be affected by the differences in habitat may change with the season. In April the Uncropped Wildlife Strips had 80% bare ground, but by July they had the most diverse and abundant vegetation cover (Table 3). A difference in activity of species in these two habitats was therefore expected. This was illustrated by the higher activity of *H. affinis*, *A. similata, A. aenea* and *A. plebeja* in arenas on bare ground than in arenas in the Uncropped Wildlife Strip in July. Whilst such differences in activity make comparisons of a species' activity in a headland treatment through a season difficult, it only applied to a few species such as *B. lampros* and *H. affinis* which were active throughout this period.

Activity also varies according to the life stage of an individual, where oviposition and mate location are associated with increased activity. These periods occur at specific times for each species and therefore give another reason for not using pitfall traps to give an index of relative abundance of species. In addition weather is important in causing temporal changes in activity of species. Ericson (1978) recorded activity of the normally nocturnal *P. melanarius* during the day when it was rainy and cloudy. Such extended periods of activity would increase their chance of capture. In this study headland treatments of 120 m x 6 m wide were arranged in a randomised block design along a single field headland, where it is possible that animals can move between treatments and between treatments and the adjacent habitats. It is therefore likely that temporal peaks of activity due to weather or life cycle will affect all treatments simultaneously allowing catches of individual species between treatments to be compared.

Temporal changes in activity make the use of pitfall traps unsuitable as a means of following changes in population density through time (Topping & Sunderland 1992), since changes in activity can mask changes in density. Where results from pitfall traps and mark-release-recapture were compared over a few days at different times in the season, few relationships were found between the two sets of data for most species. Indeed the mean catch week^{-1} of *B. lampros* on 17 May and 3 July were similar at 3.3 and 2.8 trap^{-1} week^{-1} in the Uncropped Wildlife Strips, but density estimates were 100 times higher in May than July. This suggests that changes in activity were obscuring changes in density. Baars (1979) considered pitfall trap data to only reflect density when data were used over the whole activity period of the animals, thus incorporating all alterations in activity. Results of the comparison of the density estimates from population enclosures and pitfall traps suggested that pitfall catches only weakly reflected density. However those habitats which had the highest density estimates, coincided with the habitats with the highest pitfall trap catches, therefore whilst it was unrealistic to gain an index of carabid density from pitfall traps in these headland treatments, it would appear that pitfall traps could identify crude differences in abundance between habitats. In this study the Uncropped Wildlife Strip was identified as the preferred habitat for most species by pitfall trap, mark-release-recapture and population enclosure studies.

In conclusion pitfall traps should not be used to estimate the relative abundance of different species in a habitat because species differ in the efficiency with which they are captured. Even where species do not differ in the efficiency with which they are captured pitfall traps should not be used to identify changes in relative abundance of species throughout a season, because such data are more likely to reflect changes in activity, than density.

Pitfall traps are useful for comparing the importance of headland treatments for individual species, when there are no significant differences in activity of species in the different habitat. Where habitats vary greatly in the complexity of vegetation throughout a season, care must be taken when comparing pitfall catches from the different times during a season in case the changes in complexity also influence activity. Where data are available over a whole season, some general comparisons between habitats can be made.

Acknowledgements

This work was funded by a CASE studentship between the Ministry of Agriculture, Fisheries and Food and The Game Conservancy Trust. I wish to thank J.J.H. Wilson for the provision of the study sites; P. Pask and the staff of Frederick Hiam Ltd at Ixworth Thorpe for their assistance, and Mark Hassall for the helpful comments made on this manuscript.

References

Baars, M.A. 1979. Catches in pitfall traps in relation to mean densities of carabid beetles. *Oecologia (Berl.)* 41: 25-46.

Bailey, N.T.J. 1951. On estimating the size of mobile populations from capture-recapture data. *Biometrika* 38: 293-306.

Chiverton, P.A. & Sotherton, N.W. 1991. The effects on beneficial arthropods of the exclusion of herbicides from cereal crop edges. *J. Appl. Ecol.* 28: 1027-1039.

Ericson, D. 1978. Distribution, activity and density of some Carabidae (Coleoptera) in winter wheat fields. *Pedobiolgia* 18: 202-217.

Greenslade, P.J.M. 1965. On the ecology of some British Carabid Beetles with special reference to life histories. *Transactions of the society for British Entomology* 16: 151-179.

Halsall, N.B. & Wratten, S.D. 1988. The efficiency of pitfall trapping for polyphagous predatory Carabidae. *Ecol. Entomol.* 13: 293-299.

Honek, A. 1988. The effect of crop density and microclimate on pitfall trap catches of Carabidae, Staphylinidae (Coleoptera), and Lycosidae (Aranae) in cereal fields. *Pedobiologia* 32: 233-242.

Jolly, G.M. 1965. Explicit estimates from capture-recapture data with both death and immigration - stochastic model. *Biometrika* 52: 225-247.

Luff, M.L. 1975. Some features influencing the efficiency of pitfall traps. *Oecologia (Berl.)*19: 345-357.

Luff, M.L. 1990. Spatial and temporal stability of carabid communities in a grass/arable mosaic. In: Stork, N.E. (ed.) *The role of ground beetles in ecological and environmental studies,* pp. 191-200. Intercept, Andover, Hampshire.

Ministry of Agriculture, Fisheries and Food 1993. Breckland ESA, Guidelines for farmers. Ref BD\ESA\2 Her Majesty's Stationery Office, London. 12 pp.

Murdoch, W.M, Evans, F.C, & Peterson, C.H. 1972. Diversity and pattern in plants and insects. *Ecology* 53: 819-828.

Powell, W., Dean, G.J, & Dewar, A. 1985. The influence of weeds on polyphagous arthropod predators in wheat. *Crop Protection* 4: 298: 312.

Sotherton, N.W., Boatman, N.D, & Rands, M.R.W. 1989. The 'Conservation Headland' experiment in cereal ecosystems. *The Entomologist* 108: 135-143.

Southwood, T.R.E, 1978. The components of diversity. In: Mound, L.A. & Waloff, N. (eds.) *Diversity of insect faunas.* Symposia of the Royal Entomological Society 9: 19-40. Blackwell.

Sunderland, K.D, & Vickerman, G.P. 1980. Aphid feeding by some polyphagous predators in relation to aphid density in cereal fields. *J. Appl. Ecol.* 17: 389-396.

Topping, C.J, & Sunderland, K.D. 1992. Limitations to the use of pitfall traps in ecological studies exemplified by a study of spiders in a field of winter wheat. *J. Appl. Ecol.* 29: 485-491.

A comparison of pitfall trap catches and absolute density estimates of carabid beetles in oilseed rape fields

B. Ulber & G. Wolf-Schwerin

Institute of Plant Pathology and Plant Protection, University of Göttingen, Grisebachstr. 6, D-37077 Göttingen, Germany.

Abstract
Carabid beetles were collected in oilseed rape fields by pitfall trapping within enclosures and by unfenced pitfall trapping from April until July in 1988 and 1989. An improvement of the enclosed pitfall trapping method was used for absolute density estimates. Enclosed areas of 1 m^2 each were subdivided by a grid of guiding walls. 36 small pitfall traps were set up in the corners of the subareas within each enclosure. The number of carabid beetles caught within 27 enclosures was compared with corresponding catches in 27 unfenced pitfall traps.

The composition of the carabid fauna assessed by the two sampling techniques differed considerably. A higher number of species was collected by unfenced pitfalls than by enclosed pitfalls. The catches of unfenced pitfalls were dominated by large species, whereas small species were predominant in enclosed pitfalls. Obvious differences between unfenced and enclosed pitfall catches were found between consecutive sampling dates and different species.

Key words: Carabids, pitfall traps, enclosed pitfalls, density, oilseed rape

Introduction

Pitfall trapping has been extensively used in many studies on carabid beetles and other soil surface predators. Although it is a general conclusion that the catches provide only data on the degree of activity rather than the actual population density (Greenslade 1964, Thiele 1977), the results obtained in short-term pitfall trap field studies very often are described in terms of abundance. Baars (1979) found that pitfall trap catches only gave reliable estimates of the population density of some species if sampling was done over the whole season. As catches are influenced by both abundance and activity, Heydemann (1953) introduced the concept of "activity density". The shortcomings of unfenced

***Arthropod natural enemies in arable land* · *I** *Density, spatial heterogeneity and dispersal*
S. Toft & W. Riedel (eds.). *Acta Jutlandica* vol. 70:2 1995, pp. 77-86.
 ISBN 87 7288 492

pitfall trapping have been reviewed in greater detail by Adis (1979), Halsall & Wratten (1988) and Topping & Sunderland (1992).

Absolute density estimates of carabid beetles have been assessed by ground searching (Greenslade 1964, Sunderland et al. 1987), quadrat sampling (Desender & Alderweireldt 1988), vacuum sampling (Sunderland et al. 1987), soil flooding (Brenøe 1987, Basedow et al. 1988), removal trapping (Scheller 1984) and mark and recapture methods (Ericson 1977, Luff 1982). The enclosed pitfall trapping technique used in this study for the estimation of population densities of carabid beetles represents an improvement of a technique, which had been applied primarily by Desender & Maelfait (1986), Sunderland et al. (1987) and Basedow & Rzehak (1988). The main aim of this investigation was to study the suitability of the modified method for density estimates of carabids in a two year field experiment and to collect initial results on the relationship between the absolute density assessed by this method and the number caught by commonly used pitfall trapping.

Methods

The investigation was carried out in 1988 (14 April to 29 July) and 1989 (11 April to 25 July) on 25 ha oilseed rape fields on clay soil in the region of Göttingen, Lower Saxony. The fields were sown on 22 August 1987 and 16 August 1988, respectively. Common farming practices were performed, but apart from seed treatment, no insecticides were applied. Nine plots of 96 m x 96 m each were arranged in a block design. Three unfenced pitfall traps were placed in a row in the centre of each plot, at a distance of 10 m from each other. Three enclosures were set up at regular intervals in a strip around these pitfall traps at a distance of 20 m from the traps and 20 - 30 m from the edge of the plot. After each sampling period of nearly four weeks, the enclosures were moved clockwise to a new position 10 m away from the former place.

The *enclosed pitfall traps* were set up within a 25 cm high, wood and metal square frame of 1 m^2, which was rapidly sunk 8 cm deep into the ground to prevent the escape of beetles. The vegetation within the frame was removed above ground level. The enclosed area was subdivided into 12 areas of different sizes by a system of metallic guiding plates : 4 x 0.18 m^2, 4 x 0.06 m^2 and 4 x 0.01 m^2 (Fig. 1). On the basis of catches within the different-sized subareas, the capture efficiency of the enclosures was supposed to be evaluated. 36 small pitfall traps (glass jars, 25 mm diameter x 50 mm deep) were sunk into the soil in almost every corner of the guiding system. The pitfalls were half-filled with a solution of 2% picric acid and a detergent. The dense grid of guiding plates

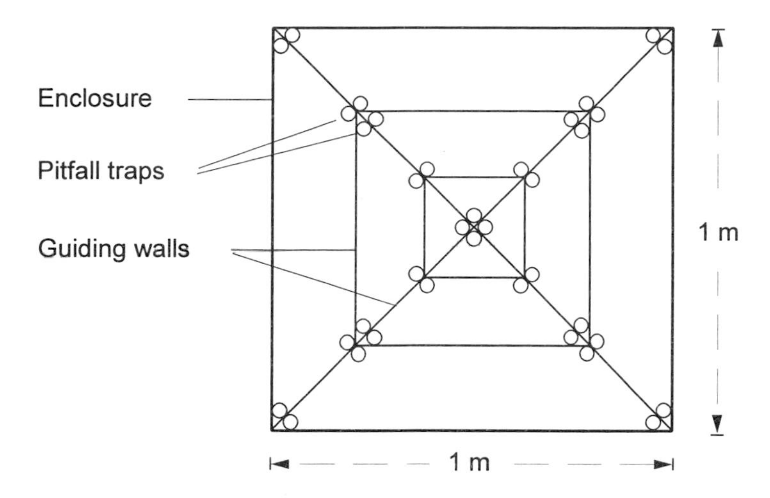

Fig. 1. Schematic outline of the enclosure with 36 pitfall traps and guiding walls

and the large number of pitfall traps within the enclosures was used to achieve a complete catch of all carabid beetles within the isolated areas. The top of the enclosure was sealed tightly with fine-mesh nylon material to prevent immigration and emigration of beetles during the sampling period.

The *unfenced pitfall traps* (plastic cups, 105 mm diameter x 88 mm deep) were half-filled with 2% picrid acid plus a detergent. A plain circular disc made of artificial sealing cement was fitted around the rim of the pitfalls to standardize the ground around the trap.

The unfenced pitfall traps were emptied fortnightly. The numbers of carabid beetles caught by these traps in two successive trapping periods were compared with the numbers of beetles caught by the enclosures in one simultaneous sampling period of nearly four weeks. For comparison of unfenced and enclosed pitfall trapping only unfenced pitfall catches just before and just following the setting up of the enclosures were considered.

Results

The composition of the adult carabid fauna caught in unfenced pitfall traps and in enclosed pitfall traps in 1988 and 1989 differed greatly (Table 1). Unfenced pitfall catches were dominated by larger carabid species, particularly *Pterostichus* spp., *Loricera pilicornis*, *Carabus* spp. and *Nebria brevicollis*, which formed up to 81% and 90% of the total catch in 1988 and 1989, respec-

Table 1. Species composition of carabid beetles in unfenced pitfall traps and in enclosed pitfall traps in 1988 and 1989 (percent frequency of each species of the total number of individuals caught by each sampling method).

	1988		1989	
	Unfenced pitfalls	Enclosed pitfalls	Unfenced pitfalls	Enclosed pitfalls
Pterostichus melanarius	46	5	80	10
Loricera pilicornis	21	1	-	-
Nebria brevicollis	-	-	7	2
Carabus cancellatus	7	< 1	-	-
Carabus granulatus	7	< 1	-	-
Pterostichus niger	-	-	3	< 1
Bembidion tetracolum	3	1	-	-
Amara spp.	-	-	2	< 1
Amara familiaris	2	11	1	5
Trechus quadristriatus	2	57	< 1	78
Agonum mülleri	2	< 1	-	-
Platynus dorsalis	1	< 1	-	-
Clivina fossor	1	21	-	-
Trechus secalis	< 1	1	-	-
Bembidion quadrimaculatum	-	-	< 1	4
Other species (< 1%)	8	3	7	4
Total	100	100	100	100
Total no. of individuals	4686	2019	3797	1687
No. of species	42	30	38	31

tively. In contrast the proportion of large species in enclosed pitfalls was very small. Small carabid species, i.e. *Trechus quadristriatus*, *Amara familiaris, Clivina fossor* and *Bembidion* spp., constituted 90% and 84% of the total catch of the enclosures in 1988 and 1989, respectively. In each year the number of carabid species caught in unfenced pitfalls was higher than the number of species caught in enclosed pitfalls.

The differences between the catches of *P. melanarius*, *T. quadristriatus*, *A. familiaris* and *C. fossor* in unfenced pitfall traps and enclosed pitfall traps in corresponding sampling periods are shown in Table 2. These species accounted for 94% and 93% of all carabids caught in the enclosures in 1988 and 1989, respectively. The highest numbers of *P. melanarius* in the enclosed pitfall traps were found on the third sampling date in each year (24 June 1988, 9 June 1989), indicating the main time of emergence of young adults from the soil. However, in unfenced pitfalls the maximum number of *P. melanarius* was caught about three weeks later on the fourth sampling date, when the reproductive activity and possibly the emigration of this species began. It is obvious that the time of hatching and maximum abundance will be misinterpreted using only unfenced pitfall traps.

Table 2. Mean number (± 95% confidence limits) of selected carabid species caught by enclosed pitfall traps (n = 27 enclosures) and by unfenced pitfall trapping (n = 27 pitfall traps) in 1988 and 1989.

Sampling period	*Pterostichus melanarius*		*Trechus quadristriatus*		*Amara familiaris*		*Clivina fossor*	
	Ind./m^2	Ind./9 traps/day	Ind./m^2	Ind./9 traps/day	Ind./m^2	Ind./9 traps/day	Ind./m^2	Ind./9 traps/day
1988								
19 April	0	0.09 ± 0.06	0	0.03 ± 0.04	0.20 ± 0.15	0.08 ± 0.11	6.44 ± 3.18	0.27 ± 0.16
1 June	1.17 ± 0.46	0.26 ± 0.15	13.37 ± 2.28	0.07 ± 0.06	1.00 ± 0.79	0.21 ± 0.18	2.87 ± 1.06	0.15 ± 0.13
24 June	2.06 ± 1.73	9.20 ± 4.86	18.67 ± 3.10	0.29 ± 0.17	4.47 ± 3.11	0.44 ± 0.35	6.71 ± 0.65	0.08 ± 0.11
19 July	0.79 ± 0.42	28.19 ± 7.16	27.70 ± 2.21	0.30 ± 0.14	6.12 ± 2.80	0.52 ± 0.48	3.42 ± 1.88	0.06 ± 0.08
1989								
26 April	0.03 ± 0.05	0.26 ± 0.10	0.03 ± 0.04	0	1.70 ± 1.07	0.20 ± 0.11		
17 May	0.20 ± 0.05	0.32 ± 0.28	0.87 ± 0.56	0	0.59 ± 0.54	0.21 ± 0.15		
9 June	2.60 ± 0.45	0.74 ± 0.70	8.53 ± 1.69	0.02 ± 0.04	0.15 ± 0.22	0.02 ± 0.04		
27 June	1.80 ± 0.58	41.40 ± 14.36	17.80 ± 0.98	0.22 ± 0.41	0.22 ± 0.31	0.06 ± 0.07		

The relationship between enclosed pitfall and unfenced pitfall catches of *T. quadristriatus* was completely different from that of *P. melanarius* (Table 2). Within the enclosures very high numbers of *T. quadristriatus* appeared in June and July whereas very small numbers of this species were caught during the same period in unfenced pitfalls.

The numbers of *A. familiaris* caught with time increased in 1988 and declined in 1989 in both the enclosed and the unfenced pitfall traps, indicating a positive correlation between the results obtained by both methods (Table 2). The peak density occurred in 1988 at the end of June and into July when tenerals hatched. In 1989 the number of beetles/m^2 decreased from April onwards. This decline possibly reflects the emigration of this phytophagous species from the field that might be initiated by a lower weed density, particularly of *Stellaria media*, which was estimated in 1988 at 26%, but only at 5% in 1989.

In 1988 *C. fossor* attained a relatively high abundance (Table 2), but in 1989 the mean density was only 0.1 ind./m^2. *C. fossor* was the only species caught which hibernates in the adult stage in the field (Desender 1983) and it attained a high density early in the season. Tenerals appeared from June onwards with a peak on 24 June. This species prefers to live below ground level. The higher humidity within the enclosures may have accelerated the epigaeic activity of this hygrophilic species and hence the catching rate. The relationship between the numbers caught in enclosed pitfalls and in unfenced pitfalls in different sampling periods was inconsistent. The higher catches in unfenced traps in April and June might be caused by reproductive activity of this spring breeder.

Discussion

In recent years enclosed pitfall trapping has been used for absolute density estimates of carabid beetles by several authors, but the size of the enclosed area, the number of pitfall traps and the period of sampling used in these studies varied considerably (Desender and Maelfait 1986, Sunderland et al. 1987, Desender & Alderweireldt 1988, Basedow & Rzehak 1988). Generally, only one to three pitfalls, connected with each other and with the edges of the enclosure by guiding plates, have been placed for one to four weeks in mesh-covered enclosures measuring between 0.1 m^2 and 1 m^2 .

In a preliminary experiment, enclosures of 0.5 m^2 with one pitfall trap (6.5 cm diameter x 9 cm deep) in each corner proved to be insufficient to trap out all carabids (unpublished data). Only 73% of the carabids were caught in the enclosed pitfalls over a period of two weeks while the remaining, mainly *Bembidion lampros* and *Notiophilus biguttatus*, were subsequently sampled by a soil flooding technique (Brenøe 1987). As a result, the enclosed area was subdivided into small plots by a dense system of guiding walls in order to improve the capture of species which have a very small range of movement. Carabid beetles are known to walk preferably along vertical edge structures when they reach a barrier (Adis 1976) and will be captured in the traps situated in the corners of the guiding system very easily. The small pitfall traps used within the enclosures proved to catch even larger species, i.e. *P. melanarius* and *Carabus* spp., efficiently, as there was a high coincidence of beetles and traps. However, the subdivision of the enclosure was not appropriate for assessing the capture efficiency within subareas of varying size, because the segregation produced by the guiding walls proved to be inadequate and the number of beetles caught per subarea was too low.

The sampling efficiency of this modified system of enclosed pitfall trapping has not yet been tested rigorously. Only one specimen of *T. quadristriatus* was detected within three enclosures by ground searching and soil flooding when these were carried out on one occasion subsequent to a four week sampling period. The capture efficiency of the enclosures is dependent to some extent on the behaviour and locomotory activity of the species involved, although this effect is minimized by the very short distances between traps and by the relatively long sampling period. The sampling period of nearly four weeks, however, will bias the estimation of actual population densities at one particular moment if tenerals emerge during that period.

In comparison with vacuum sampling and ground search methods, even carabid species which remain inactive and buried deeper in the ground (due to nocturnal activity or unfavourable microclimatic conditions) can be caught by the continuous sampling of enclosed pitfall trapping, if the beetles become active in the course of the sampling period (Luff 1978). This is important as most carabid species are predominantly nocturnal (Thiele & Weber 1968). In an arable field Greenslade (1964) sampled diurnal and nocturnal species using quadrat counts and recorded similar numbers, but in pitfall traps the catch of nocturnal species was significantly higher than that of diurnal.

Using vacuum sampling, ground search and then enclosed pitfall trapping on the same unit of area in cereals, Sunderland et al. (1987) caught high proportions of carabid beetles in enclosed pitfalls even if this was applied following the other techniques. The density of Carabidae in these cereal fields varied according to site and date between 2.0 and 53.0 ind./m^2, compared with 1.6 and 47.7 ind./m^2 in this investigation. Desender and Maelfait (1986) found in pasture plots that the relative abundance of some species, including *C. fossor*,

T. quadristriatus and *P. melanarius*, as estimated by enclosures, were of the same order of magnitude as the densities assessed by soil samples. On the other hand the mean population density of *B. lampros* was estimated by Scheller (1984) at 61 ind./m^2 by removal trapping, but only 0.3 and 1.3 ind./m^2 when sampled by quadrat counts and soil samples, respectively. In maize fields the maximum density of *C. fossor*, *T. quadristriatus* and *P. melanarius* was estimated by enclosed pitfall trapping at 15 ind./m^2, 7.5 ind./m^2 and 2.8 ind./m^2, respectively (Desender & Alderweireldt 1988). In the oilseed rape fields studied in this investigation, the numbers of these species are of the same order of magnitude (*C. fossor* 6.7 ind./m^2, T. *quadristriatus* 27.7 ind./m^2 and *P. melanarius* 2.6 ind./m^2).

When comparing the results obtained by both methods it is obvious that in unfenced pitfall catches the abundance of large mobile carabid species, e.g. *P. melanarius*, *L. pilicornis* and *Carabus* spp., will be overestimated; on the other hand the abundance of small and less soil surface-active species, e.g. *T. quadristriatus*, *C. fossor* and *Amara* spp., will often be underestimated. The population densities of many of the carabid species in these rape fields estimated by enclosed pitfall trapping, however, were too low for comparisons to be made with the numbers caught in unfenced pitfalls. In contrast to unfenced pitfalls some species were not collected at all in enclosed pitfalls although the total area sampled on each occasion was quite large. Thus unfenced pitfall trapping gave more reliable information on the qualitative species composition than did enclosed pitfall trapping, as the results of Table 1 show.

Compared with common pitfall trapping, the applicability of this modification of enclosed pitfall trapping might be limited to intensive studies on the absolute densities of carabid beetles as it is very labour-intensive. The setting of one complete enclosure at a new position took between two and three man-hours. This effort possibly could be reduced by decreasing the number of pitfalls per enclosure.

Much variation occurred between the catches of single enclosures as well as single unfenced pitfall traps at any sampling date, resulting in a high standard deviation of the means. Accordingly the results of this preliminary study are not adequate to make a clear decision whether the numbers of individuals caught in unfenced pitfall traps can be correlated with the actual population density of single species. Since the relationship between the results of each method varied much between species, consecutive sampling dates and years, it seems unlikely that reliable conclusions can be drawn from short-term pitfall trap catches in respect to the actual population density of carabid species.

Acknowledgements

The authors are grateful to Dr. W. Powell for correcting the English and Mrs. C. Trabert for technical assistance. This work was partly financially supported by the Federal Ministry of Research and Technology.

References

Adis, J. 1976. *Bodenfallenfänge in einem Buchenwald. Ökologie.* Arbeiten, Berichte, Mitteilungen. Ulm, Göttingen.

Adis, J. 1979. Problems of interpreting arthropod sampling with pitfall traps. *Zool. Anz. Jena* 202: 177-184.

Baars, M.A. 1979. Catches in pitfall traps in relation to mean densities of carabid beetles. *Oecologia (Berl.)* 41: 251-259.

Basedow, T. & Rzehak, H. 1988. Abundanz und Aktivitätsdichte epigäischer Raubarthropoden auf Ackerflächen - ein Vergleich. *Zool. Jb. Syst.* 115: 495-508.

Basedow, T., Klinger, K., Froese A. & Yanes, G. 1988. Aufschwemmung mit Wasser zur Schnellbestimmung der Abundanz epigäischer Raubarthropoden auf Äckern. *Pedobiologia* 32: 317-322.

Brenøe, J. 1987. Wet extraction - a method for estimating populations of *Bembidion lampros* (Herbst)(Col., Carabidae). *J. Appl. Ent.* 103: 124-127.

Desender, K. 1983. Ecological data on *Clivina fossor* (Coleoptera, Carabidae) from a pasture ecosystem. 1. Adult and larval abundance, seasonal and diurnal activity. *Pedobiologia* 25: 157-167.

Desender, K. & Maelfait, J.-P. 1986. Pitfall trapping within enclosures: a method for estimating the relationship between the abundances of coexisting carabid species (Coleoptera: Carabidae). *Holarct. Ecol.* 9: 245-250.

Desender, K. & Alderweireldt, M. 1988. Population dynamics of adult and larval carabid beetles in a maize field and its boundaries. *J. Appl. Ent.* 106: 13-19.

Ericson, D. 1977. Estimating population parameters of *Pterostichus cupreus* and *P. melanarius* (Carabidae) in arable fields by means of capture - recapture. *Oikos* 29: 407-417.

Greenslade, P.J.M. 1964. Pitfall trapping as a method for studying populations of Carabidae (Coleoptera). *J. Anim. Ecol.* 33: 301-310.

Halsall, N.B. & Wratten, S.D. 1988. The efficiency of pitfall trapping for polyphagous predatory Carabidae. *Ecol. Entomol.* 13: 293-299.

Heydemann, B. 1953. *Agrarökologische Problematik - dargetan an Untersuchungen über die Tierwelt der Bodenoberfläche der Kulturfelder.* Diss. Univ. Kiel.

Luff, M.L. 1982. Population dynamics of Carabidae. *Ann. Appl. Biol.* 101: 164-170.

Luff, M.L. 1985: Diel activity patterns of some field Carabidae. *Ecol. Ent.* 3: 53-62.

Scheller, H.V. 1984. The role of ground beetles (Carabidae) as predators on early populations of cereal aphids in spring barley. *J. Appl. Ent.* 97: 451-463.

Sunderland, K.D., Hawkes, C., Stevenson,J.H., McBride, T., Smart, L.E., Sopp, P.I., Powell, W., Chambers, R.J. & Carter, O.C.R. 1987. Accurate estimation of invertebrate density in cereals. *Bulletin SROP/WPRS* 1987/X/1: 71-81.

Thiele, H.U. 1977. *Carabid beetles in their environments.* Springer, Berlin.

Thiele, H.U. & Weber, F. 1968. Tagesrhythmen der Aktivität bei Carabiden. *Oecologia (Berl.)* 1: 315-355.

Topping, C.J. & Sunderland, K.D. 1992. Limitations to the use of pitfall traps in ecological studies exemplified by a study of spiders in a field of winter wheat. *J. Appl. Ecol.* 29: 485-491.

Estimating beneficial arthropod densities using emergence traps, pitfall traps and the flooding method in organic fields (Vienna, Austria)

B. Kromp, Ch. Pflügl, R. Hradetzky & J. Idinger

L. Boltzmann-Institute for Biological Agriculture and Applied Ecology
A-1110 Vienna, Rinnböckstraße 15, Austria

Abstract

From April 27 - July 12, 1993, the arthropod fauna of seven organic fields with different crops was investigated using three different sampling methods: emergence traps, the flooding method and pitfall traps.

In the emergence traps, total arthropod densities (excluding Collembola and Acari) differed widely in the different crops, ranging from 16,000 individuals/m^2 in lucerne (mostly aphids und aphidiid wasps), 6000 - 8000 in rye and 4000 - 5000 in wheat, to 2400 in root crops. Potentially beneficial arthropods (predators and parasitoids) accounted for approximately one third of the total number of arthropods. The lack of information on the actual contribution of beneficials to the suppression of pest populations is discussed.

In comparing the results of the three methods for carabids, small carabids (e.g. *Bembidion lampros, Trechus quadristriatus*) predominated in emergence traps and flooding-samples, whereas the larger-sized *Poecilus cupreus* was strongly predominant in pitfall trap samples, most likely due to its high mobility.

Key words: Beneficial arthropods, Carabidae, crop rotation, organic agriculture, sampling methods, emergence traps, flooding method, pitfall traps

Introduction

Beneficial arthropods are considered to be a major component in the self-regulation of pests in organic crops. Organic cultivation seems to enhance beneficial arthropod populations in arable fields, as has been proved so far for carabids (e.g. Kromp 1990) and lycosid spiders (Ingrisch et al. 1989). According to Booij & Noorlander (1992), however, the effects of crop structure and crop-related factors on carabids, staphylinids and spiders outweigh effects of the farming system applied. This was shown by means of pitfall trapping.

***Arthropod natural enemies in arable land · I** Density, spatial heterogeneity and dispersal*
S. Toft & W. Riedel (eds.). *Acta Jutlandica* vol. 70:2 1995, pp. 87-100.
 ISBN 87 7288 492

A broader spectrum of field arthropods can be investigated by using emergence traps (photoeclectors). Again, decreasing arthropod densities were found with increasing cultivation intensity (in terms of amounts of pesticides and mineral fertilizers applied), both in sugar beet (Büchs 1991) and winter wheat crops (Büchs 1993).

To date, no photoeclector studies have simultaneously compared different crops in a single cultivation system. Therefore, we selected this method to investigate the influence of crops and crop rotation on the arthropod fauna of organically cultivated arable fields, emphasizing the potentially beneficial predatory and parasitoid arthropods.

Another objective of this study was to compare different sampling methods with regard to their suitability in providing density-related data on epigaeic predators. Thus, pitfall traps and the soil flooding method supplemented the photoeclectors. Based on this methodological approach, this study provides preliminary results for carabids; the data on staphylinids and spiders will be presented in a future paper.

The general aim of this and related studies carried out by our working group (e.g. Kromp & Nitzlader 1995) is to contribute to a concept of optimal augmentation of naturally occurring beneficials in organic agriculture.

Materials and methods

Study sites

The sampling was done in seven organically cultivated fields at the riverside nature reserve "Obere Lobau", Vienna (48°10'N and 16°30'E; 152 m above sea level; annual average temperature 9.6°C, annual average total precipitation 510 mm).

Table 1 summarizes abbreviations of fields, crops, previous crops, acreage, cultivation specifics and sampling methods applied. The fields are between 4 and 10.3 ha in size and, with the exception of Lu (distance: 0.8 km), situated adjacent to each other. They are surrounded by deciduous riverside forest, forest strips, hedges and tracks. The soil type in all fields is similar (greyish alluvial soil); the differences within and between the fields are due to soil depth and particle size distribution, which ranges from silt to silty loam.

"Biotonne" compost (Amlinger 1993), originating from separately collected organic components of household garbage, was used for fertilization. No pesticides were applied, except wettable sulphur in wheat and rye against mildew. Soil cultivation (ploughing, seedbed combination) was basically the same in all fields, the only differences being due to crops and time of sowing.

Table 1. Specifics of the research fields (Obere Lobau, Vienna,1993).

Field	Crop	Previous crop	Acreage (ha)	Compost fertilization (autumn 92) (1)	Spraying with wettable sulfur (April-May 93) (2)	Sampling method ET	PT	FM
So	Soya	Rye	4	-	-	x	x	
Su	Sunflower	Potatoes	10.3	+	-	x	x	x
W-W	Wheat	Wheat	9	+	+	x	x	x
W-P	Wheat	Potatoes	7	+	+	x	x	
R-W	Rye	Wheat	6.8	+	+	x	x	x
R-R	Rye	Rye	7.5	-	+	x	x	x
Lu	Lucerne	Rye	4.3	-	-	x	x	x

ET: Emergence traps, PT: Pitfall traps, FM: Flooding method
(1) 30t/ha, (2) 5kg/ha

Weed management was done by curry comb and ring-roller in grainfields, cross harrowing/ridging in sunflower, and hoeing/curry comb in soybean.

Since the cultivation method, soil type, field size and surroundings are similar for all research fields, crop type and crop-related factors are considered to be the main differentiating parameters in this study. Nevertheless, a severe drought from early May to mid-June led to an increased heterogeneity in crop development in wheat fields and lucerne.

Sampling methods

In order to evaluate the emergence rates of arthropods from the fields, emergence traps (Funke 1971) were used: Square metallic frames of 0.25 m^2, with removable upper part covered by green raincoat material; a single pitfall trap was placed within each frame at ground level, containing 2% formaldehyde; upper trap with transparent cover, containing 1% formaldehyde.

Pitfall traps (transparent plastic beakers, diameter 85 mm, 2% formaldehyde; lids of acrylic glass 20 x 20 cm) were used to evaluate activity densities. Additionally, the flooding method (FM) as described by Basedow et al. (1988) was applied. Here, the metallic frame of a photoeclector was pushed into the ground quickly, and the vegetation inside the frame removed. After a first search, water was poured into the frame and all emerging arthropods (of collectable size) were collected by hand. This process was repeated twice (up to 10 litres per frame, depending on the water capacity of the soil), until no further arthropods appeared; finally, the wet soil inside the frame was trampled and searched again. The total time required to complete one flooding sample ranged between 15-35 min.

In each field, five pitfall traps (PT) and three emergence traps (ET) were placed, alternating in one line (distance between pitfall traps and emergence traps 7.5 m, between successive pitfall traps 15 m), parallel to seed rows. Each trapping line was set at a distance of at least 50 m from the nearest field margin. The sampling period of emergence traps was from 27 April to 12 July, 1993; traps were emptied fortnightly (6 sampling dates); furthermore, each trap was repositioned in the immediate vicinity monthly.

Pitfall traps were set from 19 April to 29 June; they were kept open every second fortnight (4 sampling dates).

On three occasions (late April, late May/early June, early July), 10 flooding samples were taken in each field (except for So and W-P, see Tab.1); they were placed at a distance of approximately 5 m from both sides of each pitfall trap.

Results and discussion

Arthropod emergence rates in different crops

Emergence trapping yielded a total of 32,387 individuals representing 108 taxa (mostly identified to family level; Collembola and Acari were not evaluated), listed in the Appendix. The most abundant group in all fields was Thysanoptera (nearly 13% of total catch), followed by Cecidomyidae (9.8%), beetle larvae (mostly carabids and staphylinids; 9%), Sciaridae (7.8%), Phoridae (5.4%), Chalcidoidea (5.3%) and Staphylinidae (4%). Other dominant groups included arachnids, proctotrupoid wasps (mostly Ceraphronidae and Scelionidae), cryptophagid and lathridiid beetles, chironomid midges and Cicadina. Carabid beetles represented only 1% of total catches.

The above-mentioned groups were more or less abundant in all fields. Aphids, aphidiid wasps and Psyllina were highly abundant only in the lucerne field, where total arthropod emergence rates exceeded 16,000 individuals/m^2 (Fig. 1). The second largest emergence rate was found in rye field R-R (8150 ind/m^2), followed by R-W (5900 ind/m^2), wheat field W-P (4700 ind/m^2) and W-W (3700 ind/m^2). Lowest emergence rates occurred in sunflower and soya fields (2400 and 2300 ind/m^2, respectively).

The grain fields were quite similar with regard to dominance distribution (Fig. 1). The exceptions were Nematocera and epigaeic predators, which showed highest emergence rates in the rye field R-R. In addition to differences in the lucerne field, differing arthropod coenoses were also found in the root crops: in soybean very low numbers of parasitic hymenopterans and coleopterans (other than carabids and staphylinids) occurred, while in sunflower the lowest emergence rates of nematocerans were recorded. Julidae were abundant only in rye fields, as was the case with Polydesmidae (Appendix 1).

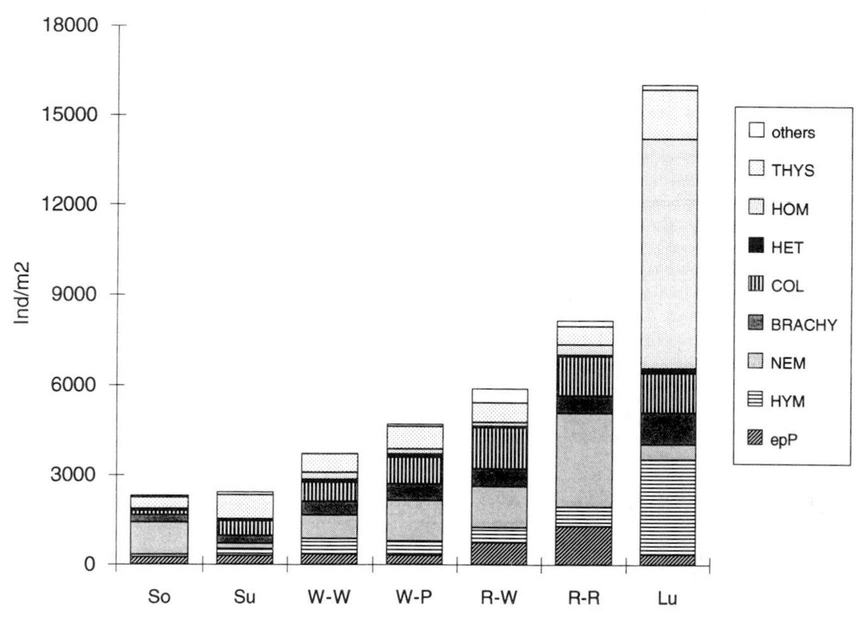

Fig. 1. Total emergence rates (m^{-2}) and dominances of arthropod groups (excluding Collembola and Acari) in 7 organic fields at Obere Lobau, Vienna; April 27 - July 12, 1993.
(For abbreviations of crops see Table 1; THYS = Thysanoptera, HOM = Homoptera, COL = Coleoptera other than carabids and staphylinids, epP = epigaeic predators, i.e. carabids, staphylinids and spiders, HET = Heteroptera, HYM = Hymenoptera, BRACHY = Brachycera, NEM = Nematocera).

Concerning the seasonal dynamics of arthropod emergence rates at Obere Lobau (Fig. 2), the fewest arthropods emerged in the root crops So and Su from late April to early June; this is possibly due to black fallow during the winter and still bare soil in spring. The highest emergence rates in spring occurred in rye field R-R, followed by R-W. Wheat fields showed medium catches in spring, increasing later in the season, as did catches in the root crops. Arthropod emergence in the lucerne field increased steeply during the season, especially in late June; this was mainly due to aphids, aphidiid wasps and Psyllina. Since aphidiids occurred in the lucerne field exclusively together with the highly abundant aphids, they probably migrated into the field from the surroundings; thus, they followed aphid infestation rather than being "ground-related parasitoids", as Jensen (1992) refers to parasitoids that stay in the field all year round.

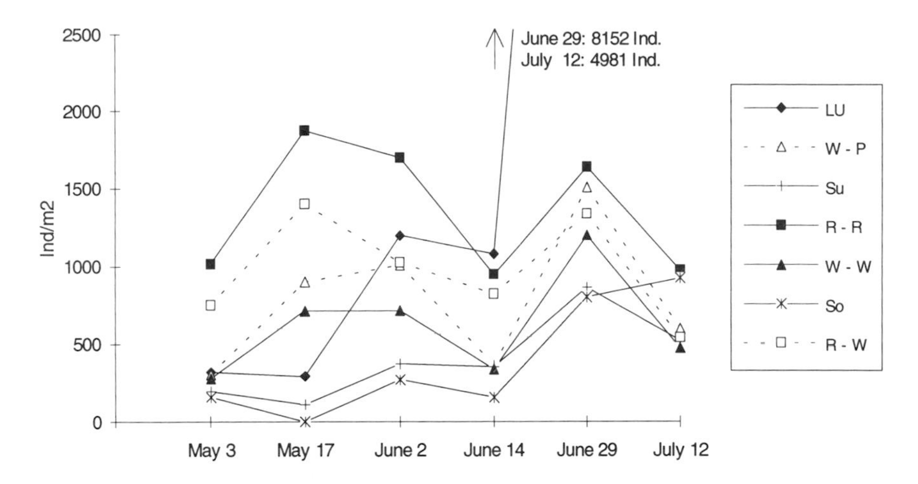

Fig. 2. Seasonal dynamics of arthropod emergence rates (excluding Collembola and Acari) in 7 organic fields (Obere Lobau, Vienna 1993; for abbreviations of crops see Table 1).

Beneficial arthropods

Fig. 3 depicts the proportions of parasitoids and predators in the total arthropod catches in six organic fields at Obere Lobau (the lucerne field is omitted here because of the above-mentioned deviating catches).

Parasitoids consisted mainly of chalcidoid and proctotrupoid wasps with a few braconids (totalling 9% of total arthropods); predators (a total of 23.9%) were dominated by beetle larvae (mainly carabids and staphylinids, 11.3%), staphylinid and carabid imagines (5.8 and 1.2%, respectively), arachnids (4.9%) and small numbers of lithobiids, predatory brachycerans and heteropterans (for further details, see Appendix 1). The above-mentioned groups, totalling one third of the field arthropod fauna collected by emergence traps, can be considered to be potential beneficials based on a rough classification of their biology.

Their actual agroecological significance in terms of pest control, however, is more difficult to assess, when their feeding biology is examined more closely. Among the parasitoids (according to Jacobs & Renner 1988), presumed pest antagonists are to be found among the chalcidoid wasps (parasitizing Lepidoptera, Diptera, Hymenoptera, Thysanoptera and Homoptera; some species, though, hyperparasitize other ichneumonids and, e.g. tachinid flies), the proctotrupoid families Scelionidae (egg-parasites on different arthropod groups like Lepidoptera, Coccina and spiders) and Platygasteridae (parasitizing harmful gall-midges), as well as the braconid wasps (parasitizing

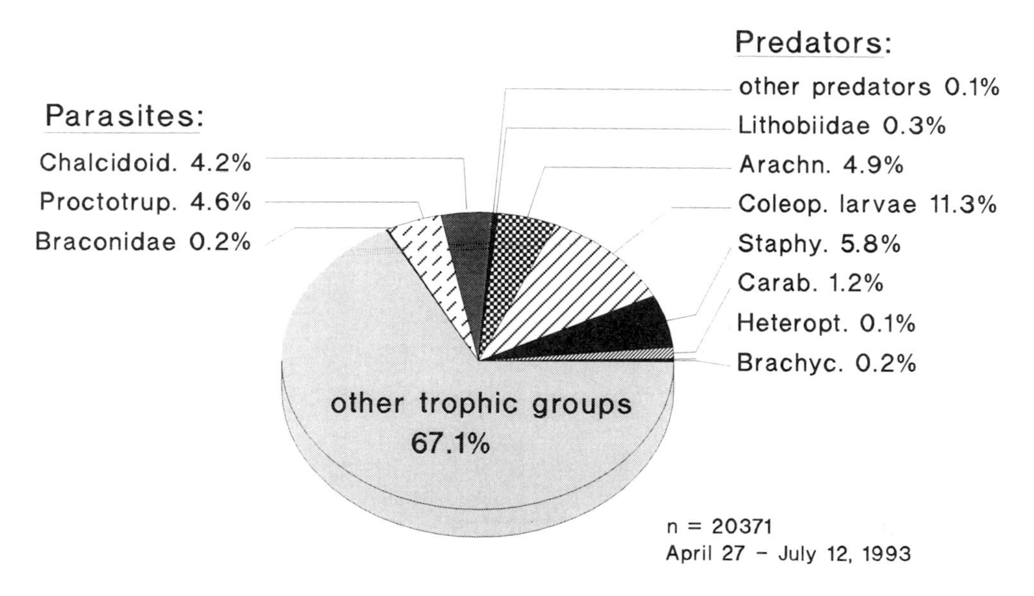

Fig. 3. Proportions of potential beneficials from total arthropod fauna (excluding Collembola and Acari), sampled with emergence traps in six organic fields at Obere Lobau, Vienna.

a broad spectrum of arthropods, among them some important pests). The prevailing proctotrupoid family Ceraphronidae, on the other hand, hyperparasitize either parasitic wasp larvae in aphids and coccinids or parasitize syrphid larvae. In terms of plant protection, this group could have a detrimental rather than favourable effect on pest regulation.

Only a taxonomic evaluation at the species level could reveal the parasitoid reservoir of relevant pests. Since the evaluation of ground-related parasitoids did not reflect the overall composition of arthropods in a Danish barley field (Jensen 1992), only an evaluation of the parasitoid index of major pests could estimate the actual contribution of parasitoids in controlling pest populations at Obere Lobau.

The latter type of information is lacking for predatory arthropods as well. Most carabid species at Obere Lobau (for dominance distribution in emergence trap samples, see Fig. 4) are well known for feeding on different stages of certain pest species. According to a review by Luff (1987), aphids, for instance, are fed upon by *Harpalus rufipes, Platynus dorsalis, Demetrias atricapillus* and *Bembidion lampros*. The latter, together with *Trechus quadristriatus*, has been shown to be a valuable antagonist in consuming the eggs of root- and stem-mining Diptera (e.g. cabbage root fly and frit fly, the latter occurring locally at Obere Lobau). The pest control status of the other abundant species

Brachinus explodens, Syntomus obscuroguttatus and *Asaphidion flavipes* is still unknown.

Due to the polyphagous feeding habits of most carabids — they also feed on other beneficial arthropods such as spiders — their impact on pest populations is difficult to evaluate (Luff 1987). *Zabrus tenebrioides* (6.1% of total carabids in the emergence traps), the only pest species among carabids, causes local damage to grainfields in Eastern Austria by larval feeding.

Although the larvae of many carabid species are known to be predatory, their role in useful predation on pest stages in the soil is largely unknown (Luff 1987). Even less is known about the feeding habits of staphylinid larvae. According to Koch (1989), staphylinid imagines are predatory in general. Certain species such as *Tachyporus hypnorum, T. chrysomelinus* and *Philonthus cognatus,* which are also abundant at Obere Lobau, are known as aphid-feeders (Dennis & Wratten 1991).

The role of spiders as biological control agents is reviewed by Riechert & Lockley (1984) and Nyffeler & Benz (1987). Like other beneficials, their actual impact on harmful arthropod populations at Obere Lobau is unknown.

Comparison of carabid data from different sampling methods

Fig. 4 reveals a basic difference in dominance distributions of carabids sampled by pitfall traps on the one hand and emergence traps and the flooding method on the other.

Pitfall trap catches (PT) were strongly dominated by *Poecilus cupreus* (70%), a relatively large, very mobile species; the medium-sized *Platynus dorsalis* and *Brachinus explodens* together reached 10% dominance, the small *Bembidion lampros* and *Asaphidion flavipes* 14%. These five species together comprised 93% of total carabids.

The catches from the other two sampling methods were clearly dominated by smaller carabids.

In flooding samples (FM), *Trechus quadristriatus, Syntomus obscuroguttatus, B. lampros* and *A. flavipes* together totalled over 60%. Two medium-sized *Amara* species comprised 20%. The FM samples contained only a few specimens of *P. cupreus, P. dorsalis* and *B. explodens.*

Emergence traps (ET) captured more *P. dorsalis* (3.8%) and *B. explodens* (2.7%). But here again, the four small carabids mentioned above that already dominated the flooding catches, together with *Demetrias atricapillus,* comprised 69%.

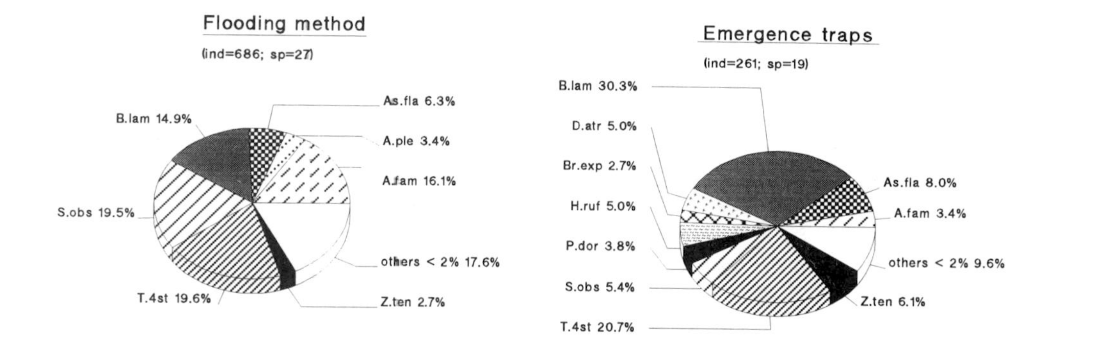

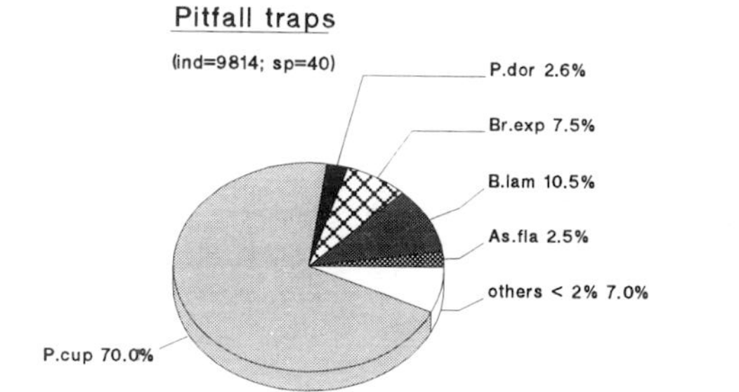

Fig. 4. Carabid dominances, obtained by three different sampling methods (for details see Materials and methods) in five organic fields at Obere Lobau, Vienna (May - July, 1993).

As.fla: *Asaphidion flavipes*; A.fam: *Amara familiaris*; A.ple: *Amara plebeja*; B.lam: *Bembidion lampros*; Br.exp: *Brachinus explodens*; D.atr: *Demetrias atricapillus*; H.ruf: *Harpalus rufipes*; P.dor: *Platynus dorsalis*; P.cup: *Poecilus cupreus*; S.obs: *Syntomus obscuroguttatus*; T.4st: *Trechus quadristriatus*; Z.ten: *Zabrus tenebrioides*.

Pitfall trap catches are clearly strongly affected by the high activity levels of larger carabid species. Among the smaller carabids, especially *T. quadristriatus* abundances are highly underestimated by pitfall traps due to this beetle's low mobility, although high population densities are indicated by flooding catches and emergence traps. *B. lampros*, though small, is an active species and therefore better represented in pitfall traps.

It is concluded that emergence trapping and the flooding method provide more reliable measures of absolute carabid densities than pitfall traps. Nevertheless, the former do have certain disadvantages. In flooding sampling, large, quickly moving arthropods like *P. cupreus* and lycosid spiders are possibly disturbed by the approaching experimenter and flee the sampling area before the sampling-frame is fully in place. Emergence traps contained rather low numbers of carabid individuals and species; some specimens were found still alive inside the eclector frame after two weeks of sampling, possibly because additional pitfall traps were not placed in the corners of the frame. Thus, thorough research in measuring and enhancing the efficacy of both the flooding method and the emergence traps remains to be done.

In comparing data from emergence traps and pitfall traps, Büchs et al. (1993, unpubl.) found very similar differences in the dominance distributions of trapped carabids: *T. quadristriatus* was underestimated, and the large, active *Pterostichus melanarius* strongly overestimated by the pitfall traps. The same pattern was found for other epigaeic active species such as *Calathus, Amara, Harpalus* and *Platynus*. Büchs et al. attempted to derive an "activity factor" for epigaeic activity; such an approach may enable recalculations of pitfall trap data resulting in more accurate density estimations.

Conclusions

This study presents background information on arthropod composition and abundances in a range of organic crops near Vienna. Major differences in emergence rates and dominance distributions of the field arthropod fauna are revealed between crop types, although all fields were cultivated without pesticides and mineral fertilizers. Black fallow during the winter, preceding root crops with bare soil in spring, appears to have a detrimental effect on the development of the within-field arthropod fauna. In winter crops like rye, favourable conditions for soil-borne arthropod populations are assumed to exist. These findings should be considered in planning crop rotations in order to augment natural pest control in organic agriculture.

The actual contribution of parasitoid and predatory arthropods in pest suppression can only be fully revealed by combining a thorough taxonomic analysis of potentially beneficial groups with an evaluation of the parasite index and predation rates of relevant pest populations.

Concerning the methodological approach, sampling by emergence traps and the flooding method is considered to provide reliable density-related data on field arthropods, whereas pitfall traps yield biased dominance distributions due to different trappabilities (Sunderland et al. 1995) of epigaeic species.

Acknowledgements

This paper is part of a research programme funded by the Austrian Ministry of Science and Research. Thanks are due to A.Velimirov for drawing the graphs.

References

Amlinger, F. 1993. *Biotonne Wien - Theorie und Praxis.* MA 48 - Stadtreinigung und Fuhrpark, Wien. 385p.

Basedow, Th., Klinger, K., Froese, A. & Yanes, G. 1988. Aufschwemmung mit Wasser zur Schnellbestimmung der Abundanz epigäischer Raubarthropoden auf Äckern. *Pedobiologia* 32: 317-322.

Booij, C.J.H. & Noorlander, J. 1992. Farming systems and insect predators. *Agric. Ecosystems Environ.* 40: 125-135.

Büchs, W. 1991. Einfluß verschiedener landwirtschaftlicher Produktions-intensitäten auf die Abundanz von Arthropoden in Zuckerrübenfeldern. *Verh. Ges. Ökol.* 20: 1-12.

Büchs, W. 1993. Auswirkungen unterschiedlicher Bewirtschaftungsintensitäten auf die Arthropodenfauna in Winterweizenfeldern. *Verh. Ges. Ökol.* 22: 27-34.

Büchs, W., Kleinhenz, A. & Zimmermann, J. 1993. Barberfallen und Boden-photoeklektoren: Aussagepotential beider Methoden bei kombinierter Anwendung. Unpublished manuscript of a lecture held at the DGaaE-conference in Jena,1993.

Dennis, P. & Wratten, S.D. 1991. Field manipulation of populations of individual staphilinid species in cereals and their impact on aphid populations. *Ecol. Entomol.* 16: 17-24.

Funke, W. 1971. Food and energy turnover of leaf-eating insects and their influence on primary production. *Ecol. Studies* 2: 81-93.

Ingrisch, S., Wasner, U. & Glück, E. 1989. Vergleichende Untersuchungen der Ackerfauna auf alternativ und konventionell bewirtschafteten Flächen. In: *Alternativer und konventioneller Landbau.* Schriftenreihe der LÖLF Nordrhein-Westfalen, Bd. 11: 113-271.

Jacobs, W. & Renner, M. 1988. *Biologie und Ökologie der Insekten* (2nd edition). G.Fischer, Stuttgart/New York, 690p.

Jensen, P.B. 1992. Ground related parasitoids in a Danish barley field. *Deutsche Ges. allgem. angew. Entomol. Nachrichten (Ulm)* 6: 150-151.

Koch, K. 1989. *Die Käfer Mitteleuropas. Ökologie.* Bd.1. Goecke & Evers, Krefeld.

Kromp, B. 1990. Carabid beetles (Coleoptera, Carabidae) as bioindicators in biological and conventional farming in Austrian potato fields. *Biol. Fertil. Soils* 9: 182-187.

Kromp, B. & Nitzlader, M. 1995. Dispersal of ground beetles in a rye field in Vienna, Eastern Austria. In: Toft, S. & Riedel, W. (eds.) *Arthropod natural enemies in arable land · I*, pp. 269-277. Aarhus.

Luff, M.L. 1987. Biology of polyphagous ground beetles in agriculture. *Agric. Zool. Rev.* 2: 237-278.

Nyffeler, M. & Benz, G. 1987. Spiders in natural pest control: A review. *J. Appl. Ent.* 103: 321-339.

Riechert, S.E. & Lockley, T. 1984. Spiders as biological control agents. *Ann. Rev. Entomol.* 29: 299-320.

Sunderland, K.D., De Snoo, G.R., Dinter, A., Hance, T., Helenius, J., Jepson, P., Kromp, B., Lys, J.-A., Samu, F., Sotherton, N.W., Toft, S. & Ulber, B. 1995. Density estimation for beneficial predators in agroecosystems. In: Toft, S. & Riedel, W. (eds.) *Arthropod natural enemies in arable land · I*, pp. 133-162. Aarhus.

Appendix 1. Total emergence trap catches of arthropods (excluding Collembola and Acari) in seven organic fields at Obere Lobau, Vienna. (Three emergence traps of 0.25m^2 per field; sampling period April 27 - July 12, 1993; six sampling dates). So: Soya, Su: Sunflower, W: Wheat, P: Potatoes, R: Rye, Lu: Lucerne.
+ Predators, ○ Parasitoids.

Fields/crops	So	Su	W-W	W-P	R-W	R-R	Lu	Total	D(>0.5%)
HETEROPTERA									
+Nabidae			4	1	5	11	4	25	
Miridae	1	3			2		24	30	
+Anthocoridae			1				5	6	
Lygaeidae	1	11	1		3	1	4	21	
Coreidae							1	1	
Scutelleridae				3				3	
Pentatomidae			3	2		5	4	14	
Cydnidae	3		56	65	13	5	61	203	0.63
Pyrrhocoridae	2		2				2	6	
Heteroptera nymphs	11	7	3	20	14	7	17	79	
HOMOPTERA									
Cicadina	4	11	150	91	88	248	61	653	2.02
Psyllina		1					630	631	1.95
Aphidina	29	24	29	27	9	16	5034	5168	15.96
Aleurodina				1				1	

Appendix 1, continued

Fields/crops	So	Su	W-W	W-P	R-W	R-R	Lu	Total	D(>0.5%)
DIPTERA									
Cecidomyidae	115	72	379	983	602	794	241	3186	9.84
Bibionidae		24	6	2	3	126	1	162	0.50
Sciaridae	672	41	187	23	295	1206	94	2518	7.77
Mycetophilidae					1			1	
Chironomidae	19		12	2	130	220	40	423	1.31
Culicidae							1	1	
+ Stratiomyidae					1			1	
+ Rhagionidae	1							1	
Phoridae	77	171	156	332	265	172	569	1742	5.38
Tabanidae	1							1	
Empididae	1			2	2	4	1	10	
Hybotidae	63	12	53	11	29	107	11	286	0.88
+ Dolichopodidae	21	1	4		7	9		42	
Syrphidae							27	27	
○ Pipunculidae			1		1			2	
○ Tachinidae				2			2	4	
+ Scatophagidae						5		5	
Anthomyidae	10		13			1	7	31	
Muscidae							1	1	
Pallopteridae			22	32	15	42		111	
Heleomyzidae						1		1	
Lauxaniidae							1	1	
+ Otitidae	1							1	
Sphaeroceridae	4	1	10	19	47	48	5	134	
Chloropidae	4	1	36	7	12	40	15	115	
Agromyzidae	2	7	2	5	12	4	93	125	
Drosophilidae	1	2	1	4	42		28	78	
Asilidae						1		1	
Diptera larvae	2	5	54	4	3	7	42	117	
HYMENOPTERA									
○ Ichneumonidae	1	3	1			1	15	21	
○ Braconidae	7	5	6	3	7	9	86	123	
○ Aphidiidae	2				1		824	827	2.55
○ Chalcidoidea	14	83	144	191	197	226	866	1721	5.31
Cynipoidae							9	9	
○ Platygasteridae	2	2	34	7	51	24	15	135	
○ Scelionidae	5	29	74	21	31	60	136	356	1.10
○ Diapriidae	3	6	48	45	33	16	6	157	
○ Proctotrupidae			1			3		4	
○ Ceraphronidae	24	39	84	88	58	141	388	822	2.54
Bethylidae	2		1	1		3		7	
○ Chrysididae			1					1	
Formicidae	1				3	2	2	8	
Andrena sp.							1	1	
Halictus sp.							22	22	
Symphyta larvae		1		1				2	

Appendix 1, continued

Fields/crops	So	Su	W-W	W-P	R-W	R-R	Lu	Total	D(>0.5%)
COLEOPTERA									
+Carabidae	34	24	39	19	78	70	50	314	0.97
+Staphylinidae	150	154	132	160	232	342	142	1312	4.05
Scarabaeidae	3	3			1			7	
Elateridae others	2			4	1	2	2	11	
Adrastus pallens	23	13	10	8	2	37	5	98	
Agriotes sputator		1	1	10				12	
Agriotes brevis		1	1	5				7	
Nitidulidae		4		1	4	6	4	19	
+Coccinellidae				2	1	1	54	58	
Ptiliidae				1	2			3	
Anthicidae others		3		1	2		12	18	
Anthicus antharinus	19	16	12	5	2	2	6	62	
Chrysomelidae others					1			1	
Oulema melanopus			36	25	7	8		76	
Cassida sp.	1	4					1	6	
Halticinae		45	11	9	1		6	72	
Curculionidae	2	41	3	4	6	3	103	162	0.50
Malachiidae			2			2		4	
Cryptophagidae	2	3	34	80	156	87	73	435	1.34
Cucujidae	1							1	
Lathridiidae	1	6	46	127	47	56	49	332	1.03
Scaphidiidae	1	3			1		2	7	
Erotylidae	5	147	2	15	1	1	2	173	0.53
Phalacridae	2	1	8	21	14	18	23	87	
Pselaphidae				1				1	
Histeridae					1		1	2	
Dermestidae			1					1	
Byrrhidae			1		1			2	
Dryopidae			1					1	
Scotylidae	1							1	
+Cantharidae			13	1	1	1		16	
+Coleoptera larvae	41	69	294	341	784	762	648	2939	9.07
INSECTA OTHERS									
Orthoptera	4		1					5	
Psocoptera	2	3		53	8	28	69	163	0.50
Thysanoptera	285	599	454	564	490	456	1250	4098	12.65
Chrysopidae		1	1	1		1	2	6	
+Chrysopidae larvae	2	1	1		1	1	3	9	
Lepidoptera				1			5	6	
Lepidoptera larvae	9	32			1	2	15	59	
MYRIAPODA									
Julidae	16	5	2	1	298	82	8	412	1.27
Polydesmidae	2				25	23	2	52	
+Lithobiomorpha	13	24	16		13	4	11	81	
+Geophilomorpha	1	1				1		3	
ARACHNIDA	5	46	86	59	242	553	72	1067	3.29
Total no. individuals	1735	1814	2787	3514	4407	6114	12016	32387	
Total no. of taxa	75	70	84	82	94	86	94	108	

Rate and local scale spatial pattern of adult emergence of the generalist predator *Bembidion guttula* in an agricultural field

Juha Helenius

Department of Applied Zoology, University of Helsinki, Finland
Present address: Institute of Crop Protection, Agricultural Research Centre, FIN-31600 Jokioinen, Finland

Abstract

Emergence rates of new generation adults of the univoltine ground beetle *Bembidion guttula* and activity density of the parent generation were estimated by trapping in a spring cereal field in two successive years. A 1.0 ha area divided into a 4 x 4 grid was used as a sampling framework. An emergence trapping method was developed, and its uses in agroecological studies is discussed.

Estimated net recruitment rates were 600,000 tenerals/ha/year in the first and 800,000 tenerals/ha/year in the second year. Spatial aggregation at the 25 m resolution was evident and the pattern was consistent over years. Peak eclosion was c. 600 day degrees (°C) after the peak activity density of the parent generation in spring. Activity density of the parents was less aggregated and not spatially correlated to the emergence rate of the offspring.

It is concluded that emergence rates can be reliably sampled for by a relatively cheap trapping design, and that the method, unlike pitfall trapping, provides population density estimates that are valuable, for example in studies that assess the effects of agricultural practices on the population dynamics of beneficial predatory ground beetles.

Key words: Carabidae, ground beetles, tenerals, sampling method, emergence trapping, population density estimates, spring cereals

Introduction

A diverse fauna of typically 40-60 species of Carabidae exists in agricultural fields throughout the temperate region. Many are generalist predators and contribute to natural biological control of pest insects (Luff 1987).

Many species breed in crop fields, and the fields can be a source of recruitment to the local populations. For example, in Scandinavia *Bembidion lampros* (Herbst) is an abundant, univoltine adult-overwinterer breeding in

Arthropod natural enemies in arable land · I *Density, spatial heterogeneity and dispersal*
S. Toft & W. Riedel (eds.). *Acta Jutlandica* vol. 70:2 1995, pp. 101-111.
 ISBN 87 7288 492

annual crops such as cereals (Wallin 1989). This species is also one of the key members of the guild of generalist predators of *Rhopalosiphum padi* Linnaeus, the major pest of spring cereals (e.g. Ekbom et al. 1992).

To answer questions like what farming practices (pesticide use, cultivation, rotations etc.) are detrimental or enhancing to ground beetles, what is the trend, if any, in population development etc., adequate sampling methods are needed.

Pitfall trapping is the most commonly used method, but the data are very difficult to interpret. In the review by Sunderland et al. (1995) it was concluded that pitfall catches are a product of abundance, activity and species-specific trappability. Absolute estimates of instantaneous densities would be most useful, but are difficult to obtain (Sunderland et al. 1995). Sunderland et al. (1987) proposed a sequential procedure of vacuum suction sampling, ground searching, isolating and removal trapping with pitfalls of a sample quadrat. This method, with variations, has been applied by others (e.g. Helenius 1991, Ulber & Wolf-Schwerin 1995) and has proven very time consuming, and even unreliable in certain conditions. In this paper, a method of emergence trapping of teneral ground beetles is described, and its uses as an alternative to instantaneous density estimates discussed.

In our study field, *Bembidion guttula* (Fabricius) (Col., Carabidae) is the most abundant species of *Bembidion* in emergence trap catches, and the most abundant among the species that are regarded as effective predators of *R. padi* (Helenius & Tolonen 1994, Holopainen & Helenius 1992). Hence, this species was chosen for developing the emergence trapping method, and for estimating the rate and temporal and spatial pattern of adult eclosion. The emphasis in this paper is in demonstrating the potential of the new method, and in providing data of recruitment rates per unit area, in order to gain deeper insight into the agroecology of beneficial ground beetles.

Materials and methods

Material was collected in 1990-1991 from a one hectare block (100 m x 100 m square) in the middle of a larger field at Viikki Experimental Farm, 61°12' N in Helsinki. The block was divided into a 25 m x 25 m grid, giving 16 plots. Originally, this layout was used for comparing four management regimes for spring cereals in a 4 x 4 Latin square in terms of their effect on ground beetles. The methods and some results of the comparisons of monocrops and undersowing-green manuring regimes were explained previously (Helenius &

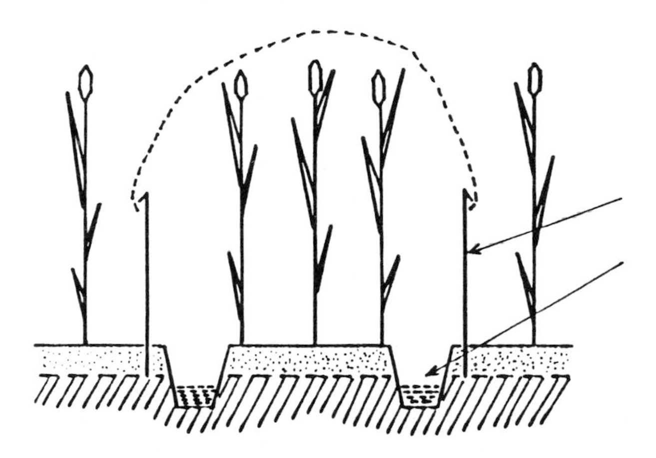

Fig. 1. Illustration of an enclosure for trapping carabids during adult eclosion. Vertical barriers (shown by the upper arrow) pushed 5cm into soil enclose 0.5m x 0.5m square, covered by insect net. One pitfall trap in each corner (lower arrow). (Modified from Helenius 1991).

Tolonen 1994). The cereal crops were spring barley in 1990 and spring wheat in 1991.

Carabid data were obtained from pitfall-trapping and from emergence trapping in the centre of each of the 16 plots (n=16). Two pitfall traps, 100 mm in diameter, were placed in each plot 2 m from each other. The trap medium was water, with some coarse salt and detergent added. The traps were emptied once every week. The trapping periods from placing the pitfalls immediately after cereal sowing to last emptying were 9/5-16/9 1990 and 13/5-14/7 1991.

Emergence trapping: the method

The emergence trap was developed from the principle of using isolators for removal trapping (Sunderland et al. 1987), following the notion that newly emerged adults may bias the estimate of instantaneous density (Helenius 1991).

The trap was a square enclosure of 0.5 m x 0.5 m, isolated from the surrounding by a frame of metal flashing, height 0.2 m, and covered with insect net to prevent escape by flying (Fig. 1). The frame was pushed into soil through the tilth (c. 5 cm). Four pitfall traps (100 mm in diameter, water with some coarse salt and detergent as a medium) were placed within the enclosure, one in each corner. This gave a net emergence trapping area of 0.2 m^2.

The vegetation within traps was left as undisturbed as possible. Pitfalls were emptied once every week, the old generation ("mature" or "spent") and new generation ("teneral" or "immature") beetles counted, sexed and aged. Softness, pale colour and undeveloped genitalia were used as criteria for identifying new generation adults (see Wallin 1989).

Microclimate (Southwood 1978) and other factors, even predation on pupae etc., are changed to an unknown extent within an emergence trap. In order to avoid accumulation of bias due to such causes, the trap was repositioned at regular intervals. There were five successive trapping periods, each of four weeks duration, overlapping by two weeks. These were 12/6-10/7, 26/6-24/7, 10/7-7/8, 24/7-21/8 & 7/8-4/9 1990 and 20/6-18/7, 4/7-1/8, 18/7-15/8, 1-29/8 & 15/8-12/9 1991. Weekly catch rates of new generation adults/m^2/week were integrated to give an estimate of net recruitment rate as numbers per unit area over the season (Helenius & Tolonen 1994).

Spatial aggregation of the emergence trap catches was described using Taylor's power law (Taylor 1961). Day-degree accumulation was calculated as DD°C above 5°C threshold from peak pitfall catches to peak emergence rates as a rough estimate of heat units requirement from egg stage to adult eclosion. The 5°C base is officially used for thermal growing season in Finland by the Finnish Meteorological Institute, and their standard data from Kaisaniemi station 10 km away was used here as no other data or further information about temperature requirements for development of the species is available.

Results

The pitfall traps caught 486 (58% females) and 1648 (55% females) specimens of *B. guttula* in 1990 and 1991 respectively. The numbers of new generation adults caught by emergence traps were 316 (51% females) and 504 (61% females) in 1990 and 1991.

Peak pitfall catch of the overwintered adults occurred on 6/6 1990 and 10/6 1991, after which the pitfall catches declined rapidly (Fig. 2). Emergence rates peaked 7-8 weeks later, when the mean instantaneous rate reached up to 40-50 new generation adults per m^2 per week (Fig. 3). Cumulative temperature between these events was 540°C DD in 1990 and 630°C DD in 1991, and accumulation from the onset of the thermal growing season until peak emergence was 830°C DD and 790°C DD, respectively.

From the 16 plots, the mean recruitment rates were 57 per m^2 and 80 per m^2 of new generation *B. guttula* in 1990 and 1991, respectively. These are equal to an annual production of c. 0.6 to 0.8 million/ha of this species alone.

Parent generation adults were caught in emergence traps in relatively low

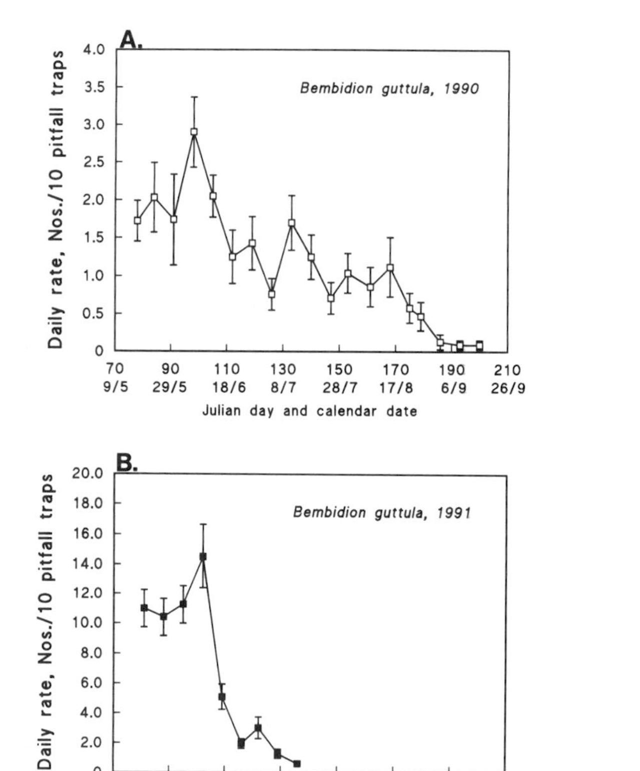

Fig. 2. Pitfall catch rates as numbers of *Bembidion guttula*, expressed per 10 pitfalls per day in 1990 (A, open squares) and 1991 (B, solid squares). Vertical bars: ±SE.

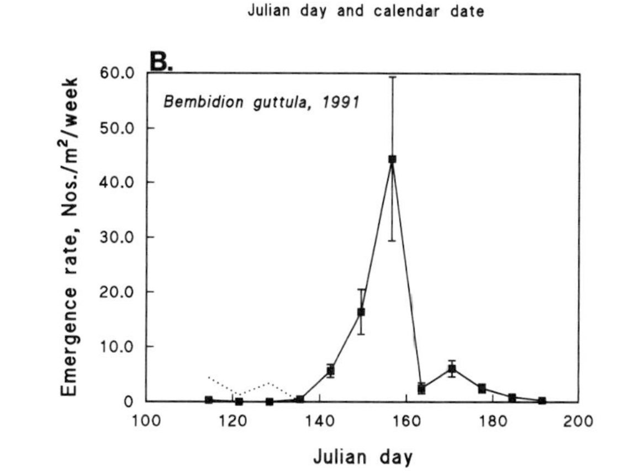

Fig. 3. Catch rates from emergence traps of new generation *Bembidion guttula* adults as numbers per square meter per week in 1990 (A, open squares) and 1991 (B, solid squares). Dashed lines show the catch rate of old generation adults. Vertical bars: ±SE.

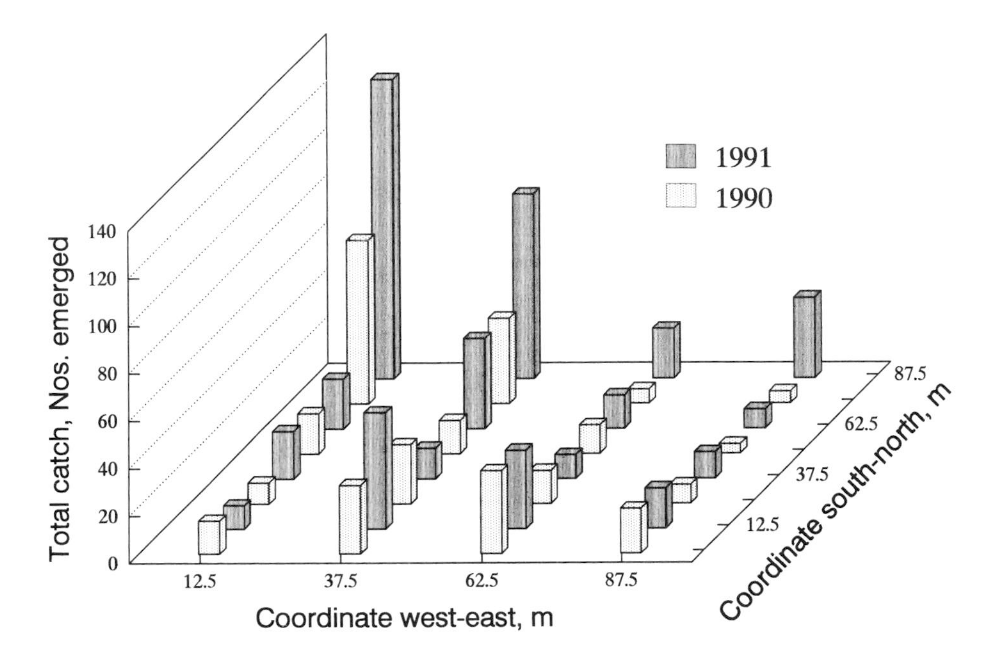

Fig. 4. Spatial pattern of the emergence trap catches of *Bembidion guttula* in the 4 x 4 grid of the 1 ha block in 1990 (dotted bars, front) and 1991 (shaded bars, back).

numbers, decreasing as the season progressed. Peak rates as numbers/m^2/week were 6.1 (SE 2.25) in 1990 and 4.4 (SE 1.32) in 1991 (Fig. 3).

Emergence was spatially aggregated to the NW corner of the 16 plot grid (Fig. 4). Temporal constancy of the pattern over the two years was marked and significant (χ^2 = 62.6, df 15, p<0.001) (Fig. 5). Power "b" of Taylor's Power Law was 1.4 and deviated significantly from unity (F(1,20)=49.3, p<0.001), demonstrating a clumped sampling distribution (Fig. 6).

The spatial distribution of the pitfall trap catches did not follow the pattern of the emergence trap catches in 1990 (χ^2 = 73.3, df 15, p<0.001) nor in 1991 (χ^2 = 234.1, df 15, p<0.001). Pitfall catches of the parent generation were less clumped and spatial trends within the grid were not evident (Fig. 7).

Discussion

Vickerman (1978) used emergence traps to study the arthropod fauna in cereal fields, but he only reported details of the beetles pooled as Coleoptera. Kromp et al. (1995) used emergence traps for ground beetles (among other groups) in

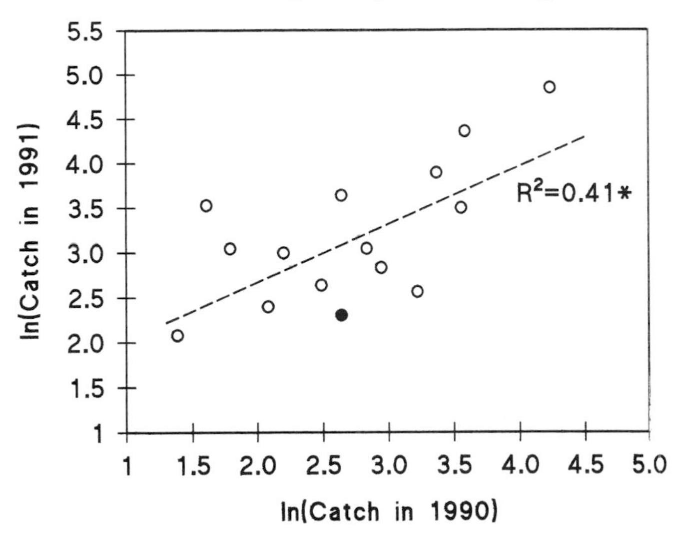

Fig. 5. Consistency of spatial pattern: season-to-season correlation of emergence trap catches of univoltine *Bembidion guttula* in a 4 x 4 grid of plots of a 1,0 ha block of spring cereal. Each data-point represents total catch in a plot, over the eclosion period from a sequence of 5 traps, 4 weeks each, overlapping by 2 weeks. Solid dot: two ovelapping data-points; star for value of R^2 means $p<0.05$.

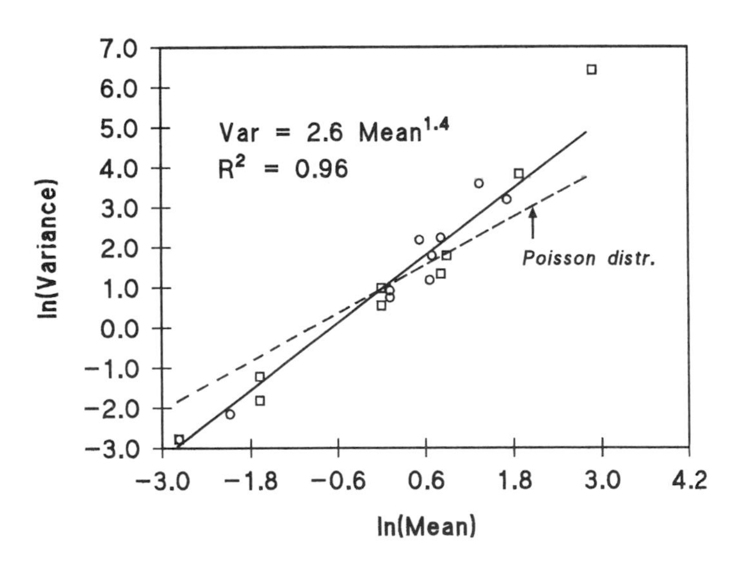

Fig. 6. Regression of variance against mean of the emergence trap catches of *Bembidion guttula* from the 4 x 4 grid. Taylor's Power Law (Taylor 1961) shows significant deviation from a random distribution ($b=1.4$). Circles: 1990; squares: 1991.

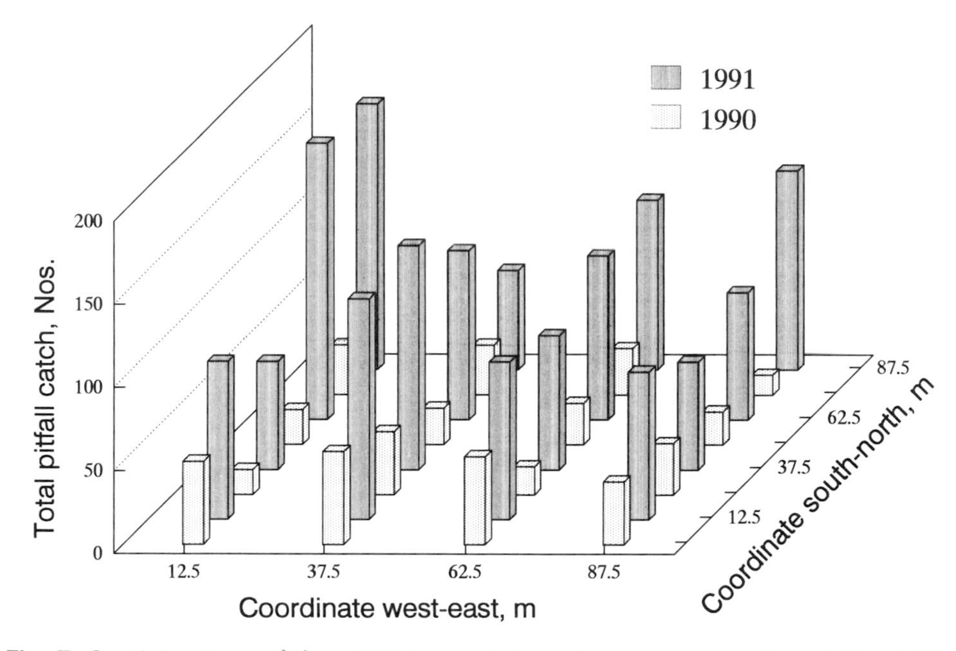

Fig. 7. Spatial pattern of the pitfall trap catches of parent generation of *Bembidion guttula* in the 4 x 4 grid of the 1 ha block in 1990 (dashed bars, front) and 1991 (shaded bars, back).

a cropping system study. Trap designs that attempt estimates for a range of taxonomic groups suffer from compromises: e.g. dark canvas often used as a tent to force flying insects into collectors seriously affects the conditions inside the trap: microclimate, crop and weed growth etc., are altered. In this respect, focusing on one species or on a "homogeneous" group of species allows a trap design that maximizes reliability.

In an agricultural field, habitat quality for soil-dwelling larvae is likely to vary along with edaphic factors. The scale of spatial contagion that could be detected by the experimental layout was 25 m. Constancy over years in the pattern indicates true habitat effects within the grid. The emergence rates may reflect suitability of soil habitat in the seemingly homogeneous, topographically flat and even field. This suitability in terms of successful development of carabid larvae may vary also on a much finer spatial scale, for which a trap as large as 0.20 m^2 may be too coarse. If the pattern was generated by egg-laying females finding most suitable microsites is not revealed by the activity-density data: the absence of correlation between activity density of parents with emergence rates of the offspring may be due to higher density (aggregation) being compensated by lower activity (slowing down) in "good sites".

Sampling from fixed coordinate points in successive years removes "noise" that would result in confounding annual variation with fine-grained spatial variation.

In life-table analysis, the emergence trapping method would be invaluable. Soil cores obtained when drilling holes for the pitfalls within the trap can be taken to the laboratory for extraction of larvae, as reported by Helenius et al. (1995). This produces a time series of the larval densities, with minimal additional sampling effort and maximal spatial relatedness of the data of larvae and the data of tenerals. The problem of estimating absolute density of, e.g. gravid females, from the often very dynamic (rapid colonization, rapid decline) parent generation remains as a major problem.

The estimates for per-unit-area production of *B. guttula* may be conservative, as the whole eclosion period was not covered. Especially in 1990, some beetles were still emerging when trapping was stopped. Also, it is unlikely that all the specimens emerging into the traps were caught. On the other hand, it is not known what factors (e.g. abiotic factors, food and predation) affecting survival were altered by the eclosures, and to what extent, in comparison with the open field. If survival was improved, the estimates may be too high. The accuracy of the estimates of day-degree accumulation from peak egg-laying up to peak eclosion was also subject to possible alteration in microclimate by the enclosures.

B. guttula was the third most abundant species of newly emerged adults from emergence traps, coming after *Trechus discus* (Fabricius) and *Clivina fossor* (Linnaeus) (Helenius & Tolonen 1994). For any given level of statistical accuracy, recruitment rates for these abundant species can be estimated by a smaller trapping effort than is the case for less abundant species. Other factors affecting sampling effort are duration of eclosion period (trapping time) and spatial sampling distribution (sample size).

The experiment was originally designed to compare cereal monocropping with three different kinds of undersowing-green manuring regimes, with respect to enhancement of predatory carabids (Helenius et al. 1995, Helenius & Tolonen 1994). A statistical model was used to predict an increase in recruitment rate among the five most common species as an exponential function of the amount of green manure given to the crop during the previous autumn (Helenius & Tolonen 1994). However, the differences in management regimes between plots in the 4 x 4 grid do not explain the spatial pattern (Helenius, in preparation).

Estimation of net recruitment rates is an advance on pitfall data (see Sunderland et al. 1995), because it contributes to an understanding of the constraints on population growth of beneficial carabids. The next step would be to sample for egg, larval and pupal densities and even for key mortality factors during these stages in the soil. However, such life table studies would require much greater resources than those needed for an emergence trapping study.

The method, as employed here, does not give estimates of recruitment rates in terms of offspring produced per gravid female. For those, absolute population estimates of the parent generation would also be required. Nevertheless, net recruitment rates per unit area provide valuable information about the quality of the arable habitat for the species under study. The method should prove valuable in comparing crop management regimes, such as the effects of pesticides, rotations and farming systems, especially when focusing on species for which field habitats are sources, rather than sinks, for the local populations.

Acknowledgement

The Carabidae were determined, sexed and aged by Timo Tolonen. Jarmo Holopainen gave valuable advice, and Keith Sunderland constructively commented on the manuscript. The study was funded by the Finnish National Research Council for Agriculture and Forestry (grants no. 1011093 and no. 1011619). Part of the work was done at South Savo Research Station, ARC.

References

Ekbom, B.S., Wiktelius, S. & Chiverton, P.A. 1992. Can polyphagous predators control the bird cherry-oat aphid (*Rhopalosiphum padi*) in spring cereals? A simulation study. *Entomol. Exp. Appl.* 65: 215-223.

Helenius, J. 1991. Integrated control of *Rhopalosiphum padi*, and the role of epigeal predators in Finland. *IOBC/WPRS Bull.* 14(4): 123-130.

Helenius, J., Holopainen, J., Muhojoki, M., Pokki, P., Tolonen, T. & Venäläinen, A. 1995. Effect of undersowing and green manuring on abundance of soil arthropods and on recruitment of ground beetles (Coleoptera, Carabidae) in cereals. *Acta Zool. Fenn.* in press.

Helenius, J. & Tolonen, T. 1994. Enhancement of generalist aphid predators in cereals: effect of green manuring on recruitment of ground beetles (Col., Carabidae). *IOBC/WPRS Bull.* 17(4): 201-210.

Holopainen, J.K. & Helenius, J. 1992. Gut contents of ground beetles (Col., Carabidae), and activity of these and other epigeal predators during an outbreak of *Rhopalosiphum padi* (Hom., Aphididae). *Acta Agric. Scand.* 42: 57-61.

Kromp, B., Pflügl, Ch., Hradetzky, R. & Idinger, J. 1995. Estimating beneficial arthropod densities using emergence traps, flooding method and pitfall traps in organic fields (Vienna, Austria). In: Toft, S. & Riedel, W. (eds.) *Arthropod natural enemies in arable land · I*, pp. 87-100. Aarhus.

Luff, M.L. 1987. Biology of polyphagous ground beetles in agriculture. *Agric. Zool. Rev.* 2: 237-278.

Southwood, T.R.E. 1978. *Ecological methods*. 2nd ed. Methuen, London.

Sunderland, K.D., Hawkes, C., Stevenson, J.H., McBride, T., Smart, L.E., Sopp, P.I., Powell, W., Chambers, R.J. & Carter, O.C.R. 1987. Accurate estimation of invertebrate density in cereals. *IOBC/WPRS Bull.* 10(1): 71-81.

Sunderland, K.D., De Snoo, G.R., Dinter, A., Hance, T., Helenius, J., Jepson, P., Kromp, B., Lys, J.-A., Samu, F., Sotherton, N.W., Toft, S. & Ulber, B. 1995. Density estimation for beneficial predators in agroecosystems. In: Toft, S. & Riedel, W. (eds.) *Arthropod natural enemies in arable land · I*, pp. 133-162. Aarhus.

Taylor, L.R. 1961. Aggregation, variance and the mean. *Nature* 189: 732-735.

Ulber, B. & Wolf-Schwerin, G. 1995. A comparison of pitfall trap catches and absolute density estimates of carabid beetles in oilseed rape fields. In: Toft, S. & Riedel, W. (eds.) *Arthropod natural enemies in arable land · I*, pp. 77-86. Aarhus.

Vickerman, G.P. 1978. The arthropod fauna of undersown grass and cereal fields. *Scient.Proc.R. Dublin Soc. A*, 6: 273-283.

Wallin, H. 1989. Habitat selection, reproduction and survival of two small carabid species on arable land: a comparison between *Trechus secalis* and *Bembidion lampros*. *Holarct. Ecol.* 12: 193-200.

Relationships between aphid phenology and predator and parasitoid abundances in maize fields

Thierry Hance

Université Catholique de Louvain, Unité d'Écologie et de Biogéographie
Place Croix du Sud, 5, B-1348 Louvain-la-Neuve, Belgium

Abstract
Tritrophic relationships between host plant, aphids and predators are extremely important in determining population fluctuations. In this context, a Latin square design was used in order to quantify the influence of four maize varieties on aphid and predator population growth. Aphid numbers and a leaf occupation index were recorded weekly on 10 maize plants per area. Simultaneously, the number of aphid mummies, coccinellids and chrysopids were counted, taking into account species and stages.

Maize variety influences aphid development and thus aphid density. The leaf occupation index, however, was quite homogenous among varieties. Percentages of parasitism are related to aphid density and the kind of relationship depends on the variety. The main species of Microhymenoptera were *Aphidius rhopalosiphi* and *Praon volucre*. Coccinellid and chrysopid egg and larval densities per plant were also established for each variety. Maximum abundance of predators and parasitoids coincided with the peak of aphid density and was probably one of the factors responsible for their decline in August.

Key words: Tritrophic relationships, aphid phenology, maize, Microhymenoptera, Coccinellidae, Carabidae, window traps, leaf occupation index

Introduction

In Belgium the area of maize cultivation has grown so that maize has become the second largest crop. Aphids are currently only a minor pest in maize fields, because aphids do not transmit viruses to maize and because maize is a fodder crop; thus, a high level of infestation is tolerated by the farmers. Maize fields are therefore seldom sprayed with insecticides. There is, however, a second reason for their minor pest status, as aphid phenology in maize crops exhibits quite an odd pattern. Maize is sown in April. Aphid infestation begins at the end of May and populations grow during June and July. The peak of abundance

 ISBN 87 7288 492

is recorded at the end of July. Then the populations decrease abruptly and disappear in August, despite the fact that the plants are still suitable for aphid nutrition. Indeed, at this time of the year, aphid rearing is still done in the laboratory on plants at the same stage as those in the field. This pattern is observed for the more abundant species, *Sitobion avenae* (Fabricius) and *Metopolophium dirhodum* (Walker). The purpose of our research program is to understand why aphid populations decline although the host plant is still suitable for them.

In order to answer that question, aphid phenology and the phenology of their main predators and parasitoids were recorded at the same time. Moreover, the influence of host plant on aphids, predators and parasitoids was estimated by comparing the number found on four maize varieties.

Materials and methods

Two experimental fields were sown with maize on May 13. A herbicide (Capsolane®, 10 l/ha) had been applied on May 7. The manure was composed of 120 kg/ha of Nitrogen, 90 kg/ha of Phosphorus and 180 kg/ha of Potassium. In field A (400 m^2), four varieties were sown in a Latin square design (Fig. 1). In each replicate (20 m^2), 10 plants were chosen and the presence of aphids and their predators (coccinellids and chrysopids), as well as the number of aphid mummies, was recorded weekly. A "leaf occupation index" was also determined each week. This index consisted of the number of leaves occupied by at least one aphid, divided by the total number of leaves. Thus, it takes into account the growth of the plant. Maize varieties were chosen according to previous experiments in which their susceptibility to aphids was determined (Hance et al. 1993). The varieties are classified by decreasing susceptibility: Topaze, DEA, Solida, and Magda.

In field B (4000 m^2), four window traps were placed as shown in Fig. 2. Each window trap consisted of two transparent plexiglass screens perpendicular to collecting water-pans. The first pan is situated 20 cm above ground level and the second pan at a height of 1.2 m. The area of each screen is 1 m^2. The traps caught mainly the individuals that entered the field because of the position of the traps at the field edge. Our aim was to quantify the importance of predator immigration into the field from the surrounding habitats (orchard or fallow). In addition, two yellow traps (water pan: Ø 21 cm) were placed in field A and in field B. The trapping fluid was water with liquid soap. Window traps and yellow traps were emptied weekly. Coccinellids, carabids and chrysopids were identified in the laboratory and counted.

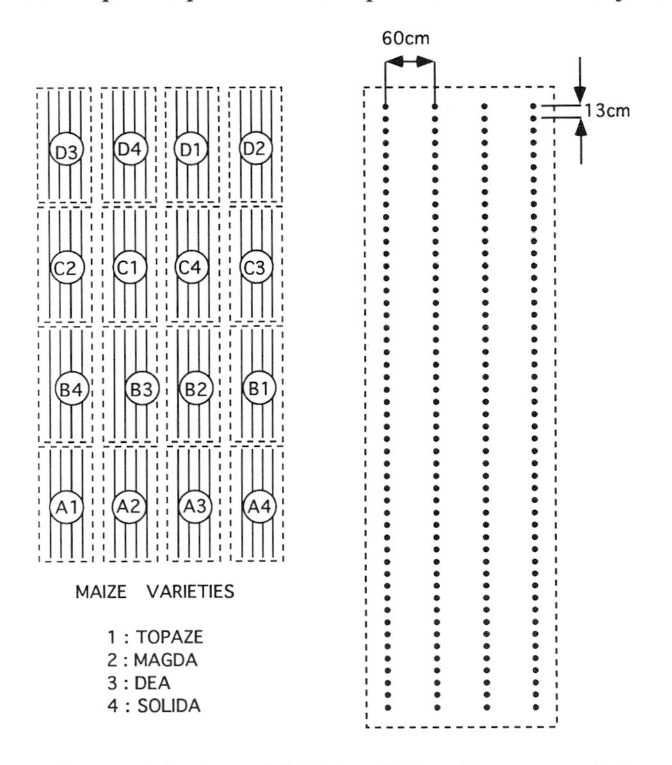

Fig. 1. Left: Experimental design of field A. Right: Structure of plots. Each dot is a maize plant.

Results and discussion

Relationship between abundance and leaf occupation

The development of the leaf occupation index by aphids is given in Fig. 3A. Testing the heterogeneity of the slopes, the ascending parts of the curves do not differ between the four varieties (N = 35, $F_{(1,28)} = 0.06$ P= 0.980), despite their differences in susceptibility (Hance et al. 1993). However, absolute aphid numbers per plant varied according to variety (Fig. 3B). This suggests that the distribution of aphids on the plants relies not on their densities but rather on the emigration or colonizing behaviour of aphids. The total aphid number at the peak of abundance is four times smaller for Magda than for Solida. This confirms the previous results concerning susceptibility of these varieties (Hance et al. 1993). Fig. 3B shows the development of the total aphid population (the two species mixed). *Sitobion avenae* was the most abundant and exhibited the same phenology as *Metopolophium dirhodum*.

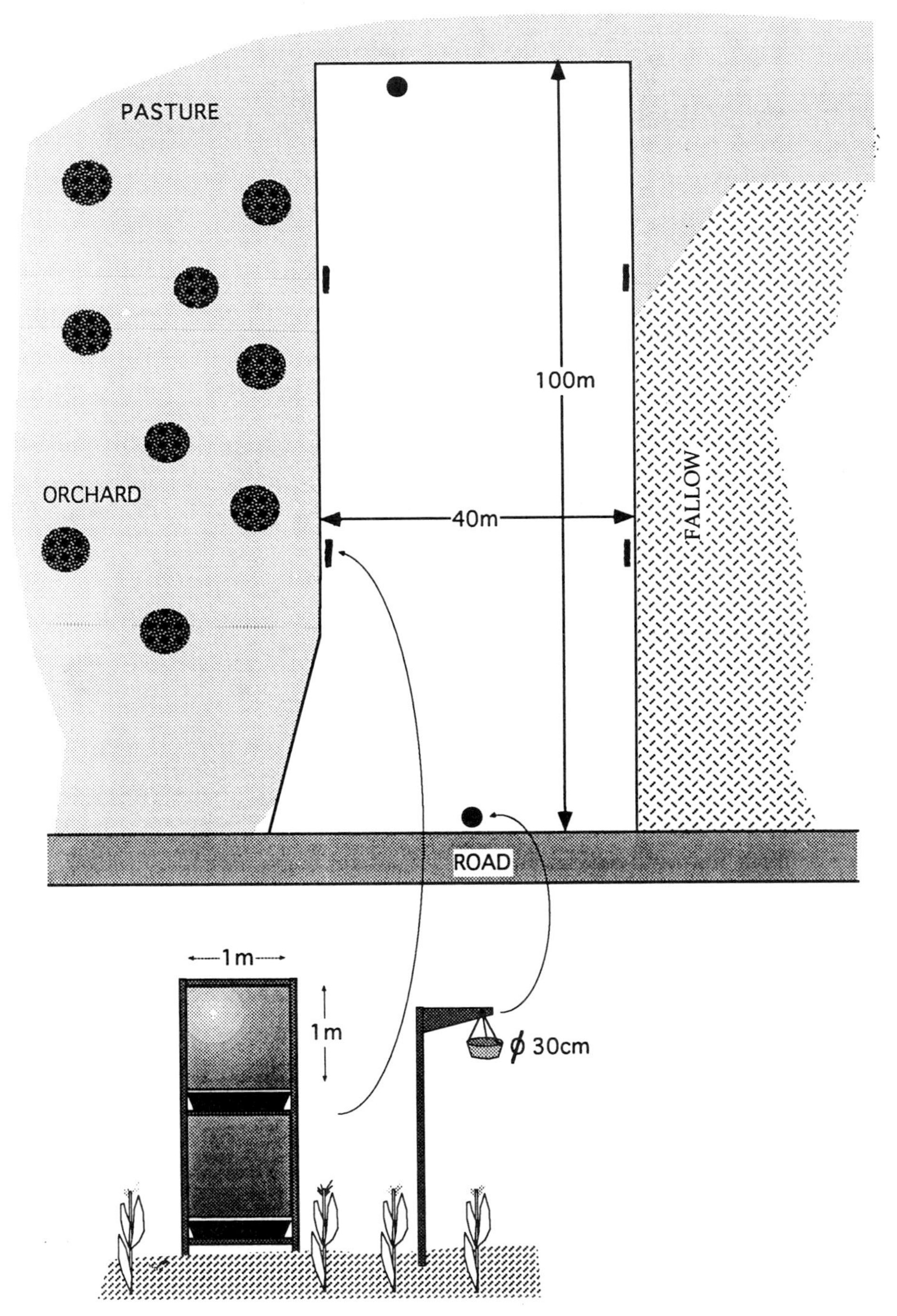

Fig. 2. Diagram of field B and its surroundings, indicating the positions of the traps. Below: design of window trap (left) and yellow trap (right).

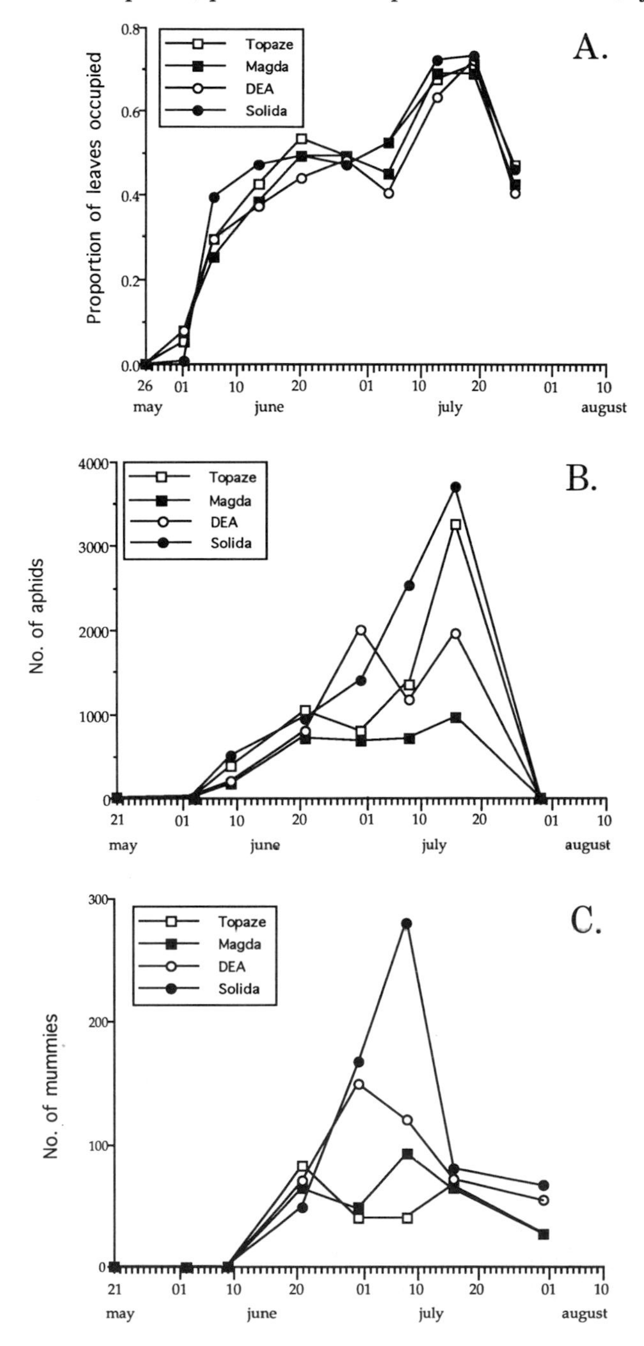

Fig. 3. Development of the leaf occupation index (A), total aphid population (B), and the number of aphid mummies (C) recorded on four maize varieties (40 plants per variety).

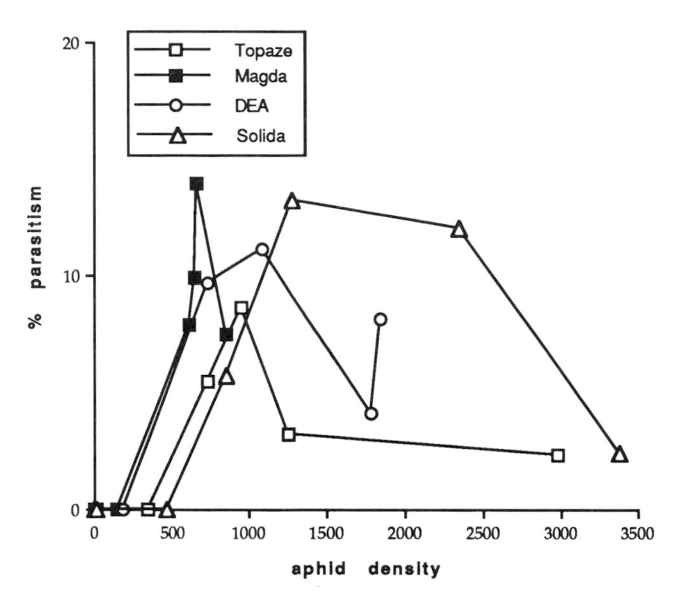

Fig. 4. Relation between the percentage parasitism and the aphid density recorded on 40 plants per variety.

Relationship between the level of parasitism and aphid densities

Microhymenopteran species were mainly *Aphidius rhopalosiphi* De Stefani-Perez and *Praon volucre* Haliday. The hyperparasitoid *Aloxysta victris* (Westwood) was also found. The first aphid mummies were recorded at the end of June when the aphid population had reached an average of 15 individuals per plant. The maximum number of mummies was observed from 10 to 15 days before the peak of aphid density, except for DEA where the two peaks coincided. Differences in numbers of mummies were observed between varieties (Fig. 3C). The greatest number of mummies was reached for Solida which also showed the highest number of aphids per plant. The lowest level of parasitism was observed for Topaze despite the second highest level of aphid infestation.

A composite graph of percentage of parasitism on aphid density shows that after an initial increase, this percentage decreased markedly (Fig. 4). This could be due to a functional response of type II by the parasitoids (Holling 1966). This response, however, varies with variety of maize, probably because the distribution of aphids differs between variety and thus influences the searching time of the parasitoids. A direct effect of the plant architecture or biochemistry on the microhymenopterans' searching behaviour could also be involved.

Abundances of coccinellids and chrysopids

Four species of Coccinellidae were regularly encountered in the fields: *Coccinella septempunctata* (L.), *Adalia bipuncata* (L.), *Propylea quatuordecim-punctata* (L.) and less often *Thea vigintiduopunctata* (L.). Fig. 5A shows the development of the *C. septempunctata* population. It is rather similar to the phenology described by Bonnemaison (1964) for the region of Paris. In our case, it also coincides with the maximum aphid density. The numbers of individuals observed were unfortunately too low to enable a comparison between maize varieties. Concerning Chrysopidae, *Chrysopa carnea* was the main species observed, but few adults and larvae were recorded. The number of eggs was larger and was also clearly related to aphid densities, as more eggs were observed on the more infested maize variety (Fig. 5B).

Window trap and yellow trap sampling

The purpose of using the window traps was to estimate the rate of immigration of predators into the field. Table 1 gives the numbers of coccinellids and carabids caught in each kind of trap from May 21 until August 30, and Fig. 5C gives the development of the catches for *Cocccinella septempunctata* and *Chrysopa carnea* in each kind of trap. Nine species of coccinellids were caught, the most abundant being the species already found in the fields. The numbers of individuals found in the window traps were low. Window traps corresponded to 4 metres of the field edge, while the perimeter of the field was 280 metres. If we assume that immigration into the field was constant on each side and limited to a height of 2 m, the number of *C. septempunctata* that entered the field was probably 2590 individuals for an area of 4000 m^2. This estimate needs to be checked by more intensive trapping, but it indicates the potential immigration of predators. More *C. septempunctata* were caught in the upper part of the traps (71%). Orientation of the traps towards adjacent crops may also have some importance, as 63% of individuals were trapped on the side of the orchard and only 27% on the side of the fallow. The same applies to *Adalia bipunctata,* as 67% of the individuals were trapped in the upper part of the window and 63% on the orchard side. According to Hodek (1973), coccinellids like *A. bipunctata* commonly breed and hibernate in orchards, but they move to other habitats in the course of the breeding season. This observation matches our results and shows the importance of spatial heterogeneity on a large scale for improving pest control. For *C. carnea*, 79% of individuals were trapped in the upper part of the window traps, but only 33 % on the orchard side. Window traps were also used by Katsoyannos et al. (1989) for monitoring aphids and

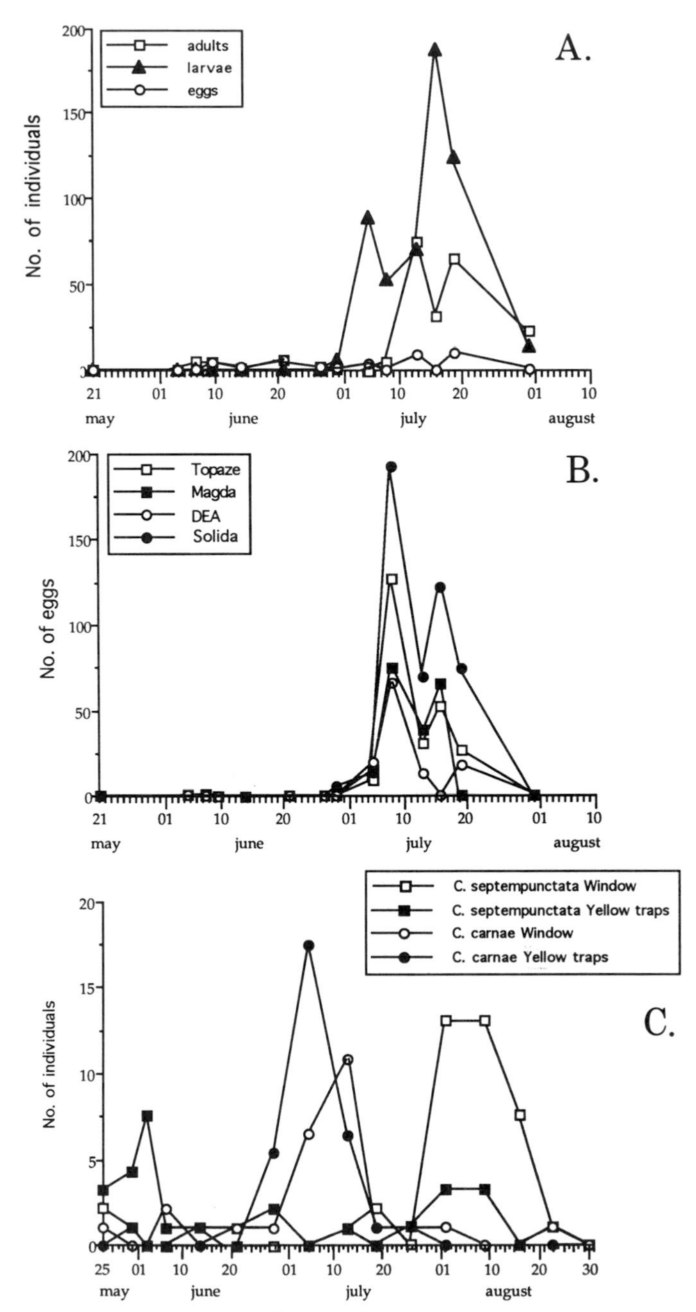

Fig. 5. Development of the *Coccinella septempunctata* population in four maize fields (totals on 160 plants)(A), the number of eggs of *Chrysopa carnea* on four varieties of maize (totals on 40 plants per variety) (B), and the captures of *Coccinella septemputctata* and *Chrysopa carnea* in window traps and yellow traps.

their aphidophagous predators in maize fields in Greece. In their case, the number of coccinellids trapped were very low, probably because of the small size of their traps.

The use of window traps provides new and interesting information concerning carabid beetles (Table 1). For example, *Demetrias atricapillus* had never before been collected in pitfall traps in the crops of this region. At the same time, *Bradycellus verbasci*, while well represented in the window traps, was found only a few times, although an intense pitfall trapping study was carried out from 1977 to 1986 (see Hance et al. 1989 for the species list found in that particular agricultural region).

Window trap data may thus supplement pitfall trap data in the study of carabid fauna. This is particularly important for small macropterous species or in the determination of flight periods of more common species such as *Trechus quadristriatus*. Window traps could also give new information on the role of adjacent habitats for the colonization of agricultural fields by predators. This could be of a great importance in the context of landscape management.

Table 1. List of species and numbers of individuals caught in the window and yellow traps.

Species	Window traps	Yellow traps
Coccinellidae:		
Coccinella septempuctata	37	19
Adalia bipunctata	11	34
Propylea quatuordecimpuctata	7	8
Thea vigintiduopunctata	6	1
Stethorus punctillum	4	0
Pullus auritus	1	1
Pullus haemoroidalis	1	1
Pullus melanarius	1	0
Pullus subvillosus	1	0
Carabidae:		
Trechus quadristriatus	23	5
Bradycellus verbasci	20	0
Demetrias atricapillus	9	1
Bembidion quadrimaculatum	5	4
Amara nitida	3	0
Harpallus cordatus	2	0
Dromius quadristriatus	1	1
Asaphidion flavipes	1	0
Amara plebeja	1	1
Amara strenua	1	1
Bembidion harpaloides	1	0
Amara aulica	1	0
Agonum dorsale	1	1

Conclusions

This first survey of aphids and their natural enemies in maize fields in Belgium shows that the decline of the aphid population at the end of August coincides with the maximum abundance of their main predators and parasitoids. Indeed, microhymenopterans, coccinellids and chrysopids are at their peak densities a little before or at the same time as aphids. The pressures exerted by this complex of predators and parasitoids (sensu Frazer et al. 1981) may cause a marked reduction in numbers of aphids. However, before it is possible to conclude that they are responsible for that decline, a lot of work has still to be done to improve the estimation of abundance of these predators and to analyse the effect of predator activity on their prey populations. More data are also needed on aphid phenology and life cycles and on their interactions with host plants. The present work has also indicated the importance of host-plant variety for the pest and for the behaviour of its predators and parasitoids. For instance, the host-plant could enhance or compromise the success of an IPM control programme by increasing or delaying the activities of biological control agents.

Acknowledgements

The author is indebted to Dr. G. Van Impe for his advice and for comments on the manuscript. He is also grateful to L. Renier and H. Vanderlinden for their technical help.

References

Bonnemaison, L. 1964. Observations écologiques sur la coccinelle à 7 points (*Coccinella septempunctata* L.) dans la région parisienne (Col.). *Bull. Soc. Entomol. France* 69: 64-83.

Frazer, B.D, Gilbert, N., Nealis, V. & Raworth, D.A, 1981. Control of aphid density by a complex of predators. *Can. Ent.* 113: 1035-1041.

Hance, Th, Delannoy, O. & Foucart, G. 1993. The screening of maize resistance to aphids as a contribution to integrated pest management. In: Struik, P.C., Vredenberg, W.S., Renkema, J.A. & Parleviet, J.E. (eds.) *Plant production on the threshold of a new century*, pp. 407-409. Kluwer Academic Publisher, Dordrecht.

Hance, Th., Gregoire-Wibo, C., Stassart, P. & Goffart, F. 1989. Etude de la taxocénose des Carabidae dans les écosystèmes agricoles du Brabant-Wallon. In : Institut Royal des Sciences Naturelles de Belgique (ed.) *Invertébrés de Belgique*, pp. 315-324. Bruxelles.

Hodek, I. 1973. *Biology of Coccinellidae.* Junk, N.V. Publishers, The Hague, Holland.

Holling, C.S. 1966. The functional response of invertebrate predators to prey density. *Mem. Ent. Soc Can.* 48: 1-86.

Katsoyannos, P., Mellidis,V., Katsadonis, N. & Sfakianakis, I. 1989. Aphid monitoring on Maize in two areas in Nothern Greece. In: Cavaloro, R. (Ed.) *'Euraphid' Network : Trapping and aphid prognosis*, pp. 271-284. Commission of the European Communities, Luxembourg.

An improved suction sampling device to collect aphids and their predators in agroecosystems

A. MacLeod[1], *N.W. Sotherton*[2], *R.W.J. Harwood*[1] & *S.D. Wratten*[3]

[1]Department of Biology, School of Biological Sciences, University of Southampton, Bassett Crescent East, Southampton SO9 3TU, England
[2]The Farmland Ecology Unit, The Game Conservancy Trust, Fordingbridge, Hampshire, SP6 1EF, England
[3]Department of Entomology & Animal Ecology PO Box 84, Lincoln University, New Zealand

Abstract
A petrol driven machine, normally used to collect leaf litter, was modified to be used as a suction sampler for polyphagous predators in cereals and grassy habitats and aphids in cereals. Recovery efficiency of Araneae, Carabidae and Staphylinidae did not differ significantly during winter sampling between various stands of single species grassland, although the structure of these grasses was very different. Summer sampling of aphids and their predators in winter wheat showed that the new sampler captured significantly more target organisms per unit area than did a traditional suction sampler. The machine was also lighter, cheaper and much easier to use than a traditional machine.

Key words: Arthropods, cereals, grass, suction sampler, sweeper vac

Introduction

Estimating pest and beneficial arthropod populations constitutes a major part of studies of their spatial and temporal dynamics. There are a variety of sampling methods available for terrestrial fauna (Southwood 1978). In arable crops, suction sampling and especially the use of the Dietrick vacuum insect sampler or D-vac (Dietrick, Schlinger & Bosch 1959) and related designs, e.g. the Thornhill vacuum sampler (Thornhill 1978) have become standard sampling methods. These machines are heavy and can become uncomfortable if used for extended periods. Ideally field workers require a lightweight, affordable and efficient machine which generates high air speeds and is able to sample adequate areas of vegetation (Southwood 1978, Holtkamp & Thompson 1985, Hand 1986).

Arthropod natural enemies in arable land · I *Density, spatial heterogeneity and dispersal*
S. Toft & W. Riedel (eds.). *Acta Jutlandica* vol. 70:2 1995, pp. 125-131.
 ISBN 87 7288 492

The Ryobi RSV3100E sweeper-vac described in this paper is one of a number of similar hand-held machines designed to collect leaf litter and light-weight garden debris by sucking it into a collecting bag. Such machines are readily available from retail outlets and generally cost less than £200.00 in the UK. With slight modification, the leaf-gathering suction devices can be used to sample aphids from grassy habitats (De Barro 1991). In this paper we assess the effectiveness of one such vac at collecting polyphagous predatory arthropods from grassy habitats and cereals and aphids from cereals. If the machine's efficiency for these groups is shown to be acceptable, it could prove to be a cheaper, lighter and more comfortable alternative to larger conventional samplers.

Materials and methods

The Ryobi weighs 5.2 kg and is 110 cm tall. The engine housing is 40 cm x 25 cm. The fan is driven by a two stroke 0.31 cc, air-cooled engine operating at 6800-7000 rpm (manufacturers' data). The bottom end of the air intake nozzle provided by the manufacturer had an oblique opening. This was cut to form a circular sampling area of 0.01 m^2. When used to collect arthropods the vac was held vertically. A fine weave nylon/cotton mix voile collecting bag was attached to an inner galvanised steel sleeve which was slightly narrower than the inner diameter of the modified air intake nozzle. The open end of the bag was attached to the outside of the sleeve and fastened with a large hose-clip. Once modified, the Ryobi vac and a suction sampler based on the design of Thornhill were weighed and air intake flows and noise levels at maximum revolutions measured.

Vac efficiency

Duration of suction time was calculated using the vac to collect epigeal predators from grassland in winter. Optimum duration was calculated as 30 seconds at which time over 75% of all epigeal predators were collected. Efficiency was not improved by increasing the suction time by up to 160 seconds (MacLeod, Wratten & Harwood 1994). Immediately after the last period of sampling, the vegetation and soil to a depth of 3 cm was removed from the sampled area using a soil corer. Cores were broken down in the laboratory and any polyphagous predators collected using a pooter.

Collecting polyphagous predators from single-species grass stands

Two study farms in southern England were used for this study where various replicated plots of single grass species had been sown for other experimental purposes. In Oxfordshire the grasses sampled were (n= number of replicates) *Holcus lanatus* (L.) (n=15), *Festuca rubra* (L.) cv. Dawson (n=18) and *Dactylis glomerata* (L.) (n=30). In Hampshire, the grasses sampled were *Agrostis stolonifera* (L.) (n=12), *Lolium perenne* (L.) (n=6), *H. lanatus* (n=12) and *D. glomerata* (n=6).

In the laboratory, the above-ground vegetation was removed from the soil cores and dried at 65°C for 24h, then weighed. The range of polyphagous predatory species varied considerably between and within the winter grass samples. Predators were therefore pooled into three groups: Araneae, Carabidae and Staphylinidae. The number of individuals in each group, recovered by the vac at each sampling point, was converted to proportions of the total of each group recovered from the vac and soil core together. These proportions were transformed using the arcsine transformation. Linear regressions were calculated between this transformed proportional recovery of each predatory group and grass dry weight for each grass species.

Comparison of a conventional suction sampler and a vac sampling for aphids and polyphagous predators in winter wheat

During May 1992, in two fields of winter wheat both at G.S. 36 (Zadoks, Chang & Konzak 1974) on a farm in Hampshire, the vac and a suction sampler based on a design of Thornhill were used to collect aphids. Samples were taken across the fields with the suction sampler and vac approximately 3 m apart. In the first field (cv. Mercia), seven samples of twenty cumulative 5s extractions were taken with both devices. The Thornhill-type suction device sampled 0.1 m^2 per 5s sampled from a combined total area of 14 m^2. The vac sampled a total area of 1.4 m^2. In the second field (cv. Maris Huntsman), eleven samples of ten cumulative single 5s extractions were taken. The suction sampler sampled a total of 22 m^2, the vac sampled a total area of 2.2 m^2.

At a later growth stage (G.S. 69) polyphagous predators were collected from cv. Maris Huntsman using both the sampler based on Thornhill's design and the vac. Both samplers took 10 samples of 10 cumulative 5s extractions. Statistical analysis of the catches was by Student's t-test after log(n+1) transformation.

Results

The vac with modifed nozzle weighed 6.1 kg. The Thornhill type suction sampler weighed 27.2 kg. Both samplers were weighed with full fuel tanks. The mean air flow for the vac was 9.6 m^3 min^{-1} over a sampling area of 0.01 m^2 (air speed 16 ms^{-1}) that for the suction sampler was 34.2 m^3min^{-1} over a sampling area of 0.1 m^2 (air speed of 5.7 ms^{-1}). Maximum noise level measured for the operator using the vac was 98dB and for the suction sampler 102dB which was a noticeable reduction in noise level for the operator.

Vac samples contained mainly Araneae and Coleoptera, but Diptera and Isopoda were also retrieved. All fauna were captured intact with no noticeable damage caused by the sampling procedure. Some loose soil and small stones were collected by the vac, the heaviest stone collected weighing 3.9 g. The heaviest carabid beetle collected was *Agonum dorsale* (Pont.) at 15.6 mg.

Collecting polyphagous predators from single-species grass stands

There were no significant ($P<0.05$) relationships between extraction efficiencies and grass dry weights up to a weight of 19.7 g per 0.01 m^2. Numbers of polyphagous predators collected by the vac and vac plus soil cores combined from the single grass species stands at both study sites are shown in Table 1.

There were no significant differences in the percentages of Carabidae, Staphylinidae or Araneae captured between grass species at the Hampshire site. There were no significant differences extracting Staphylinidae or Araneae at the Oxfordshire site. However there were differences extracting Carabidae. A higher proportion of Carabidae were captured from *D. glomerata* than from *F. rubra*. This was due to the higher proportion of Carabidae that were actually buried in the soil under the *F. rubra* than in the soil under the *D. glomerata* during sampling.

Table 1. % efficiency of the Ryobi vac suction sampler to collect epigeal predators from five single-species stands of grass, Southern England, Winter 1991/92.

	Total Carabidae			Total Staphylinidae			Total Araneae			Total no. of Invertebrates		
	Vac	Total	%	Vac	Total	%	Vac	Total	%	Vac	Total	%
Dactylis glomerata	222	366	60.6	194	331	58.6	223	250	89.2	639	947	67.5
Holcus lanatus	311	750	41.3	47	368	12.7	198	218	90.8	376	661	56.9
Festuca rubra	44	74	59.4	90	234	38.5	114	121	94.2	248	429	57.8
Agrostis stolonifera	5	11	45.4	18	44	40.9	27	36	75.0	50	91	54.9
Lolium perenne	6	15	40.0	18	44	40.9	27	36	75.0	51	95	53.7
Mean % efficiency			49.3			38.3			84.8			58.2
± SE			4.45			7.35			4.10			2.44

Table 2. Mean densities (95% confidence limits) per 0.1 m^2 of *Tachyporus* spp. adults and larvae, and of cereal aphids collected from wheat fields in May using the Ryobi vac and a Thornhill-type suction sampler, Southern England, 1992.

	Ryobi Vac		Thornton-type Sampler	
Tachyporus				
Adults	2.10	(1.01- 3.19)	0.04	(0.02- 0.06)
Larvae	16.3	(9.0 - 23.6)	2.4	(1.7 - 3.1)
Aphids	2.06	(0.83 - 3.29)	0.16	(0.09 - 0.22)

Overall sampling efficiency was about 50%. Sampling efficiency was highest for the Araneae and lowest for the Staphylinidae. Invertebrates were extracted most easily from the fine leaved *F. rubra* and less easily from the densely tussocked *D. glomerata* (Table 1).

Comparison of a conventional suction sampler and a vac sampling for aphids and polyphagous predators in winter wheat

Although the conventional suction sampler sampled an area 10 times greater than that of a vac, the vac captured significantly more cereal aphids per unit area (Table 2).

When used in winter wheat during G.S. 69, the most abundant predatory group recovered were adults and larvae of *Tachyporus* spp. The vac captured significantly more of these larvae per unit area than did the suction sampler (Table 2).

Discussion

The vac has a number of advantages over the traditional suction samplers. For fieldworkers, the weight, reduced noise level and ease of use will appeal; the lower cost is also an important factor. However, unless the new types of suction device are shown to be as effective or better than the traditional type of suction sampler, they will not be adopted as sampling tools. Used in grasses of different structure in the winter, the vac was able to extract substantial proportions of the arthropods present. Within the limits of the weight of the vegetation sampled, increasing weight of vegetation did not significantly reduce capture efficiency, suggesting that a threshold weight, where the mass of vegetation reduces extraction efficiency, had not been reached. However experiments to discover the upper weight limit and the effects of grass weight or structure on extraction efficiency have not been conducted.

When comparisons were made with the suction sampler during summer sampling, the vac collected significantly greater numbers of aphids and *Tachyporus* spp. adults and larvae from cereal fields, probably due to its greater rate of air flow per unit area over the sampling site. However more care has to be taken when placing the vac over cereals for sampling than when using the traditional suction sampler because of the narrower collecting area of the vac. We also have no information regarding the impact of using such a narrow nozzle on the sampling efficiency of species with aggregated distributions. Taking fewer samples with a narrow nozzle could decrease sampling efficiency of aggregated species. The trade off between nozzle size and spatial distribution is currently under investigation using mathematical models. However the Ryobi's narrower nozzle may also make it more susceptible to errors caused by the "edge effect" compared to machines using wider nozzles. The "edge effect" is where invertebrates are sucked into the machine under the sides of the nozzle from areas outside the sampling head. In other words from a greater area than that of the nozzle itself. This phenomenon has been previously reported by Duffey (1974).

Large suction samplers carried on the back of field workers have been used over many years and have established themselves as a standard piece of field-work equipment when sampling insects on herbage. The "densities" recorded via the vac compared with the larger suction sampler were, however, much higher. This has implications for the many studies in which field populations of these and other arable invertebrates have been assessed using the D-vac and related machines. Such studies often compare densities before and after pesticide use (Greig-Smith et al. 1992) or use the densities obtained for modelling the effects of predation on pest populations (Winder et al. 1994). If the underestimates reported here are typical of the larger suction samplers, then calculation of the populations of such predators in arable land and possibly of their role in reducing pest numbers, may have been severely underestimated. The fact that significant proportions of the predator populations in the shallow soil cores were not extracted (Table 1) adds to the extent of this probable underestimation in published studies.

The new vac described here is not a panacea for sampling terrestrial predatory arthropods. Sunderland et al. (1987) suggested an "absolute" sampling regime which combined a range of methods to accurately determine predator numbers. In this regime the limitations of using conventional suction samplers were overcome by using pitfall traps and ground searching techniques. This sampling approach is however very labour intensive and involves some habitat destruction.

This paper forms an initial attempt to describe the efficiency of a new, more user-friendly suction sampler. It is not the definitive work and much more detailed, comparative quantification is necessary.

Acknowledgements

This work was carried out during two SERC CASE awards between Southampton University (SU) and The Game Conservancy Trust and SU and Willmot Industries Ltd. (Conservation Department). We thank M. Lower & M. Wright for help with measuring air speeds and noise levels, fieldwork was assisted by G. Cox, P. De Barro, J. Holland, F. Rothery, & L. Sijun. Thanks are also due to Prof. J. Hirst for comments on earlier work.

References

De Barro, P.J. 1991. A cheap lightweight, efficient vacuum sampler. *J. Aust. Ent. Soc.* 30: 207-208.

Dietrick, E.J., Schlinger, E.I. & Bosch, R. Van Den 1959. A new method for sampling arthropods using a suction collecting machine and modified Berlese funnel separator. *J. Econ. Ent.* 52: 1095-1091.

Duffey, E. 1974. Comparative sampling methods for grassland spiders. *Bull. Br. Arachnol. Soc.* 3: 34-37.

Greig-Smith, P.W., Frampton, G.K. & Hardy, A.R. (eds.) 1992. *Pesticides, cereal farming and the environment. The Boxworth Project.* London.

Hand, S.C. 1986. The capture efficiency of the Dietrick vacuum insect net for aphids on grasses and cereals. *Ann. Appl. Biol.* 108: 233-241.

Holtkamp, R.H. & Thompson, J.I. 1985. A lightweight, self-contained insect suction sampler. *J. Aust. Ent. Soc.* 24: 301-302.

MacLeod, A., Wratten, S.D. & Harwood, R.W.J. 1994. The efficiency of a new lightweight suction sampler for sampling aphids and their predators in arable land. *Ann. Appl. Biol.* 124: 11-17.

Southwood, T.R.E. 1978. *Ecological Methods with Particular Reference to the Study of Insect Populations*. 2nd Edition. London.

Sunderland, K.D., Hawkes, C. Stevenson, J.H., McBride, T., Smart, L.E., Sopp, P.I., Powell, W., Chambers, R.J. & Carter, O.C.R. 1987. Accurate estimation of invertebrate density in cereals. *Bulletin SROP* 10:71-81.

Thornhill, E.W. 1978. A motorised insect sampler. *PANS* 24:205-207.

Winder, L., Hirst, D.J., Carter, N., Wratten, S.D. & Sopp, P.I. 1994. Estimating predation of the grain aphid *Sitobion avenae* (F.) by polyphagous predators. *J. Appl. Ecol.* 31: 1-12.

Zadoks, J.C., Chang, T.T. & Konzak, C.F. 1974. A decimal code for the growth stages of cereals. *Weed Res.* 14: 415-421.

Density estimation for invertebrate predators in agroecosystems

K.D. Sunderland[1], *G.R. De Snoo*[2], *A. Dinter*[3], *T. Hance*[4], *J. Helenius*[5], *P. Jepson*[6], *B. Kromp*[7], *J.-A. Lys*[8], *F. Samu*[9], *N.W. Sotherton*[10], *S. Toft*[11] & *B. Ulber*[3]

[1]Horticulture Research International, Littlehampton, West Sussex, BN17 6LP, England
[2]Centre of Environmental Science, Leiden University, P.O. Box 9518, NL-2300 RA Leiden, The Netherlands
[3]Institute for Plant Diseases and Plant Protection, University of Hannover, Herrenhäuserstrasse 2, D-30419 Hannover, Germany
[4]Université Catholique de Louvain, Unité d'Écologie et de Biogéographie, Place Croix du Sud, 5, 1348, Louvain-la-Neuve, Belgium
[5]Institute of Crop Protection, Agricultural Research Centre, FIN-31600 Jokioinen, Finland
[6]Department of Biology, Medical & Biological Sciences Building, Bassett Crescent East, Southampton SO9 3TU, England
[7]Ludwig Boltzmann Institute for Biological Agriculture and Applied Ecology, Rinnböckstraße 15, A-1110 Vienna, Austria
[8]Zoological Institute, University of Bern, Baltzerstraße 3, CH-3012 Bern, Switzerland
[9]Plant Protection Institute of the Hungarian Academy of Sciences, Budapest II, Herman Ottó út 15, Hungary
[10]The Game Conservancy Trust, Fordingbridge, Hampshire SP6 1EF, England
[11]Department of Zoology, University of Aarhus, Building 135, DK-8000 Århus C, Denmark

(2nd to 12th author in alphabetical order)

Abstract
A review of the principal methods available for the estimation of predator density or abundance (suction apparatus, habitat search, mark-release-recapture, fenced pitfall traps, ground photoeclectors, soil flooding, microhabitat removal, catch per unit effort, distance method, trap stones and unfenced pitfall traps) indicated that good data on the efficiency of these methods, and the causes of variation in efficiency, are generally lacking. The relative advantages, disadvantages and limitations of each of these methods are discussed. Since no single sampling method is appropriate for all circumstances, methods appropriate for specific requirements are recommended here, taking account of the type of predator (aphid-specific predators, spiders, carabid and staphylinid beetle adults and larvae) and

Arthropod natural enemies in arable land · I *Density, spatial heterogeneity and dispersal*
S. Toft & W. Riedel (eds.). *Acta Jutlandica* vol. 70:2 1995, pp. 133-162.
 ISBN 87 7288 492

crop involved and the scale of the investigation. Gaps in our ability to assess the abundance of some species reliably are revealed, especially in extensive sampling programmes, where effort per sample must be limited.

Key words: Density, abundance, polyphagous predators, agroecosystems, sampling methods, sampling efficiency, Araneae, Carabidae, Staphylinidae, Coleoptera larvae, suction apparatus, habitat search, mark-release-recapture, fenced pitfall traps, ground photoeclectors, soil flooding, microhabitat removal, catch per unit effort, distance method, trap stones, visual count sampling, trapping webs, unfenced pitfall traps

Introduction

Reliable estimates of population density are vital to studies of the population dynamics of invertebrate predators in agroecosystems. For other purposes (such as pest control modelling, ecotoxicological investigations and assessments of the effectiveness of measures taken to augment the numbers of predators) indices of predator abundance may be adequate. However, to have any value, these must relate reliably to population density. Indices of abundance must therefore be without bias, or, failing this, any biases should be quantified and accounted for. For example, indices of predator abundance in a range of crops may suggest differences in catch between sites. It cannot be concluded safely from these indices that predator density varies from crop to crop until variation in sampling efficiency caused, for example, by the crop structure, is accounted for. Unfortunately, quantitative investigations of the causes of variable sampling efficiency have only rarely been made. In some cases, even basic efficiency data (e.g. from measurements made under standard laboratory conditions) are lacking. The provision of efficiency data might enable re-interpretation of some results from the published literature. It would also improve the reliability of results emanating from simple, uncomplicated apparatus; an important consideration in countries where expensive equipment is not available.

The first section of this paper is a review of the principal methods currently used for density estimation and abundance assessment. It draws attention to any associated advantages, disadvantages, limitations and significant deficiencies of information. Aspects such as the optimal size and number of sample units to achieve a given precision for a given effort, and the pattern, timing and frequency of sampling in relation to predator dispersion pattern will not be treated extensively here, since this is adequately covered elsewhere (e.g. Southwood 1978). In the second section an attempt is made to indicate which methods (or combinations of methods) are most appropriate for specific requirements, taking account of the species or higher taxonomic group of predator involved, the crop type, the scale and intensity of investigation, and the resources available. Perennial crops and non-crop habitats are excluded from detailed consideration here, although aspects of the review will be relevant to sampling problems in these habitats. The review focusses on epigeal invertebrate predators and detailed treatment of methods applicable to soil and air are not included.

Methods for density estimation and abundance assessment

Suction apparatus

The use of suction apparatus for collecting invertebrates in the field appears to date back to McGinnis in 1923 (quoted in Hills 1933). The Dietrick vacuum insect net (D-vac) (Dietrick 1961) is perhaps the most widely used type of suction sampler. It has a sampling nozzle diameter of 33 cm and a nozzle air velocity of about 40 m sec^{-1} (De Barro 1991). Duffey (1974) compared the D-vac catch of spiders with *in situ* searching in calcareous grassland; 9 spiders per 0.5 m^2 were collected by D-vac, compared with 6 per 0.5 m^2 by searching. In a later paper he measured D-vac efficiency by subjecting pasture turves, that had been sampled by D-vac, to Tullgren funnel extraction (Duffey 1980). D-vac efficiency was lower for Araneae (7-49%) and Coleoptera (7-27%) than for Hemiptera (22-65%) and Hymenoptera (83-92%) and it was reduced by high and dense vegetation. Overall efficiency was only 14-18% in May and 33-58% in August. Duffey (1974) also found that significantly more spiders were caught in 99 x 0.09 m^2 D-vac samples (arranged into 11 subsamples, 3 m apart, along a set of transects) than for D-vac collection within 9 random 1 m^2 quadrats (using a frame), even though the same area was sampled in both cases. This might have been due to the disturbance caused by placement of the frame. Alternatively, it is possible that spiders were sucked into the D-vac nozzle from just beyond its 0.09 m^2 area; if this is the explanation then this source of error could be reduced by suction sampling within larger quadrats. Pruess, Lal Saxena & Koinzan (1977) encountered the same phenomenon when suction sampling insects in alfalfa and they described a mathematical conversion to correct for this effect. Other workers have avoided the problem by sampling within 30-60 cm high metal cylinders sealed into the ground (Johnson, Southwood & Entwistle 1957, Southwood & Pleasance 1962, Törmälä 1982, Summers, Garrett & Zalom 1984, Toft, Vangsgaard & Goldschmidt 1995). Although D-vac efficiency is low for large, heavy predators (Bayon et al.

1983), it has also been found to vary with species of predator within size groups and in relation to vertical stratification (Sunderland & Topping 1995). These authors also found that efficiency varied according to site/year, season and degree of weed cover. Density of vegetation as a factor reducing D-vac efficiency has also been noted by Hand (1986) and Dewar, Dean & Cannon (1982) in relation to sampling cereal aphids. Dense vegetation may affect efficiency by reducing airflow and also by forming a filter under which predators can hide and avoid capture. The Thornhill suction sampler has a nozzle airspeed of 5.7 m sec^{-1} and an airflow of 34.2 m^3 min^{-1} (Macleod et al. 1995). It is similar in design to the D-vac and capture rates of various invertebrates, including Staphylinidae and Araneae, were also very similar (Thornhill 1978).

Other types of machine are characterised by narrower suction hoses, giving higher intake velocities for sampling over a very localised area. Such samplers are often used to systematically remove predators within quadrats in habitats with short vegetation but are not usually suitable for sampling large crop plants. The apparatus of Arnold, Needham & Stevenson (1973) has a nozzle diameter of 15 cm. Henderson & Whitaker (1976) found that the efficiency of this machine in pasture varied most according to taxonomic group and height of vegetation. Johnson, Southwood & Entwistle (1957) designed a machine with a 3 cm diameter nozzle that had an efficiency (determined by handsorting vegetation and soil in the laboratory) in grass of 67% for Chilopoda, 70% for Coleoptera larvae and 92% for Staphylinidae. The machine of Heikinheimo & Raatikainen (1962) also had a nozzle diameter of c. 3 cm. They added known numbers of leafhoppers to grass, which was sampled two hours later, at which time they recorded a sampling efficiency of 75% for nymphs and 88% for adults. The Burkard Univac, with a 19.6 cm^2 nozzle opening, caught only 2.5 spiders per 0.5 m^2 in grassland, compared with 6 per 0.5 m^2 by *in situ* search (Duffey 1974). Using the same machine in grass heathland, Workman (1978) collected only 3 *Trochosa terricola* (Thorell) (Araneae: Lycosidae) per m^2 compared with 12 per m^2 by *in situ* search and 31 per m^2 by heat extraction of removed turves. In very dense grassland, invertebrates can remain inaccessible to suction samplers below a thick mat of vegetation, but these machines remain useful for estimating densities of the field layer fauna in such habitats. Törmälä (1982) compared a Burkard suction sampler with four other sampling methods and concluded that suction gave the most unbiased representation of the field layer community in dense grassland.

New machines continue to be designed and assessed. Summers, Garrett & Zalom (1984) compared an Echo PB-400 Power Blower with venturi attachment (or UC-Vac) with a D-vac and found the sampling efficiency of the two machines to be similar, except that the UC-vac captured more Heteroptera and Araneae than the D-vac. De Barro (1991) evaluated a modified McCulloch Eager Beaver Blower/Vac R, which was an improvement on the machine described by Holtkamp & Thompson (1985). The De Barro machine had a sampling efficiency of 60-93% for the cereal aphid *Rhopalosiphum padi* (L.) in grassland. Efficiency was reduced in tall grassland, compared with short, but did not appear to be affected by aphid density. Wright & Stewart (1992) compared a Blowvac with a D-vac in natural grassland; the Blowvac had greater suction power near to the ground and caught more Araneae and Coleoptera. The Ryobi RSV3100E sweeper-vac has a nozzle airspeed of 16 m sec^{-1} and an airflow of 9.6 m^3 min^{-1} and is capable of collecting invertebrates of up to 3.9 g. Its efficiency seemed to be little affected by vegetation density (MacLeod et al. 1994, 1995). The Gravesen machine sucks animals into a glass jar, the tube between nozzle and jar being lined with fine cloth to reduce damage to the sampled animals (Toft, Vangsgaard & Goldschmidt 1995). Machines such as this, with high intake velocities, may sometimes rupture spider eggsacs and produce an inflated estimate of the density of immature spiders. Toft, Vangsgaard & Goldschmidt (1995) occasionally obtained samples with small numbers of spiders of an instar that occurs only within the eggsac.

Various modifications can be made to suction samplers to alter their performance. Kennedy, Evans & Feeney (1986) used adjuster bars attached to the D-vac head to sample adults and larvae of *Tachyporus hypnorum* F. (Coleoptera: Staphylinidae) at various vertical distances from the ground in cereal fields. Topping & Sunderland (1994) used a steel extension tube attached to the D-vac head to reduce spider mortality during sampling of mature winter wheat and thus permit live-sorting of the catch. Yeargan & Cothran (1974) demonstrated that addition of a flange, set at an angle of forty degrees within the D-vac head, reduced escape rates of large predators from the D-vac during sampling in alfalfa. Estimates of the density of adult Nabidae (Heteroptera) were nearly twice as great in flanged compared with unflanged D-vacs. Summers, Garrett & Zalom (1984) attached a narrow hose suction adaptor to the proximal end of their UC-vac to increase efficiency.

Suction samplers have also been used in novel ways, or in conjunction with other methods, to obtain better estimates of density. Haas (1980) used repeated suction sampling within a 0.25 m^2 biocenometer (or gauze covered isolator, see Mühlenberg 1993) in the vegetation layer of grassland and calculated the efficiency of this method for different arthropod taxa on the basis of decreasing numbers caught in successive samples. The sampling efficiency for spiders was estimated at 85%. Marston, Davis & Gebhardt (1982) caged part of a soybean crop, sprayed permethrin inside to knock down invertebrates from the crop, and then used a D-vac to collect the invertebrates from inside the cage 15 minutes later. This method collects predators only from the vegetation layer. Hills (1933) attached a cylindrical cage to the end of a 1.4 m long pitchfork handle and used it for rapid caging of sugar beet plants, in an attempt to incarcerate highly active species that normally escape from the sampling area when the experimenter approaches. He then removed insects from the cage using an electric vacuum collector. Following the same principle, Turnbull & Nicholls (1966) devised a "quick trap" for sampling invertebrates in grassland. A net was attached to a ring on top of a tripod and was held above the crop for 24 hours to allow active invertebrates to re-distribute naturally below. It was then triggered from a distance of c. 10 m and fell to the ground trapping the invertebrates below. A narrow nozzle suction sampler was then used to remove invertebrates from within the cage. This technique revealed the true density of active invertebrates to be very much greater than that recorded by other methods, e.g. the density of adult Diptera was 1.5 per m^2 by suction sampler, but 30 per m^2 by the "quick trap" (Turnbull & Nicholls 1966). A modification of the "quick-trap" technique was employed by Gromadzka & Trojan (1967) who sampled invertebrates in meadows using a 0.25 m^2 biocenometer (or mesh-covered pyramidal isolator) that was dropped quickly onto the grass from a small hand-held crane. Mason & Blocker (1973) used the same principle, with improvements, in the "drop trap". In this apparatus, the trap was held vertically on supports. When it was pushed over, the cage isolated an area of ground 5 m distant from the experimenters. Metal blades below the cage allowed it to cut into the turf when it dropped, thus improving the seal. It has the advantage of eliminating the 24 hour delay, and it can be used repeatedly in the same field to achieve the required number of sample units per sample. The authors did not provide data on the sampling efficiency of their "drop trap".

Accurate density estimates of specific growth stages of predators, rather than general indices of abundance, are usually required in population dynamics studies. Suction samplers form one component of the methods that have been designed to provide these estimates. An attempt is made to remove a very high proportion of the target predators from the vegetation and ground surface of a unit area of habitat. Simultaneous removal from vegetation and ground reduces the problem of differential sampling efficiency related to diel cycles in the vertical distribution of predators (Vickerman & Sunderland 1975, Loughridge & Luff 1983, Leathwick & Winterbourn 1984, Chiverton 1988). Dinter & Poehling (1992) used an intensive D-vac technique to sample spiders in winter wheat. Shoots within 0.25 m^2 cylinders were shaken then excised 10 cm above ground, the area was then sampled for 3 minutes using a D-vac with a 20 cm diameter nozzle. The collecting bag was emptied after 1 minute to guarantee a strong suction pressure for the remaining 2 minutes. The efficiency of this method varied from 75% to 100% (estimated from recapture rates of marked spiders), depending on the species and sex of spider. Repeated D-vac sampling of the same area yielded little increase in efficiency (Dinter 1995), as was also found by Pruess, Lal Saxena & Koinzan (1977). Another approach, which has been used for spiders in winter wheat, is to collect predators within a defined area using a D-vac, with a 33 cm diameter nozzle, and then immediately make a careful search of the ground in the area just sampled by the D-vac; density is then estimated by numbers collected by D-vac plus ground search (Topping & Sunderland 1994, Sunderland & Topping 1995). Sunderland et al. (1987) tested this method for total polyphagous predators in winter wheat, but with the added component of pitfall trapping within the area already sampled by D-vac and ground search; pitfall traps were enclosed in a tightly sealed isolator. Pitfalls contributed a significant proportion of the the total catch (especially for adults and larvae of Carabidae and Staphylinidae), suggesting that some predators were below ground during the period when D-vac and ground search collections were made. The carabid beetle, *Harpalus rufipes* (Degeer), for example, can occur at depths of 25 cm below ground during the daytime (Luff 1978). Similar findings have been made for some other species of carabid (Desender & Maelfait, 1983, Desender, van den Broeck & Maelfait, 1985) and it is mainly the large, nocturnal, larval-overwintering species of carabid (Luff 1978, Kegel 1990) that are to be found below ground during the daytime. Wallin & Ekbom (1988) observed the carabid *Pterostichus melanarius* (Ill.) burrowing into the soil of a cereal field and *Pterostichus niger* (Schall.) squeezing into crevices in the soil. The carabid *Clivina fossor* (L.) is structurally adapted for burrowing into soil (Forsythe 1987). Some small linyphiid spiders, such as *Porhomma microphthalmum* (O.P.-Cambridge) also appear to be partly subterranean (Topping & Sunderland 1995) and staphylinid beetles of the genus *Philonthus* can be found at depths of up to 15 cm in light soils (Frank 1967).

A significant limitation to the use of techniques involving most suction samplers is that the habitat to be sampled must normally be dry (e.g. Törmälä 1982), otherwise invertebrates become stuck to the collecting nozzle, hose and net and sampling efficiency is drastically reduced. Thus sampling may not be possible at night or at dawn, when there is a heavy dew, or during the daytime after a period of rain. If sampling conditions are marginal, it may be advantageous to use a Blower/Vac sampler with a high nozzle air velocity, so that free water in the sample is expelled through the blower tube. De Barro (1991) used such a machine and claimed that its performance was not adversely affected when vegetation was wet.

Habitat search

Sessile life stages of predators are not effectively sampled by suction machines. Hance (1995) used *in situ* plant searches to record the abundance of a range of life stages of aphid-specific predators on maize. Chambers & Adams (1986), and Cowgill, Wratten & Sotherton (1993), found that eggs and larvae of Syrphidae and Coccinellidae could be numerous on weeds below winter wheat, as well as on the crop plants. Chambers & Adams (1986) counted the number of these predators on 100 to 400 shoots per sample, then excised the wheat shoots in 10 x 0.1 m^2 quadrats and counted eggs and larvae on the stubble, weeds, soil surface and under stones. Data on predators per shoot and per quadrat were combined with counts of shoot density to provide total habitat predator density estimates.

Nyffeler (1982) counted spiders on the vegetation of grassland and cereals within 1 m^2 quadrats; smaller quadrats (0.04 - 0.16 m^2) were used for ground searching. He noted that Lycosidae were too active to be assessed adequately by this method. Searching confined to the ground zone (e.g. Sunderland, Crook, Stacey & Fuller 1987) omits any predators that remain attached to the plant during the searching period. Sunderland (1987) considered that, where ground search and D-vac sampling are concurrent, better density estimates result from selecting the highest value (D-vac cf. ground search) for each species on each date, than by using either D-vac or ground search estimates separately or summed. If the study is confined to relatively large species, such as adult Carabidae, it should be possible to find the majority of individuals by *in situ* search of plant, ground surface and the soil to a depth of 15 cm (Hance 1992). Efficient search of the ground surface and upper layers of soil necessitates removal of the plant cover (Ulber & Dinter, unpublished). Winder et al. (1994) searched 0.1 m^2 quadrats *in situ* for up to 15 min each, then used a trowel to remove weeds, stones, wheat plants and roots into a plastic tray for further searching. It should be noted that habitat search is labour-intensive, and especially so in dense, complex habitats; e.g. Duffey (1974) required 60 - 90 man minutes to search 0.5 m^2 of calcareous grassland (mean height 15 cm) and remove all spiders present.

Mark - release - recapture (MRR)

When marked predators are released into a habitat the proportion of marked animals recaptured at a later date enables calculation of population size. Various assumptions are implicit in this method and there are many different equations available for calculating population size, some of which allow for dilution of the marked population by mortality and emigration. Release of marked predators should be carried out in such a way that the assumption, that marked and unmarked predators have the same probability of recapture, is not invalidated. The MRR programme is best planned in advance, to suit a particular predator and crop combination, and with prior knowledge of which equation will be used to calculate population size, since this determines the exact MRR tactics to be adopted. Extensive information on these aspects of MRR are available in Seber (1973), Southwood (1978) and Begon (1979). MRR for predators in agroecosystems usually involves the use of pitfall traps. By recording the distance moved by marked predators within a trapping grid, the area of influence of individual traps, or of the trapping grid as a whole, can be estimated (Kuschka, Lehmann & Meyer 1987, Franke, Friebe & Beck 1988). Alternatively, the area of influence can be measured by defining concentric rectangles within the grid and noting at which size of rectangle the density estimate becomes stabilised (Loreau & Nolf 1993). Density is calculated as population size divided by area of influence. Where MRR is conducted within large field enclosures (Loreau 1984), the area of influence is coincident with the area of the enclosure. The use of such enclosures has the advantage of suppressing migration, but it is unlikely that sufficient enclosures can be set up to allow for spatial variation in population density (Loreau 1984).

Large, active predators (such as adult carabid beetles and lycosid spiders), that have a high capture rate in pitfall traps, are suitable candidates for MRR. Marking can be by various types of paints (e.g. Hackman 1957, Samu & Sárospataki 1995), inks (e.g. Kromp & Nitzlader 1995), typewriter correction fluid (Sotherton unpublished), microcautery (e.g. Manga 1972, Ericson 1977) or by using a drill or medical saw to make scratches or pits on the exoskeleton (e.g. Benest 1989a, Loreau & Nolf 1993). It is also possible to mark large, surface-active beetle larvae using paint and tags (Nelemans 1986). Marked individuals do not usually suffer greater mortality than unmarked controls (Loreau 1984). MRR has been used for carabids (e.g. Ericson 1977, 1978, Samu et al. 1991, Kromp & Nitzlader 1995, Samu & Sárospataki 1995), lycosids (Hackman 1957, Greenstone 1979, Samu & Sárospataki 1995) and staphylinids (Frank 1968). A problem sometimes encountered with this method is that recapture rates can be too low to permit the calculation of population size. Samu & Sárospataki (1995), for example, marked 512 of the carabid *Harpalus rufipes* (Degeer) and released them into alfalfa in August; only 0.6% were recaptured, which was insufficient for a reliable estimation of population size. Ericson (1977) had the same problem with this species. Kromp & Nitzlader (1995) recaptured only 4.9% of 8035 marked carabids (24 species) in an organic winter rye field. Begon (1979) gives tables of sample sizes required (for given levels of accuracy) in relation to numbers marked and recaptured. Sexes should be treated separately if they have very different capture rates in pitfall traps, as is often the case for lycosid spiders (Samu & Sárospataki 1995) and some carabids (Ericson 1977).

Fenced pitfall traps

This method is suitable for species of predator that are active on the ground surface and are caught reliably by pitfall traps. It has been used mainly for carabid beetles (Basedow 1973, Bonkowska & Ryszkowski 1975, Baars 1979, Dennison & Hodkinson 1984, Desender & Maelfait 1986, Helenius 1995, Ulber & Wolf-Schwerin 1995) but occasionally for Araneae and Staphylinidae (Basedow & Rzehak 1988) or non-predatory invertebrates (Gist & Crossley 1973). An area of crop, containing pitfalls, is isolated by a barrier or fence and the top of the isolator is sealed with a fine mesh material to prevent any further entry or exit of animals from the area. The pitfall catch over several weeks provides an estimate of predator density, because it is assumed that a high proportion of predators within the fenced area will have been caught during this period. In addition, some authors (e.g. Gist & Crossley 1973, Scheller 1984) have exploited the relationship between capture rate and cumulative numbers caught to calculate (see Seber 1982, Southwood 1978) the number of predators initially present within the fenced area. Loreau (1984) used this method for carabids in a beech forest and found that the density estimates obtained were similar to estimates obtained by mark-release-recapture within the enclosures.

Unfortunately, densities of large, extremely active, predators will be underestimated if they flee from the area when the barrier is being established. Members of the guild of wandering spiders, for example, have been seen moving away from investigators engaged in quadrat sampling (Uetz & Unzicker 1976). This limitation applies to all the methods described in this review, with the exception of those employing the "drop trap" principle, and unfenced pitfall traps.

Fences are usually made of wood (e.g. Desender & Maelfait 1983, Dennison & Hodkinson 1984) or metal (e.g. Bonkowska & Ryszkowski 1975, Grégoire-Wibo 1983a,b, Scheller 1984, Helenius 1995) and internal barriers, connecting pitfalls together, are recommended to increase their capture efficiency (Durkis & Reeves 1982, Ulber & Wolf-Schwerin 1995). It is vital that the isolators are sunk securely into the ground and that the mesh covering is very tightly sealed (e.g. with stong, waterproof, sticky tape). This is necessary because experience has shown that climbing predators (e.g. the staphylinid genus *Tachyporus* and the carabid *Demetrias atricapillus* L.) are attracted to isolators, can climb even smooth, shiny materials and will squeeze through very small gaps, thus artificially inflating the density estimates. Basedow (1973), Gist & Crossley (1973), Bonkowska & Ryszkowski (1975), Baars (1979) and Scheller (1984) did not report that the enclosures used in their studies were sealed.

When marked carabids were released into fenced areas, recapture rates were usually high (e.g. 95% of *Pterostichus melanarius* Ill. within 2 weeks in pasture (Desender, van den Broeck & Maelfait 1985), 66%-100% of *Pterostichus* spp, *Abax* spp. and *Nebria* spp. within 3 weeks in mixed woodland (Dennison & Hodkinson 1984), 96% of carabids (unspecified species) in potato and rye fields (Bonkowska & Ryszkowski 1975). This suggests that density estimates by this method would be reasonably accurate, at least for some species, if sufficient fenced pitfall areas could be deployed. The number needed would depend on mean density, degree of spatial heterogeneity (Powell et al. 1995) and the level of sampling precision required. Desender & Maelfait (1986) reported that densities of carabids in a pasture, estimated from fenced pitfalls, were similar to estimates from soil samples. Similarly, Gist & Crossley (1973) found good agreement between density estimates derived from fenced pitfalls and sorting of habitat samples, for Araneae and Coleoptera in hardwood forest. Dennison & Hodkinson (1984) compared fenced pitfalls and leaf litter flotation for carabids in mixed woodland and found seasonal variation in the relative ranking of the two density estimates. Density estimates from fenced pitfalls were correlated with abundance estimates from unfenced pitfalls summed over the "activity season" of the carabids *Pterostichus versicolor* Sturm and *Calathus melanocephalus* L. in heathland (Baars 1979). Fenced pitfalls may have to be run for several weeks to obtain a density estimate (e.g. Ulber & Wolf-Schwerin 1995), especially during periods of cold weather when predator activity is reduced. It should be noted that densities derived by this method can only be attributed to the time of erection of the fence if it is known that there has been no eclosion from eggs or pupae during the trapping period (i.e. fenced pitfalls are also, effectively, emergence traps, see Helenius 1995). This aspect can be used to advantage in population dynamics studies to estimate recruitment, and the spatial distribution of variation in recruitment, provided teneral adults are easily recognised, as was the case for the carabid *Bembidion guttula* (F.) in spring cereals (Helenius 1995).

Ground Photoeclectors

An eclector is an emergence trap, and a photoeclector is a design of emergence trap that exploits the positive phototaxis of emerging invertebrates. Ground photoeclectors were originally used by Funke (1971) to assess the density of insects emerging from the ground within an enclosed area of beech forest. Insects were caught in a clear perspex collecting vial attached to the top of the dark, cloth-covered photoeclector. Ground photoeclectors, with surface areas usually between 0.25 m^2 and 1.0 m^2, have since been used fairly extensively in agroecological studies (e.g. Törmälä 1982, Bosch 1990, Büchs 1991, 1993, Wehling & Heimbach 1991, Kleinhenz & Büchs 1993). The method is very effective for collecting Diptera, Hymenoptera and Staphylinidae, but the timing of their emergence, and that of other invertebrates, may be altered in response to the changed microclimate within the photoeclector (Büchs 1993). One pitfall trap is usually established inside the photoeclector, near its wall, to remove epigeic predators and prevent predation on emerging insects (Mühlenberg 1993). Central positioning of the pitfall trap is inadequate, Kromp et al. (1995) reported many carabids still alive within the photoeclector after two weeks when a single pitfall was used in the centre of a 0.25 m^2 photoeclector. If larger numbers of pitfall traps, connected by internal barriers, were to be established within the photoeclector (as was done for fenced pitfall traps by Ulber & Wolf-Schwerin (1995)) the sampling efficiency of this method might be increased considerably.

Törmälä (1982) considered suction samplers to be more efficient than photoeclectors for estimating the density of most species of invertebrates in the field layer of grassland. However, this author used photoeclectors lacking internal pitfall traps, and a photoeclector trapping time of only 2 hours. Similar conclusions were reached by Gromadzka & Trojan (1967), but, again, the trapping time was a maximum of 30 minutes. Spiders of the vegetation layer of *Calluna* heathland were over-represented in photoeclectors compared with densities determined by extraction of samples using a Kempson, Lloyd & Ghelardi (1963) extractor (Breuer 1987). In a beech forest, Grimm, Funke & Schauermann (1975) recorded higher densities of staphylinid beetles in photoeclectors than by extraction with a Kempson, Lloyd & Ghelardi (1963) extractor.

In addition to their potential role in density estimation, when ground photoeclectors are operated concurrently with unfenced pitfall traps in the same habitat, a comparison of the numbers, species composition, sex and age ratios of invertebrates collected by the two methods can yield much useful information on the biology and behaviour of polyphagous predators in agroecosystems (Büchs, Kleinhenz & Zimmerman, unpublished).

Soil flooding

Originally devised for sampling carabid and staphylinid beetles in riparian habitats (Desender & Segers 1985), this method has since been applied to agroecosystems (Brenøe 1987, Basedow et al. 1988). 0.1 m^2 isolators are sunk into the soil, the vegetation within is examined, excised and removed, then 2 litres of water are poured into each isolator. Predators seen on the ground surface are removed and later another 2 litres of water are added and any additional predators that appear are removed. The total collection time is 10 minutes per quadrat. Basedow et al. (1988) used this method in winter wheat and sugar beet and compared it to handsorting soil cores (each 30 cm deep and requiring 3 man hours to sort); they concluded that soil flooding was an effective method for small- to medium-sized adult Carabidae, Staphylinidae and Araneae. Kromp et al. (1995), investigating a range of organic crops, found that small carabids were more abundant than large carabids in soil flooding samples, the reverse being true for unfenced pitfalls. It is likely that, in this case, soil flooding provided a more reliable indication of relative abundance and that the pitfall catch was affected by the higher relative activity levels of the larger carabids. Brenøe (1987) assessed the efficiency of soil flooding for estimating the density of the small carabid beetle *Bembidion lampros* (Herbst) in a cauliflower field on loamy soil. After soil flooding the soil was removed to a depth of 20 cm and washed in the laboratory, no further beetles being found. Beetles were, however, highly aggregated, and only 23 were extracted from 2.25 m^2 during the entire investigation.

Recent experiences with this method (Ulber & Dinter, unpublished) suggest that it is not practicable under some soil conditions. The soil surface can become blocked with silty mud, preventing the emergence of predators. Conversely, in drying clay soils with large vertical cracks, or where mice or earthworms have made extensive perforations, more than 20 litres of water can be added per 0.25 m^2 without any success in flooding predators out of the soil.

Microhabitat removal

Unit areas of habitat may be removed from the field, then hand sorted in the laboratory or put into some type of extraction apparatus (e.g. Duffey 1962, Kempson, Lloyd & Ghelardi 1963, Hassall et al. 1988) or into a soil washing and flotation apparatus (e.g. Sotherton 1984, 1985). Edwards & Fletcher (1971) provide a useful review of some of the extraction methods available. Examples of microhabitat removal and predator extraction are given below. Absolute efficiency of the techniques or apparatus used was, however, not normally reported in these publications. Kempson, Lloyd & Ghelardi (1963) suggest a means of estimating extraction efficiency. Known numbers of animals can be added to defaunated habitat to record the proportion extracted; this probably underestimates efficiency because the handled animals may be injured and the defaunated habitat (e.g. by autoclaving) may lose pore spaces and trap animals. A second estimate of efficiency can be obtained by careful search of the habitat sample after extraction is completed, looking for corpses. This may overestimate efficiency if some corpses are not found. The true value for efficiency therefore lies between these two estimates.

Franke, Friebe & Beck (1988) assessed the density of the macro- and mesofauna of beech wood litter by handsorting. The efficiency, as estimated by sorting the same sample three times by three different people, was found to be 63-100%, depending on taxonomic group. The efficiency was considered to be inadequate for density estimation of the mesofauna. Thomas et al. (1992) handsorted 0.04 m^2 x 0.1 m deep grassy turves to remove carabids, staphylinids and spiders. Chiverton (1989) used modified Tullgren funnels to remove carabids, staphylinids and spiders from samples taken from field overwintering sites. De Keer & Maelfait (1987, 1988) and Alderweireldt (1987) removed 156 cm^2 sample units from maize, ryegrass and pasture, handsorted them in the laboratory, then put them into Tullgren-Berlese extractors to remove spiders. 95% confidence limits were high because of the small size of the sample units in relation to the scale of spider aggregation in islets of high vegetation. Workman (1978), studying the lycosid spider *Trochosa terricola* Thorell in heathland, used Tullgren funnels for small sample units (0.02 m^2) and a modified bowl extractor for larger ones (0.1 m^2). The latter method yielded 31 *T. terricola* per m^2 compared with 12 m^{-2} for *in situ* handsearching. Duffey (1962) took 25 cm x 25 cm turves of limestone grassland and removed spiders from them in a horizontal platform extractor. This method was slightly more efficient than Tullgren funnels. It should be noted that extractors that raise the temperature of the sample may cause premature emergence of larvae or nymphs from eggs, and eclosion of adults from pupae, resulting in overestimation of the density of larvae, nymphs and adults, relative to their density in the field at the time of sampling (e.g. Törmälä1982, Dinter 1995). Where this is likely to be a problem, a proportion of the sample units could be subjected to flotation techniques for comparison. Other limitations to the use of extractors are that the capacity of the machine could be inadequate to process the large number of sample units per sample that might be needed when a predator species is at low density, and that the relatively long period for extraction (e.g. 11 days in a Kempson, Lloyd & Ghelardi (1963) machine) constrains the minimum inter-sample interval.

Soil samples from arable fields have been examined layer by layer and the carabids removed (Dubrovskaya 1970), or carabids were extracted by some other (unstated) method (Bonkowska & Ryszkowski 1975, Pollet & Desender 1985). Sotherton (1984, 1985) took 0.04 m^2 x 35 cm deep soil cores from fields and field boundaries in winter to extract earwigs plus carabid and staphylinid adults and larvae. The cores were coarsely sieved then mixed into saturated salt solution; floating vegetation was sieved off several times and put in a white tray to pick out predators. Processing samples on the day of collection was considered to be more efficient than delaying extraction to a later occasion, because dead animals are less easily found in the tray than moving ones. Dennison & Hodkinson (1984) also used flotation to extract carabids from 0.25 m^2 x 4.5 cm deep litter samples from mixed woodland.

Unfenced pitfall traps

Although pitfall traps do not provide density estimates, some authors have interpreted the catch data as an index of abundance. It has been known, however, for some time, that pitfall catch is affected by numerous factors (e.g. species, sex and hunger level of predator, size, shape, pattern and spacing of trap, presence of a trap cover, material from which the trap is constructed, preservative used, substrate, vegetation, climate and season; Heydemann 1962, Luff 1975, Uetz & Unzicker 1976, Thomas & Sleeper 1977, Adis 1979, Müller 1984, Niemelä et al. 1986, Benest 1989b). Inter-trap variability in pitfall catch can also be increased by weeds, ants, soil type and the pH of the topsoil (Honěk 1988, Powell et al. 1995). Carcasses of small mammals that have fallen into the traps may attract some species of invertebrates (Törmälä 1982). There is also some indication that, where traps are operated without trapping fluid (dry traps), individual carabid beetles may persistently enter or avoid traps, perhaps as the result of a learning process (Benest 1989a). This has also been reported in other Coleoptera (Thomas & Sleeper 1977). More recently it has become apparent that individual behaviours of the species being trapped can have a considerable effect on trappability. Halsall & Wratten (1988), using video techniques in the laboratory, showed that capture efficiencies of Carabidae adults in dry pitfall traps were low and markedly species-specific. Topping (1993) observed the behaviour of linyphiid spiders in a laboratory grass arena containing an array of preservative-filled pitfall traps. Again, capture efficiencies were extremely low, with the mean number of spider-trap encounters required to achieve capture varying from 16 to 57, depending on species. Spiders entering a trap attached silken drag lines to the sides and were only caught if they lost their footing. Some species built webs in the traps, which introduces the possibility of complex interspecific interactions also affecting trap catch. There is an additional problem in ecotoxicological studies because sub-lethal dosages of pesticides can directly increase locomotory activity (e.g. Jepson et al. 1987) or directly decrease it (e.g. Everts et al. 1991), depending on species and active ingredient, or they can affect mobility indirectly by altering the food supply (e.g. Chiverton 1984).

It is regrettable that the term "activity-density" has been used to describe pitfall trap catch; the more accurate term "activity-trappability-density" would alert authors to the problems inherent in interpreting pitfall catch data, where trapping efficiency is variable within and between species and habitats.

It has been found that the species composition, size distribution, sex ratio and relative abundance of spiders in winter wheat as measured by pitfall trapping, are very different from more accurate estimates obtained using intensive D-vac or suction and ground search methods (Dinter & Poehling 1992, Topping & Sunderland 1992, Dinter 1995). The situation is similar for carabid beetles. Dennison & Hodkinson (1984) found that the seasonal fluctuation in numbers of carabids caught in pitfalls in a mixed woodland was much greater than that for soil flotation or fenced pitfalls. Desender & Maelfait (1983, 1986) found poor correlations between pitfall catches of carabids in pasture and numbers in soil cores and fenced pitfalls. Similar results were reported for carabids in oilseed rape (Ulber & Wolf-Schwerin 1995), rye and potatoes (Bonkowska & Ryszkowski 1975) and winter wheat (Basedow & Rzehak 1988). Lohse (1981) found that some species of carabid (e.g. *Amara brunnea* Gyll.) and staphylinid (e.g. within the genera *Philonthus* and *Tachyporus*), although extremely numerous in forest habitats, as determined by handsearching, were absent from (or very poorly represented in) pitfalls. The author considered this to be due to low trappability of these species, resulting from their behaviour (e.g. microsite preferences, trap avoidance behaviour). It is likely that specialised designs of pitfall trap, such as those that collect beetle larvae from below the soil surface (Loreau 1987), will be subject to the same limitations as for more traditional designs.

Some authors (e.g. Baars (1979) for the carabids *Pterostichus versicolor* Sturm and *Calathus melanocephalus* L. in heathland, and Kowalski (1976) for the staphylinid *Philonthus decorus* Gr. in oak woodland) claim that pitfall catches summed over the entire seasonal activity period of a species provide an index of abundance in a habitat. Comparisons can only, however, be made within species and within habitat types and the "seasonal activity" approach cannot be used for monitoring changes in abundance within seasons. In addition, this index is not valid for all species. Loreau (1984), for example, demonstrated that for some carabids, such as *Pterostichus oblongopunctatus* (F.), annual activity was not linearly related to population density (as determined by mark-release-recapture within large woodland enclosures).

If unfenced pitfall traps are arranged in a trapping web sampling design with concentric circles of traps at fixed distances from a central point (Buckland et al. 1993) it is possible to calculate density from the resulting catch, using a negative exponential density estimation model (Parmenter & MacMahon 1989). The latter authors used this method with tenebrionid beetles in a shrub-steppe habitat and demonstrated that it produced density estimates with 95% confidence intervals that always included the "true" density value (validation was by adding known numbers of beetles to fenced trapping webs). The authors stress that various assumptions (e.g. all animals at the centre of the web must be captured, there must be no preferential direction of movement and all captures must be independent events) must be fulfilled and that the target species must have an adequately high capture rate. Prior knowledge is needed of "home range", movement patterns and trappability. Density estimates by this method will apply only to the portion of a population that is active on the ground surface during the sampling period. Parmenter & MacMahon (1989) were dealing with an artificially simplified case because their populations were marked and enclosed. There were no complications due to recruitment or migration and they further assumed that beetle mortality was negligible during the 3-6 week trapping periods.

It is clear from the foregoing that pitfall trap catch data can be used as a reliable index of abundance only in special cases, and where sufficient information is available about the biology of the species and the efficiency of capture. This conclusion has been reached previously by other authors (e.g. Lohse 1981). For certain taxonomic groups, unfenced pitfall traps may be useful as an aid for establishing a species list for a habitat. This appears to be true for Carabidae (Ulber & Wolf-Schwerin 1995) and the guild of wandering spiders (Uetz & Unzicker 1976), but not, in general, for Staphylinidae, because many species that emerge from arable soils spend little time active on the soil surface (Büchs, Kleinhenz & Zimmermann, unpublished).

Other methods

Greenstone (1979) used an *in situ* visual "catch per unit effort" method to estimate the density of lycosid spiders (*Pardosa* spp.) at the edge of a pond. The results were well correlated with density estimates from a mark-release-recapture programme. However, he stressed that the method required a very open habitat and standardised conditions of weather and collection technique during the period of density estimation. This method will not often, therefore, be applicable to agroecological studies.

With large, easily-seen predators, such as adult coccinellid beetles, it is sometimes possible to estimate density with reference to a calibrated and standardised visual count sampling method. Elliott et al. (1991) compared visual counts of coccinellids in wheat fields with densities derived from a removal sampling programme and found that visual counts were influenced by temperature and aphid density. When these variables were incorporated into regression models relating visual counts to population densities the resulting coefficients of determination ranged from 0.63 to 0.94, depending on species.

The "closest individual" or "distance method" (Southwood 1978) can be used for animal artefacts, such as spider webs, to provide an index of spider abundance. Toft, Vangsgaard & Goldschmidt (1995) used this method to estimate spider abundance in spring barley and winter wheat fields. Potato starch was used to reveal webs, a random point in the field was selected and the distances from this point to the 5-10 nearest webs were measured, and the results used to estimate the density of web-building spiders (Southwood 1978, Krebs 1989). There was a significant correlation between density estimates from the distance method and from concurrent suction samples within quadrats. The distance method has the advantage of being very rapid, but the disadvantage that it can both overestimate (because some spiders may construct webs daily and these may remain intact for several days) and underestimate (not all spiders produce webs) true spider density. Web density was also reduced dramatically, and temporarily, after a period of heavy rain.

"Trap stones" (Mühlenberg 1993) are small Plaster of Paris or concrete structures containing channels, sealed above by a cover plate, that are placed on the ground to provide daytime refuges for small invertebrates. They have potential to be calibrated as an index of abundance for appropriate nocturnal, ground-active predators.

Appropriate methodology for specific requirements

There is no single method of density or abundance estimation that is suitable for all circumstances. Suitable methods vary according to the type(s) of predators being assessed, whether the study is intensive (much effort devoted to one or two fields, e.g. for a population dynamics study) or extensive (e.g. survey of a large number of fields) and whether the crop is homogeneous (i.e. plant and ground can be sampled simultaneously by one method, as in cereals and grasses) or heterogeneous (plant and ground must be treated separately, as in mature field vegetables). Consideration is given, below, to appropriate methods for each of these situations; this is based on an examination of the literature and the authors' own experience. In a few cases it was not possible to suggest any appropriate methods that would meet the standards of reliability outlined in the Introduction.

Intensive studies

Homogeneous crops

In situ plant and ground searches are appropriate for assessing the abundance of the various life-stages of aphid-specific predators. Parasitism of aphids can be reliably quantified by collecting live aphids and determining what proportion become mummified under standardised conditions in captivity (e.g. Borgemeister, Haardt & Höller 1991), or by the use of electrophoresis (Castanera, Loxdale & Novak 1983). Coleoptera larvae are best assessed by microhabitat removal and extraction by soil washing or flotation, or in a heat and light extractor. Efficiency data should be obtained for the species under consideration and allowance should be made for premature emergence of larvae from eggs in heat and light extractors..

The majority of small spiders (epigeic, field layer, web-builders and others; e.g. Linyphiidae, Theridiidae, Araneidae) will be efficiently sampled using an intensive suction sampling technique (e.g. Dinter 1995, Toft, Vangsgaard & Goldschmidt 1995)) or a combined suction and ground search method (e.g. Topping & Sunderland 1994). Unfortunately, sampling will not normally be possible in wet weather and the technique could be problematical on very sandy soils because of the large quantity of soil picked up by the suction sampler. It is also possible that under very dry conditions, on soils that become deeply fissured, spiders will retire below the level of influence of the suction sampler (Toft, Vangsgaard & Goldschmidt 1995). The density of highly active cursorial spiders (e.g. Lycosidae) can be assessed with a "quick trap" or "drop trap" or by mark-release-recapture (MRR) or by unfenced pitfall traps in a trapping web. Abundant species of Carabidae will, in most cases, be reliably assessed using fenced pitfall traps, and microhabitat removal or soil-flooding may be viable alternatives under appropriate conditions. The abundance of highly-active species present only at low densities cannot practicably be estimated by these methods, because of the large number of sample units that would be required. Such predators, although at low density, could, nevertheless, play a significant role in pest control, because their high level of activity could result in a high encounter rate with prey. MRR or the use of a trapping web is recommended for these species. Less information is available on which to base guidelines for reliable density estimation of the Staphylinidae. Some genera (e.g. *Tachyporus*) can be highly active on the ground and in the field layer of agroecosystems, others may spend some time in burrows or crevices. The "absolute density" method of Sunderland et al. (1987) should provide accurate estimates, but most of the methodology described for carabids is probably also applicable to staphylinids. Preliminary results suggest that photoeclectors can be used to provide approximate density estimates for some staphylinid species, but allowance must be made for the altered microclimate in relation to phenological data.

Heterogeneous crops

Sampling methods for the ground component of heterogeneous crops are the same as for homogeneous crops (above). To this must be added searching of the plant. The protocol to be adopted will depend on the growth form of the plant and the type(s) of predator under consideration. If predators are entirely evident on the surface of vegetation, *in situ* surface examination will be sufficient, otherwise *in situ* plant dissection, or removal of plant parts for later dissection in the laboratory, will be required. If a small number of predator species are being counted *in situ*, the data can be entered sequentially, during sampling, into a programmable field logger to provide immediate feedback on sampling precision, thus optimising sampling effort (and avoiding examination of an unnecessarily large number of plants). Counts of the number of plants per unit area can be made to convert predator counts per plant to density estimates.

Extensive studies

Pitfall trapping has most often been used in extensive studies, in the hope of making conclusions about the relative abundance of predators in a range of crops or treatments. Unfortunately, for the reasons given above, pitfalls do not provide a reliable index of abundance, especially where different crop types are being sampled. They are therefore rejected for current purposes (although pitfall trapping remains a useful technique in certain restricted cases such as for "activity season" estimates, or as one of the indicators of site quality where sites are being differentiated by multivariate analyses, e.g. Rushton, Topping & Eyre 1987). When more is known about the scale of inaccuracy of abundance estimation inherent in pitfall methodology, this may still prove to be a useful technique in cases where inaccuracy is low compared with the scale of effect caused by the factors (e.g. pesticides, farming operations, crop types) under investigation. Although less extensively researched, water traps and sweep-netting are thought to have similar limitations to those described for pitfall traps.

Suction sampling is sufficiently rapid to provide the basis for an extensive sampling programme (e.g. 100 cereal fields sampled in a few days, Aebischer & Potts 1990), but interpretation of results would be aided by a better understanding of the causation of variation in efficiency (e.g. Sunderland & Topping 1995). With this information, it would be possible to classify fields/treatments into groups, such that within-group variation in efficiency is negligible. Then suction sampling could be calibrated in one field per group (e.g. by an "absolute density" method such as that of Sunderland et al. 1987) and accurate densities for the whole sampling programme calculated using conversion factors. The principle of combining "absolute" methods, for calibration, with easier and quicker techniques, is not new, and has been used, for example, in sampling programmes for beneficial invertebrates on cotton in the USA (Smith, Stadelbacher & Gantt 1976). The calibration data need not be collected as frequently as the main survey data, and in general, it is worth giving careful consideration to whether frequency of sampling can be "traded-off" in favour of quality of sampling, without compromising the aims of the sampling programme. It may even be possible to find ways to increase the efficiency of sample sorting and identification to free resources for better quality sampling.

There does not currently appear to be any feasible means of reliably estimating the abundance, in extensive programmes, of the "problem species" (highly-active, at low density, or underground during the daytime) of Lycosidae and Carabidae referred to above. Mark-release-recapture carried out simultaneously in a large number of fields would not be possible. Since the species concerned are relatively few in number, and all tend to be caught easily in pitfall traps, it would be a useful (and probably tractable) long-term aim to calibrate pitfall trapping (or some other inexpensive method, such as "trap stones") against reliable density estimates, just for these species. It is already known that conversion factors would need to be species-specific, and it would probably also be necessary to take into account the sex of the predator, the season, the crop type and weather variables during the trapping period.

Conclusions

Although there are a large number of methods available for estimating predator density or abundance, information on the efficiency of these methods (and variation in efficiency in the field), tends to be sparse. Attempts to devise reliable sampling protocols for a wide range of specific requirements revealed deficiencies in our current knowledge. Much work remains to be done on the development of suitable methodology before it can be claimed that methods are available for the reliable estimation of the density or abundance of all species of beneficial predators in agroecosystems.

Acknowledgement

This publication was made possible through an EU Concerted Action "Enhancement, dispersal and population dynamics of beneficial insects in integrated agroecosystems". KDS was funded by the UK Ministry of Agriculture, Fisheries & Food, FS was funded by OTKA Grant F5042.

References

Aebischer, N.J. & Potts, G.R. 1990. Long-term changes in numbers of cereal invertebrates assessed by monitoring. *Proc. 1990 Brit. Crop Prot. Conf. - Pests & Diseases* 1: 163-172.

Adis, J. 1979. Problems of interpreting arthropod sampling with pitfall traps. *Zool. Anz. Jena* 202: 177-184.

Arnold, A.J., Needham, P.H. and Stevenson, J.H. 1973. A self-powered portable insect suction sampler and its use to assess the effects of azinphos methyl and endosulfan on blossom beetle populations on oilseed rape. *Ann. Appl. Biol.* 75: 229-233.

Alderweireldt, M. 1987. Density fluctuations of spiders on maize and Italian ryegrass fields. *Meded. Fac. Landbouwwet. Rijksuniv. Gent* 52: 273-282.

Baars, M.A. 1979. Catches in pitfall traps in relation to mean densities of carabid beetles. *Oecologia (Berl.)* 41: 25-46.

Basedow, T. 1973. Der Einfluss epigäischer Rauparthropoden auf die Abundanz phytophager Insekten in der Agrarlandschaft. *Pedobiologia* 13: 410-422.

Basedow, T. & Rzehak, H. 1988. Abundanz und Aktivitätsdichte epigäischer Rauparthropoden auf Ackerflächen - ein Vergleich. *Zool. Jb. Syst.* 115: 495-508.

Basedow, T., Klinger, K., Froese, A. & Yanes, G. 1988. Aufschwemmung mit Wasser zur Schnellbestimmung der Abundanz epigäischer Rauparthropoden auf Äckern. *Pedobiologia* 32: 317-322.

Bayon, F., Fougeroux, A., Reboulet, J.N. & Ayrault, J.P. 1983. Utilisation et intérêt de l'aspirateur "D-vac" pour la détection et le suivi des populations de ravageurs et d'auxiliaires sur blé au printemps. *La Défense des Végétaux* 223: 276-297.

Begon, M. 1979. *Investigating Animal Abundance: Capture-Recapture for Biologists.* Edward Arnold, London.

Benest, G. 1989a. The sampling of a carabid community. I. The behaviour of a carabid when facing the trap. *Rev. Écol. Biol. Sol.* 26: 205-211.

Benest, G. 1989b. The sampling of a carabid community. II. Traps and trapping. *Rev. Écol. Biol. Sol.* 26: 505-514.

Bonkowska, T. & Ryszkowski, L. 1975. Methods of density estimation of carabids (Carabidae, Coleoptera) in fields under cultivation. *Pol. Ecol. Stud.* 1: 155-171.

Borgemeister, C., Haardt, H. & Höller, C. 1991. Fluctuations in relative numbers of *Aphidius* species (Hymenoptera, Aphidiidae) associated with cereal aphids. In: Polgár, L., Chambers, R.J., Dixon, A.F.G. & Hodek,I. (eds.) *Behaviour and Impact of Aphidophaga*, pp 23-28. SPB Academic Publishing bv., The Hague.

Bosch, J. 1990. Die Arthropodenproduktion des Ackerbodens. *Mitt. Biol. Bundesanst. Land-Forstwirtschaft H.* 266: 74.

Brenøe, J. 1987. Wet extraction - a method for estimating populations of *Bembidion lampros* (Herbst) (Col., Carabidae). *J. Appl. Ent.* 103: 124-127.

Breuer, M. 1987. Ein Vergleich verschiedener Erfassungsmethoden zur Untersuchung der Spinnenfauna eines Calluna-Heidebiotops. *Mitt. Dtsch. Ges. Allg. angew. Ent.* 5: 120-124.

Büchs, W. 1991. Einfluss verschiedener landwirtschaftlicher Produktionsintensitäten auf die Abundanz von Arthropoden in Zuckerrübenfeldern. *Verh. Ges. Ökol.* 20: 1-12.

Büchs, W. 1993. Auswirkungen unterschiedlicher Bewirtschaftlungsintensitäten auf die Arthropodenfauna in Winterweizenfeldern. *Verh. Ges. Ökol.* 22: 27-34.

Buckland, S.T., Anderson, D.R., Burnham, K.P. & Laake, J.L. 1993. *Distance Sampling*. Chapman & Hall, London.

Castanera, P., Loxdale, H.D. & Novak, K. 1983. Electrophoretic study of enzymes from cereal aphid populations. II. Use of electrophoresis for identifying aphidiid parasitoids of *Sitobion avenae* (F.)(Hymenoptera: Aphidiidae; Hemiptera: Aphididae). *Bull. ent. Res.* 73: 659-665.

Chambers, R.J. & Adams, T.H.L. 1986. Quantification of the impact of hoverflies (Diptera: Syrphidae) on cereal aphids in winter wheat: an analysis of field populations. *J. Appl. Ecol.* 23: 895-904.

Chiverton, P.A. 1984. Pitfall-trap catches of the carabid beetle *Pterostichus melanarius*, in relation to gut contents and prey densities, in insecticide treated and untreated spring barley. *Entomol. Exp. Appl.* 36: 23-30.

Chiverton, P.A. 1988. Searching behaviour and cereal aphid consumption by *Bembidion lampros* and *Pterostichus cupreus*, in relation to temperature and prey density. *Entomol. Exp. Appl.* 47: 173-182.

Chiverton, P.A. 1989. The creation of within-field overwintering sites for natural enemies of cereal aphids. *Proc. 1979 Brit. Crop Prot. Conf. - Weeds* 1093-1096.

Cowgill, S.E., Wratten, S.D. & Sotherton, N.W. 1993. The selective use of floral resources by the hoverfly *Episyrphus balteatus* (Diptera: Syrphidae) on farmland. *Ann. Appl. Biol.* 122: 223-231.

De Barro, P.J. 1991. A cheap lightweight efficient vacuum sampler. *J. Aust. ent. Soc.* 30: 207-208.

De Keer, R. & Maelfait, J.P. 1987. Life history of *Oedothorax fuscus* (Blackwall, 1834)(Araneae, Linyphiidae) in a heavily grazed pasture. *Rev. Écol. Biol. Sol.* 24: 171-185.

De Keer, R. & Maelfait, J.P. 1988. Observations on the life cycle of *Erigone atra* (Araneae, Erigoninae) in a heavily grazed pasture. *Pedobiologia* 32: 201-212.

Dennison, D.F. & Hodkinson, I.D. 1984. Structure of the predatory beetle community in a woodland soil ecosystem. IV. Population densities and community composition. *Pedobiologia* 26: 157-170.

Desender, K. & Maelfait, J.P. 1983. Population restoration by means of dispersal, studied for different carabid beetles (Coleoptera, Carabidae) in a pasture ecosystem. In: Lebrun, P., Andre, H.M., De Medts, A., Gregoire-Wibo, C. & Wauthy, G. (eds.) *New Trends in Soil Biology*. Dieu-Brichart, Ottignies-Louvain-la-Neuve, pp. 541-550.

Desender, K. & Segers, R. 1985. A simple device and technique for quantitative sampling of riparian beetle populations with some carabid and staphylinid abundance estimates on different riparian habitats (Coleoptera). *Rev. Écol. Biol. Sol.* 22: 497-506.

Desender, K. & Maelfait, J.P. 1986. Pitfall trapping within enclosures: a method for estimating the relationship between the abundances of coexisting carabid species (Coleoptera, Carabidae). *Hol. Ecol.* 9: 245-250.

Desender, K., van den Broeck, D. & Maelfait, J.P. 1985. Population biology and reproduction in *Pterostichus melanarius* III. (Coleoptera, Carabidae) from a heavily grazed pasture ecosystem. *Meded. Fac. Landbouwwet. Rijkuniv. Gent* 50, 567-75.

Dewar, A.M., Dean, G.J. & Cannon, R. 1982. Assessment of methods for estimating the numbers of aphids (Hemiptera: Aphididae) in cereals. *Bull. ent. Res.* 72: 675-685.

Dietrick, E.J. 1961. An improved backpack motor fan for suction sampling of insect populations. *J. econ. Ent.* 54: 394-395.

Dinter, A. 1995. Estimation of epigeic spider population densities using an intensive D-vac sampling technique and comparison with pitfall trap catches in winter wheat. In: Toft, S. & Riedel, W. (eds.) *Arthropod natural enemies in arable land · I*, pp. 23-32. Aarhus.

Dinter, A. & Poehling, H.M. 1992. Spider populations in winter wheat fields and the side-effects of insecticides. *Asp. Appl. Biol.* 31: 77-85.

Dubrovskaya, N.A. 1970. Field carabid beetles (Coleoptera, Carabidae) of Byelorussia. *Ent. Rev.* 49: 476-483.

Duffey, E. 1962. A population study of spiders in limestone grassland. *J. Anim. Ecol.* 31: 571-599.

Duffey, E. 1974. Comparative sampling methods for grassland spiders. *Bull. Br. arachnol. Soc.* 3: 34-37.

Duffey, E. 1980. The efficiency of the Dietrick Vacuum Sampler (D-VAC) for invertebrate population studies in different types of grassland. *Bull. Ecol.* 11: 421-431.

Durkis, T. J. & Reeves, R.M. 1982. Barriers increase efficiency of pitfall traps. *Ent. News* 93: 8-12.

Edwards, C.A. & Fletcher, K.E. 1971. A comparison of extraction methods for terrestrial arthropods. In: J. Phillipson (ed.) *Methods of Study in Quantitative*

Soil Ecology: Population, Production and Energy Flow , IBP Handbooks, 18: 150-185.

Elliott, N.C., Kieckhefer, R.W. & Kauffman, W.C. 1991. Estimating adult coccinellid populations in wheat fields by removal, sweep net, and visual count sampling. *Can. Ent.* 123: 13-22.

Ericson, D., 1977. Estimating population parameters of *Pterostichus cupreus* and *P. melanarius* (Carabidae) in arable fields by means of capture-recapture. *Oikos* 29: 407-417.

Ericson, D., 1978. Distribution, activity and density of some Carabidae (Coleoptera) in winter wheat fields. *Pedobiologia* 18: 202-217.

Everts, J.W., Willemsen, I., Stulp, M., Simons, L., Aukema, B. & Kammenga, J. 1991. The toxic effect of deltamethrin on linyphiid and erigonid spiders in connection with ambient temperature, humidity, and predation. *Arch. Environ. Contam. Toxicol.* 20: 20-24.

Forsythe, T.G. 1987. *Common Ground Beetles.* Naturalists' Handbooks No. 8., Richmond Publishing Company: Surrey, UK.

Frank, J.H. 1967. The insect predators of the pupal stage of the winter moth, *Operophtera brumata* (L.) (Hydriomenidae). *J. anim. Ecol.* 36: 375-389.

Frank, J.H. 1968. Notes on the biology of *Philonthus decorus* (Grav.)(Col., Staphylinidae). *Entomologist's mon. Mag.* 103: 273-277.

Franke, U., Friebe, B. & Beck, L. 1988. Methodisches zur Ermittlung der Siedlungsdichte von Bodentieren aus Quadratproben und Barberfallen. *Pedobiologia* 32: 253-264.

Funke, W. 1971. Food and energy turnover of leaf-eating insects and their influence on primary production. *Ecol. Stud.* 2: 81-93.

Gist, C.S. & Crossley, D.A. 1973. A method of quantifying pitfall trapping. *Environ. Entom.* 2: 951-952.

Greenstone, M.H. 1979. A line transect density index for wolf spiders (*Pardosa* spp.), and a note on the applicability of catch per unit effort methods to entomological studies. *Ecol. Ent.* 4: 23-29.

Grégoire-Wibo, C. 1983a. Incidences écologiques des traitments phytosanitaires en culture de betterave sucrière, essais expérimentaux en champ. I. Les Collemboles épigés. *Pedobiologia* 25: 37-48.

Grégoire-Wibo, C. 1983b. Incidences écologiques des traitments phytosanitaires en culture de betterave sucrière, essais expérimentaux en champ. II. Acariens, Polydesmes, Staphylins, Cryptophagides et Carabides. *Pedobiologia* 25: 93-108.

Grimm, R., Funke, W. & Schauermann, J. 1975. Minimalprogramm zur Ökosystemanalyse: Untersuchungen an Tierpopulationen in Wald-Ökosystemen. *Verh. Ges. Ökol.* 1974: 77-87.

Gromadzka, J. & Trojan, P. 1967. Comparison of the usefulness of an entomological net, photoeclector and biocenometer for investigation of entomocenosis. *Ekol. Polska* 15: 505-529.

Haas, V. Methoden zur Erfassung der Arthropodenfauna in der Vegetationsschicht von Grasland-Ökosystemen. *Zool. Anz. Jena* 204: 319-330.

Hackman, W. 1957. Studies on the ecology of the wolf spider *Trochosa terricola* Deg. *Soc. Scient. Fenn. Comm. Biol.* XVI: 1-34.

Halsall, N.B. & Wratten, S.D. 1988. The efficiency of pitfall trapping for polyphagous predatory Carabidae. *Ecol. Ent.* 13: 293-299.

Hance. T. 1992. Spring densities of ground beetles (Coleoptera: Carabidae) in cultivated fields. *Bull. Annls Soc. belge Ent.* 128: 319-324.

Hance, T. 1995. Relationships between aphid phenology and predator and parasitoid abundances in maize fields. In: Toft, S. & Riedel, W. (eds.) *Arthropod natural enemies in arable land · I*, pp. 113-123. Aarhus.

Hand, S.C. 1986. The capture efficiency of the Dietrick vacuum insect net for aphids on grasses and cereals. *Ann. Appl. Biol.* 108: 233-241.

Hassall, M., Dangerfield, J.M., Manning, T.P. & Robinson, F.G. 1988. A modified high-gradient extractor for multiple samples of soil macro-arthropods. *Pedobiologia* 32: 21-30.

Heikinheimo, O. & Raatikainen, M. 1962. Comparison of suction and netting methods in population investigations concerning the fauna of grass leys and cereal fields, particularly in those concerning the leafhopper, *Calligypona pellucida* (F.). *Publ. Finn. State Agric. Res. Board* No. 191.

Helenius, J. 1995. Rate and local scale spatial pattern of adult emergence of the generalist predator *Bembidion guttula* in an agricultural field. In: Toft, S. & Riedel, W. (eds.) *Arthropod natural enemies in arable land · I*, pp. 101-111. Aarhus.

Henderson, I.F. & Whitaker, T.M. 1976. The efficiency of an insect suction sampler in grassland. *Ecol. Ent.* 2: 57-60.

Heydemann, B. 1962. Untersuchungen über die Aktivitäts- und Besiedlungsdichte bei epigäischen Spinnen. *Zool. Anz. Suppl.* 25: 538-556.

Hills, O.A. 1933. A new method for collecting samples of insect populations. *J. econ. Ent.* 26: 906-910.

Holtkamp, R.H. & Thompson, J.I. 1985. A lightweight, self-contained insect suction sampler. *J. Aust. ent. Soc.* 24: 301-302.

Honěk, A. 1988. The effect of crop density and microclimate on pitfall trap catches of Carabidae, Staphylinidae (Coleoptera) and Lycosidae (Araneae) in cereal fields. *Pedobiologia* 32: 233-242.

Jepson, P., Cuthbertson, P., Downham, M., Northey, D., O'Malley, S., Peters, A., Pullen, A., Thacker, R., Thackray, D., Thomas, C. & Smith, C. 1987. A quantitative ecotoxicological investigation of the impact of synthetic pyrethroids on beneficial insects in winter cereals. *Bull. SROP/WPRS* 1987/X/1: 194-205.

Johnson, C.G., Southwood, T.R.E. & Entwistle, H.M. 1957. A new method of extracting arthropods and molluscs from grassland and herbage with a suction apparatus. *Bull. ent. Res.* 48: 211-218.

Kegel, B. 1990. Diurnal activity of carabid beetles living on arable land. In: *The Role of Ground Beetles in Ecological and Environmental Studies* (N.E. Stork, ed.) Intercept, Andover, UK, pp 65-76.

Kempson, D., Lloyd, M. & Ghelardi, R. 1963. A new extractor for woodland litter. *Pedobiologia* 8: 1-21.

Kennedy, T.F., Evans, G.O. & Feeney, A.M. 1986. Studies on the biology of *Tachyporus hypnorum* F. (Col. Staphylinidae), associated with cereal fields in Ireland. *Ir. J. agric. Res.* 25: 81-95.

Kleinhenz, A. & Büchs, W. 1993. Einfluss verschiedener landwirtschaftlicher Produktionsintensitäten auf die Spinnenfauna in der Kultur Zuckerrübe. *Verh. Ges. Ökol.* 22: 81-88.

Kowalski, R. 1976. Obtaining valid population indices from pitfall trapping data. *Bull. de l'Acad. pol. des Sciences* 23: 799-803.

Krebs, C.J. 1989. *Ecological Methodology.* Harper & Row, New York.

Kromp, B. & Nitzlader, M. 1995. Dispersal of ground beetles in a rye field in Vienna, Eastern Austria. In: Toft, S. & Riedel, W. (eds.) *Arthropod natural enemies in arable land · I*, pp. 269-277. Aarhus.

Kromp, B., Pflügl, C., Hradetzky, R. & Idinger, J. 1995. Estimating beneficial arthropod densities using emergence traps, pitfall traps and the flooding method in organic fields (Vienna, Austria). In: Toft, S. & Riedel, W. (eds.) *Arthropod natural enemies in arable land · I*, pp. 87-100. Aarhus.

Kuschka, V., Lehmann, G. & Meyer, U. 1987. Zur Arbeit mit Bodenfallen. *Beitr. Ent., Berlin* 37: 3-27.

Leathwick, D.M. & Winterbourn, M.J. 1984. Arthropod predation on aphids in a lucerne crop. *N.Z. Entomol.* 8: 75-80.

Lohse, G.A. 1981. Bodenfallenfänge im Naturpark Wilseder Berg mit einer kritischen Beurteilung ihrer Aussagekraft. *Jber. naturwiss. Ver. Wuppertal* 34: 43-47.

Loreau, M. 1984. Population density and biomass of Carabidae (Coleoptera) in a forest community. *Pedobiologia* 27: 269-278.

Loreau, M. 1987. Vertical distribution of activity of carabid beetles in a beech forest floor. *Pedobiologia* 30: 173-178.

Loreau, M. & Nolf, C.L. 1993. Occupation of space by the carabid beetle *Abax ater*. *Acta Oecol.* 14: 247-258.

Loughridge, A.H. & Luff, M.L. 1983. Aphid predation by *Harpalus rufipes* (DeGeer) (Coleoptera: Carabidae) in the laboratory and field. *J. Appl. Ecol.* 20: 451-462.

Luff, M.L. 1975. Some features influencing the efficiency of pitfall traps. *Oecologia (Berl.)* 19: 345-357.

Luff, M.L. 1978. Diel activity patterns of some field Carabidae. *Ecol. Ent.* 3: 53-62.

MacLeod, A., Sotherton, N.W., Harwood, R.W.J. & Wratten, S.D. 1995. An improved suction sampling device to collect aphids and their predators in agro-ecosystems. In: Toft, S. & Riedel, W. (eds.) *Arthropod natural enemies in arable land · I*, pp. 125-131. Aarhus.

Macleod, A., Wratten, S.D. & Harwood, R.W.J. 1994. The efficiency of a new lightweight suction sampler for sampling aphids and their predators in arable land. *Ann. Appl. Biol.* 124: 11-17.

Manga, N. 1972. Population metabolism of *Nebria brevicollis* (F.) (Coleoptera: Carabidae). *Oecologia (Berl.)* 10: 223-242.

Marston, N., Davis, D.G. & Gebhardt, M. 1982. Ratios for predicting field populations of soybean insects and spiders from sweep-net samples. *J. econ. Ent.* 75: 976-981.

Mason, C.E. & Blocker, H.D. 1973. A stabilised drop trap for unit-area sampling of insects in short vegetation. *Environ. Entom.* 2: 214-216.

Mühlenberg, M. 1993. *Freilandökologie.* Quelle & Meyer Verlag, Heidelberg.

Müller, J.K. 1984. Die Bedeutung der Fallenfang-Methode für die Lösung ökologischer Fragestellungen. *Zool. Jb. Syst.* 111: 281-305.

Nelemans, M.N.E. 1986. Marking techniques for surface-dwelling Coleoptera larvae. *Pedobiologia* 29: 143-146.

Niemelä, J., Halme, E., Pajunen, T. & Haila, Y. 1986. Sampling spiders and carabid beetles with pitfall traps: the effect of increased sampling effort. *Ann. Ent. Fennici* 52: 109-111.

Nyffeler, M. 1982. *Field studies on the ecological role of the spiders as insect predators in agroecosystems (abandoned grassland, meadows, and cereal fields).* Thesis, Swiss Federal Institute of Technology, Zurich.

Parmenter, R.R. & MacMahon, J.A. 1989. Animal density estimation using a trapping web design: field validation experiments. *Ecology* 70: 169-179.

Pollet, M. & Desender, K. 1985. Adult and larval feeding ecology in *Pterostichus melanarius* Ill. (Coleoptera, Carabidae). *Med. Fac. Landouwwet. Rijksuniv. Gent* 50: 581-594.

Powell, W., Hawthorne, A., Hemptinne, J.L., Holopainen, J.K., Den Nijs, L.J.F.M., Riedel, W. & Ruggle, P. 1995. Within-field spatial heterogeneity of arthropod predators and parasitoids. In: Toft, S. & Riedel, W. (eds.) *Arthropod natural enemies in arable land · I*, pp. 235-242. Aarhus.

Pruess, K.P., Lal Saxena, K.M. & Koinzan, S. 1977. Quantitative estimation of alfalfa insect populations by removal sweeping. *Environ. Entom.* 6: 705-708.

Rushton, S.P., Topping, C.J. & Eyre, M.D. 1987. The habitat preferences of grassland spiders as identified using Detrended Correpondence Analysis (DECORANA). *Bull. Br. arachnol. Soc.* 7: 165-170.

Samu, F., Sárospataki, M., Fazekas, J. & Bíró, Z. 1991. Density estimation of the carabid beetle *Calasoma auropunctatum* in alfalfa by a marking technique: a preliminary report. *Proc. 4th EC5/XIII.SIEEC, Gödöllö* 1991, 98-100.

Samu, F. & Sárospataki, M. 1995. Estimation of population sizes and "home ranges" of polyphagous predators in alfalfa using mark-recapture: an exploratory study. In: Toft, S. & Riedel, W. (eds.) *Arthropod natural enemies in arable land · I*, pp. 47-55. Aarhus.

Scheller, H.V. 1984. The role of ground beetles (Carabidae) as predators on early populations of cereal aphids in spring barley. *Z. ang. Ent.* 97: 451-463.

Seber, G.A.F. 1973. *The Estimation of Animal Abundance and Related Parameters.* Griffin, London.

Smith, J.W., Stadelbacher, E.A. & Gantt, C.W. 1976. A comparison of techniques for sampling beneficial arthropod populations associated with cotton. *Environ. Entom.* 5: 435-444.

Sotherton, N.W. 1984. The distribution and abundance of predatory arthropods overwintering on farmland. *Ann. Appl. Biol.* 105:423-429.

Sotherton, N.W. 1985. The distribution and abundance of predatory arthropods overwintering in field boundaries. *Ann. Appl. Biol.* 106: 17-21.

Southwood, T.R.E. 1978. *Ecological Methods.* Chapman & Hall, London.

Southwood, T.R.E. & Pleasance, H.J. 1962. A hand-operated suction apparatus for the extraction of arthropods from grassland and similar habitats, with notes on other models. *Bull. ent. Res.* 53: 125-128.

Summers, C.G., Garrett, R.E. & Zalom, F.G. 1984. New suction device for sampling arthropod populations. *J. econ. Ent.* 77: 817-823.

Sunderland, K.D. 1987. Spiders and cereal aphids in Europe. *Bull. SROP/WPRS* 1987/X/I: 82-102.

Sunderland, K.D., Crook, N.E., Stacey, D.L. & Fuller, B.T. 1987. A study of feeding by polyphagous predators on cereal aphids using ELISA and gut dissection. *J. Appl. Ecol.* 24: 907-933.

Sunderland, K.D. & Topping, C.J. 1995. Estimating population densities of spiders in cereals. In: Toft, S. & Riedel, W. (eds.) *Arthropod natural enemies in arable land · I*, pp. 13-22. Aarhus.

Sunderland, K.D., Hawkes, C., Stevenson, J.H., McBride, T., Smart, L.E., Sopp, P.I., Powell, W., Chambers, R.J. & Carter, O.C.R. 1987. Accurate estimation of invertebrate density in cereals. *Bull. SROP/WPRS* 1987/X/I: 71-81.

Thomas, D.B. & Sleeper, E.L. 1977. The use of pitfall-traps for estimating the abundance of arthropods, with special reference to the Tenebrionidae (Coleoptera). *Ann. ent. Soc. Am.* 70: 242-248.

Thomas, M.B., Wratten, S.D. & Sotherton, N.W. 1992. Creation of 'island' habitats in farmland to manipulate populations of beneficial arthropods: predator densities and species compositon. *J. Appl. Ecol.* 29: 524-531.

Thornhill, E.W. 1978. A motorised insect sampler. *PANS* 24: 205-207.

Toft, S., Vangsgaard, C. & Goldschmidt, H. 1995. The distance method used to measure densities of web spiders in cereal fields. In: Toft, S. & Riedel, W. (eds.) *Arthropod natural enemies in arable land · I*, pp. 33-45. Aarhus.

Topping, C.J. 1993. Behavioural responses of three linyphiid spiders to pitfall traps. *Entomol. Exp. Appl.* 68: 287-293.

Topping, C.J. & Sunderland, K.D. 1992. Limitations to the use of pitfall traps in ecological studies exemplified by a study of spiders in a field of winter wheat. *J. Appl. Ecol.* 29: 485-491.

Topping, C.J. & Sunderland, K.D. 1994. Methods for quantifying spider density and migration in cereal crops. *Bull. Br. arachnol. Soc.* 9: 209-213

Topping, C.J. & Sunderland, K.D. 1995. Methods for monitoring aerial dispersal by spiders. In: Toft, S. & Riedel, W. (eds.) *Arthropod natural enemies in arable land · I*, pp. 245-256. Aarhus.

Törmälä, T. 1982. Evaluation of five methods of sampling field layer arthropods, particularly the leafhopper community, in grassland. *Ann. Ent. Fenn.* 48: 1-16.

Turnbull, A.L. & Nicholls, C.F. 1966. A 'quick trap' for area sampling of arthropods in grassland communities. *J. econ. Ent.* 59: 1100-1104.

Uetz, G.W. & Unzicker, J.D. 1976. Pitfall trapping in ecological studies of wandering spiders. *J. Arachnol.* 3: 101-111.

Ulber, B. & Wolf-Schwerin, G. 1995. A comparison of pitfall trap catches and absolute density estimates of carabid beetles in oilseed rape fields. In: Toft, S. & Riedel, W. (eds.) *Arthropod natural enemies in arable land · I*, pp. 77-86. Aarhus.

Vickerman, G.P. & Sunderland, K.D. 1975. Arthropods in cereal crops: nocturnal activity, vertical distribution and aphid predation. *J. Appl. Ecol.* 12: 755-766.

Wallin, H. & Ekbom, B. 1988. Movements of carabid beetles (Coleoptera: Carabidae) inhabiting cereal fields: a field tracing study. *Oecologia (Berl.)* 77: 39-43.

Wehling, A. & Heimbach, U. 1991. Untersuchungen zur Wirkung von Pflanzenschutzmitteln auf Spinnen (Araneae) am Beispiel einiger Insektizide. *Nachrictenbl. Deut. Pflanzenschutzd.* 43: 24-30.

Winder, L., Hirst, D.J., Carter, N., Wratten, S.D. & Sopp, P.I. 1994. Estimating predation of the grain aphid *Sitobion avenae* by polyphagous predators. *J. Appl. Ecol.* 31: 1-12.

Workman, C. 1978. Life cycle and population dynamics of *Trochosa terricola* Thorell (Araneae: Lycosidae) in a Norfolk grass heath. *Ecol. Ent.* 3: 329-340.

Wright, A.F. & Stewart, A.J.A. 1992. A study of the efficacy of a new inexpensive type of suction apparatus in quantitative sampling of grassland invertebrate populations. *Brit. Ecol. Soc. Bull.* XXIII: 116-120.

Yeargan, K.V. & Cothran, W.R. 1974. An escape barrier for improved suction sampling of *Pardosa ramulosa* and *Nabis* spp. populations in alfalfa. *Environ. Entom.* 3: 189-191.

SPATIAL HETEROGENEITY

Using mark-release-recapture techniques to study the influence of spatial heterogeneity on carabids

Wilf Powell & Jim Ashby

Entomology & Nematology Department, IACR-Rothamsted, Harpenden, Herts. AL5 2JQ, England

Abstract
Two contrasting attempts to use mark-release-recapture techniques in order to investigate the effects of within-field spatial heterogeneity on carabid beetles are described. These studies are used as examples to illustrate the value and the limitations of mark-release-recapture methodologies.

Key words: Carabidae, mark-recapture, spatial heterogeneity

Introduction

Mark-release-recapture techniques can be used to investigate several different aspects of the ecology and behaviour of arthropod natural enemies, including relative abundance, locomotory activity and dispersal behaviour (Powell & Walton 1994). Information obtained from mark-release-recapture studies of ground-dwelling predators such as carabid beetles can help in the interpretation of data from pitfall trap catches, which are influenced by beetle activity and beetle trappability as well as by beetle abundance. Both the locomotory activity of beetles and their local abundance can be affected by small-scale spatial heterogeneity within the environment. Local differences in soil type, micro-topography, vegetation type and plant density are amongst the factors which contribute to small-scale spatial heterogeneity.

Mark-release-recapture techniques have been used at Rothamsted to address two different problems involving the effects of spatial heterogeneity within individual fields on carabid beetles. These are described briefly and illustrate two different ways in which mark-release-recapture can be used and also highlight a few points which need to be borne in mind when using this technique.

***Arthropod natural enemies in arable land* · *I** *Density, spatial heterogeneity and dispersal*
S. Toft & W. Riedel (eds.). *Acta Jutlandica* vol. 70:2 1995, pp. 165-173.
 ISBN 87 7288 492

Problem 1

This concerns spatial heterogeneity in a winter wheat field resulting from the use of different soil cultivation techniques on small experimental plots. These plots formed part of a multidisciplinary study of the long-term effects of different methods of straw disposal. Pitfall catches of carabid beetles in summer were frequently greater in plots which had been ploughed in the previous autumn than in plots which had received minimum cultivation (tine cultivated). In order to determine the biological significance of these data it was important to know whether differences in pitfall catches between the two treatments were due to differences in beetle abundance or in beetle activity.

Problem 2

The farmland ecosystem is a patchy mosaic of crop and semi-natural habitats in which insect populations live. The movement patterns of individual insects within this spatially heterogeneous environment have an important influence on the population dynamics and dispersion of their species. An experimental "patchy arena" has been established in two fields in order to study several aspects of insect ecology and behaviour in spatially heterogeneous environments. Carabids are one of the groups which are being studied in this system, especially their local movement patterns within and between areas of semi-natural habitat and areas of cereal crop. The main aim is to assess whether individual beetles forage throughout the whole area and whether they use both the patches of semi-natural habitat and the intervening crop areas.

Methods

Problem 1

The replicated plot experiment on which the work was done consisted of 48 plots of winter wheat, each measuring 9 m x 28 m, arranged in 4 blocks of 12 (Fig. 1). The blocks were 10 m apart and the plots within a block were 3 m apart. On individual plots the soil was either tine cultivated (non-inversion tillage) or ploughed. Four pitfall traps (white plastic beakers with a diameter of 60 mm and partly filled with a 50% methanol solution) were placed in each plot and emptied weekly from late spring until harvest in 1987, 1988 and 1989. The traps were also operated in autumn 1990.

On July 10 1989, 60 *Pterostichus melanarius,* which had been collected from outside the study area, were released onto each of four ploughed (soil inverted to a depth of 20 cm) and four tined plots (to a depth of 10 cm without soil inversion). One tined and one ploughed plot was selected at random from each block (Fig. 1). Beetles were scattered over each release plot, except in the

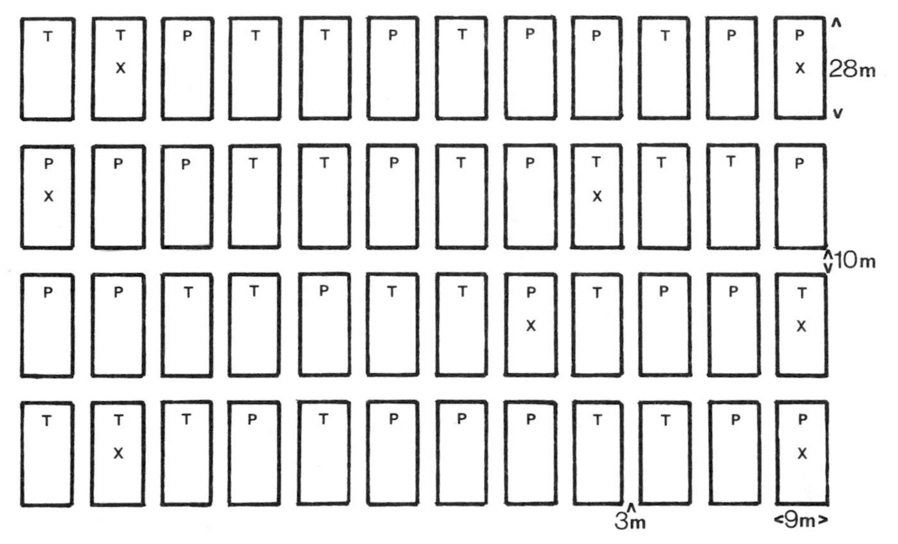

Fig. 1. Plan of field plots of winter wheat used in a mark-release-recapture study to investigate the activity of the carabid beetles *Pterostichus melanarius* and *Nebria brevicollis* on ploughed (P) and tine-cultivated (T) plots. Groups of marked beetles were released on plots marked "X".

area within one metre of the plot edge. Before release the beetles were marked using enamel paint so that the plot on which they had been released could be determined when they were recaptured. Four extra pitfall traps were placed in each release plot and all traps throughout the site were changed weekly. For each recaptured beetle, its release plot and position of recapture were recorded. A second experiment was done in autumn 1990 using *Nebria brevicollis* collected from outside the study area, 96 of which were released on each of the same 8 plots on October 31.

Problem 2

The study area was set up in 1989 and consisted of a series of semi-natural, square patches (created by sowing a mixture of four grasses and four herbaceous plants) laid out within a field of winter barley. Pitfall traps were placed in both the sown patches (81 traps) and the cereal crop areas (60 traps) (Fig. 2). A further 11 traps were placed in the hedgerow bordering the experiment. Over a three week period in July and a two week period in August 1992, the traps were run with no preservative so that captured beetles remained alive. During these two trapping periods the traps were usually checked daily. All captured individuals of large carabid species were taken to the laboratory, marked with an individual code using a surgical cautery needle and then released on the same day approximately 1 metre from the trap in which they had been caught. The date and location of all recaptures were recorded. The barley crop was harvested mid-way through the first trapping period.

Results

Problem 1

In all three years significantly more ($p<0.05$) *P. melanarius* from the unmarked, natural population were caught in pitfall traps in ploughed plots than in traps in tined plots, even though the total numbers caught varied greatly between years (Table 1). Similarly, during the mark-release-recapture experiment more marked beetles from the released population were recaptured in the ploughed plots than in the tined plots (Table 2). Of the 480 marked beetles released in the experiment 214 (46%) were recaptured, and the relative proportions caught on tined and ploughed plots were similar for the natural population and for the released population (Tables 1 & 2). When only those beetles recaptured on the same plots on which they had been released were considered, the proportion caught on each treatment remained unchanged.

Pitfall trapping during the autumn was only done in one year of the experiment. A total of 220 unmarked *N. brevicollis* were caught from the natural population and of these 57% were caught on tined plots. Recaptures from the marked, released population gave similar results. Of the 768 beetles released, 207 (27%) were recaptured, with 60% caught in tined plots (Table 3).

Problem 2

A total of 449 large carabids, belonging to eight species, were caught and marked during the two mark-release-recapture periods; of these only 34 (7.6%) were recaptured (Table 4). Half of the beetles marked (51%) were *P. melanarius* and this species formed 71% of the recaptures.

The locations of initial capture and recapture for *Pterostichus* spp. are shown in Fig. 2. Many of the individuals moved long distances across the experimental area but most appeared to spend the majority of their time within the crop rather than in the semi-natural patches, both before and after harvest. Of the 26 recaptures, 23 involved individuals which were initially caught in the crop and also recaptured in the crop, 2 were caught in the crop and recaptured in semi-natural patches and 1 was initially caught in a patch and then recaptured in the crop. The longest straight-line distance between the initial capture and recapture traps was 107 metres, and time intervals between first capture and recapture ranged between 1 and 29 days (Fig. 2).

There were very few recaptures of other species. The two recaptures of *Carabus violaceus* suggested that this species moved across both crop and patch

Table 1. Unmarked *Pterostichus melanarius* caught in pitfall traps in ploughed and tine cultivated plots in a replicated field plot experiment during summer in three consecutive years.

Year	Total number caught	% caught in Tined plots	% caught in Ploughed plots
1987	1017	42%	58%
1988	2669	35%	65%
1989	11712	46%	54%

Table 2. *Pterostichus melanarius* from a marked, released population caught in pitfall traps in ploughed and tine cultivated plots in July 1989. Sixty beetles were released in each of 4 ploughed and 4 tined plots.

Where recaptured	Total recaptured	% caught in Tined plots	% caught in Ploughed plots
Release plot	89	40%	60%
Other plots	125	38%	62%
All plots	214	39%	61%

Table 3. *Nebria brevicollis* from a marked, released population caught in pitfall traps in ploughed and tine cultivated plots in November 1990. Ninety-six beetles were released in each of 4 ploughed and 4 tined plots.

Where recaptured	Total recaptured	% caught in Tined plots	% caught in Ploughed plots
Release plot	43	58%	42%
Other plots	164	61%	39%
All plots	207	60%	40%

Table 4. Carabid beetles marked and recaptured within a spatially heterogeneous environment, consisting of a series of sown, grass/herb patches laid out in a winter barley field.

Species	Number caught and marked	Number of recaptures
Amara aulica (Panzer)	75	3
Calathus fuscipes (Goeze)	31	0
Carabus violaceus L.	9	2
Harpalus affinis (Schrank)	3	0
Harpalus rufipes (Degeer)	46	3
Nebria brevicollis (F.)	26	0
Pterostichus madidus (F.)	30	2
Pterostichus melanarius (Ill.)	229	24
Total	449	34

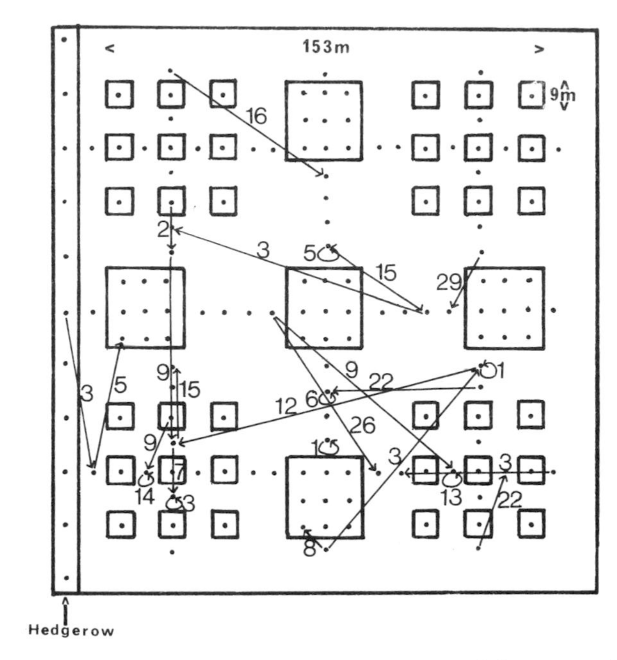

Fig. 2. Recaptures of marked carabids (*Pterostichus* spp.) in a spatially heterogeneous field study area, consisting of a series of sown grass/herb patches (squares) within a winter barley crop. Pitfall trap positions are indicated by solid circles, and the arrows show the positions of first capture and recapture for individual beetles. The number of days between first capture and recapture is shown against each arrow.

areas. In contrast, the three *Amara aulica* recaptures all occurred in the same semi-natural patch in which they were initially caught and re-released, after periods of 3, 5 and 7 days. Of the three *Harpalus rufipes* recaptured, two were recaptured in the same trap in which they had initially been caught, one in the crop (3 days after re-release) and one in a patch (9 days after re-release). The third was caught in the crop and then recaptured in an adjacent patch (9 metres away) after 4 days. *H. rufipes* may move over much smaller distances than does *P. melanarius* but more data are needed to confirm this.

Discussion

These two studies demonstrate two contrasting ways in which mark-release-recapture can be used to investigate aspects of carabid behaviour and ecology, as influenced by spatial heterogeneity. In the first case, with spatial heterogeneity created by different soil cultivation techniques, beetles were collected from outside the study area and released onto plots to provide equal populations of marked beetles on the two treatments. This technique resulted in a high recapture rate, providing sufficient data for a comparison of the pitfall catches of marked and unmarked beetles. The similarity of the data from the natural and introduced populations suggests that differences in pitfall catches between treatments (ploughed and tine cultivated plots) were caused, at least partly, by differences in beetle activity. Reasons for this difference in activity on ploughed and tined plots need to be investigated but could be due to differences in soil topography (surface roughness), plant density (weeds and volunteers) or food availability (soil fauna). Differences in soil topography are unlikely to be the cause since the soil surface would be expected to be rougher on ploughed plots.

In contrast, the study of beetle movement in the patchy environment involved the capture of individuals from the natural population present on the site, and their recapture following marking and release. In this case information was required about the movement patterns of individual beetles rather than the general levels of activity within the population. Therefore, it was important not to introduce large numbers of marked beetles from outside the study area, because their movement patterns may have been abnormal following introduction. Also, large numbers of introduced beetles would have resulted in sudden changes in beetle densities in local areas, and this could have influenced individual behaviour patterns. However, the limited number of beetles which can be marked and the low rate of recapture are obvious problems with only marking individuals as they are caught from the resident population. Capture rates may be improved by increasing the density of traps in the study area, but this could cause excessive disturbance to the beetle population and increase the probability of interference effects between neighbouring traps (Luff 1975). Nevertheless, the few data obtained did provide some indications of differences between species in their movement patterns within the spatially heterogeneous environment, both in their use of different habitats and distances travelled by individuals.

In assessing the use of mark-release-recapture techniques to study the effects of spatial heterogeneity on carabids a number of points need to be considered. Firstly, most studies are conducted on large species, as in the examples described above, principally because they are easier to handle. However, it would be naive to assume that small carabid species behave in exactly the same way as their larger relatives, particularly in their foraging behaviour. This highlights the need for efficient methods of marking small beetles. Also, if distances travelled, and therefore areas of ground over which foraging movements occur, are limited in the case of small beetles, studies of such species may require a different size and design of pitfall trap to those used in studies of larger species.

Secondly, recapture rates are often low, as in our second example, providing too few data for firm conclusions to be made, so that follow-up studies using different techniques may be necessary. A compromise must be made between increasing trap density in order to increase recapture rates and maintaining sufficient distances between traps to minimize interference. If a beetle is unable to move very far without encountering a trap, then its natural movement pattern will be disrupted, negating the value of the recapture data.

Thirdly, mark-release-recapture data only provide superficial information on the movement patterns of individuals. No information is obtained about the actual path travelled by the beetle between the release point and position of recapture, so that only the minimum distance travelled can be calculated. For example, in our second study there is no way of knowing whether a beetle which was recaptured in the same trap as its initial capture after a period of several days, remained in the close vicinity of that trap or moved widely throughout the area, returning to that location by chance. These data, therefore, need to be supplemented by tracking data such as those obtained by Baars (1979) and Wallin & Ekbom (1988), although the tracking techniques devised so far may sometimes interfere with normal movement.

Fourthly, both studies were done over short periods of a few weeks. Many carabid beetles survive as adults over much longer periods and their activity levels and movement patterns can change with time, being influenced by sex and reproductive state as well as environmental factors (Ericson 1978, Lys & Nentwig 1991).

Finally, in both our examples the study areas were not discrete so that marked and released beetles were able to move in and out of the trapping area, and this probably reduced the recapture rate. Immigration and emigration from a trapping area can be prevented, or considerably reduced by the erection of physical barriers. However, the presence of such barriers may in itself affect the movement patterns of beetles within the enclosed area, leading to erroneous data.

Mark-release-recapture remains a useful technique which can be used to address a variety of questions. However, as with the use of pitfall trapping as a sampling method for ground-dwelling predators, the limitations of the technique need to be appreciated and it is probably of most benefit when used to supplement data obtained in other ways.

References

Baars, M.A. 1979. Patterns of movement of radioactive carabid beetles. *Oecologia (Berl.)* 44:125-140.

Ericson, D. 1978. Distribution, activity and density of some Carabidae (Coleoptera) in winter wheat fields. *Pedobiologia* 18: 202-217.

Luff, M.L. 1975. Some features influencing the efficiency of pitfall traps. *Oecologia (Berl.)* 19: 345-357.

Lys, J.A. & Nentwig, W. 1991. Surface activity of carabid beetles inhabiting cereal fields. *Pedobiologia* 35: 129-138.

Powell, W. & Walton, M.P. 1994. Natural enemy populations and communities. In: M.A. Jervis & N.A.C. Kidd (eds.) *Insect Natural Enemies: Practical approaches to their study and evaluation.* Chapman & Hall, London, in press.

Wallin, H. & Ekbom, B.S. 1988. Movements of carabid beetles (Coleoptera: Carabidae) inhabiting cereal fields: a field tracing study. *Oecologia (Berl.)* 77: 39-43.

Spatio-temporal patterns of activity density of some carabid species in large scale arable fields

C.J.H. Booij, L.J.M.F. den Nijs & J. Noorlander

Research Institute for Plant Protection, P.O. Box 9060, NL-6700 GW Wageningen, The Netherlands.

Abstract
Trapping data from grid samples of *Pterostichus cupreus, Agonum dorsale* and *Bembidion tetracolum* from oilseed rape and winter wheat were analysed. The spatial activity density patterns were more homogeneous than expected. Even in early spring, numbers caught were not substantially affected by the presence of nearby field margins or shelterbelts. For *Agonum dorsale* only, spring colonisation of the fields from overwintering sites was more apparent. Favourable overwintering conditions in crops like oilseed rape and winter wheat, and rapid colonisation by flight, may explain the rather even distribution of numbers in early spring.

Key words: Spatial distribution, ground beetles, dispersal, grid sampling

Introduction

Due to the dynamic nature of the farmland habitat and the patchy distribution of favourable and unfavourable habitats, considerable variation occurs in activity density of carabid beetles. Since many carabid species can disperse rapidly either by flight or by walking (den Boer 1981), they are expected to redistribute themselves over places where food, reproduction or survival conditions are optimal.

Levels of activity density may quickly change when the conditions change due to crop rotation and crop management practices such as soil tillage and chemical treatments. The maintenance of many species can be secured only if they are well-adapted to the frequent disturbances or if they can rapidly redistribute over favourable habitat patches. In this context the role of field margins, weed strips and other non-crop environments as refuges during adverse conditions has been studied extensively (e.g. Duelli 1990, Sotherton 1984, 1985, Lys & Nentwig 1992, Riedel 1991, Thomas et al. 1991).

***Arthropod natural enemies in arable land** · I* *Density, spatial heterogeneity and dispersal*
S. Toft & W. Riedel (eds.). *Acta Jutlandica* vol. 70:2 1995, pp. 175-184.

The spatio-temporal distribution of activity density can be studied by grid sampling at regular intervals over a certain period. Grid sampling has been used at various spatial scales for problems ranging from microhabitat preference (e.g. Ericson 1978, Hengeveld 1979), and recolonisation of experimentally sprayed plots (e.g. Duffield & Moffatt 1991, Jepson 1989, Thomas 1988), to population movements at the field level (Lys & Nentwig 1991, 1992), the farmland level (Coombes & Sotherton 1986, Jensen et al. 1989), and even at the landscape scale (Duelli 1990). Current research in this field is aimed at gaining more insight into the effects of farming practices and agroecological infrastructure on the dynamics of beneficial arthropods. New frameworks and approaches in this field are currently being developed (Booij & den Nijs 1992).

To add some additional realism to population dynamics, distribution heterogeneity and large scale movement should be taken into account when highly dispersive species are studied. In this paper we analyse carabid trap catches from grid samples taken by Everts (1990), in order to detect seasonal aggregation patterns in Dutch agro-ecosystems, with emphasis on border effects in large scale arable fields.

Methods

During large scale studies on the ecotoxicological effects of deltamethrin and fenitrothion, several fields with autumn-sown oil-seed rape and winter wheat were studied in the agricultural area of Oostelijk Flevoland in the Netherlands in 1985 and 1986 (Everts 1990). Widely spaced grids of pitfall traps were used to monitor carabids and spiders in this study. Collected carabids and some processed data were kindly supplied by J. Everts for further analysis.

An analysis was made of the data from 5 fields: Nz27 and Oz80 (winter wheat 1985), Oz81 and Oz82 (oilseed rape 1985) and Cz36 (winter wheat 1986), each being 50 to 100 ha. Each grid consisted of 4 * 5 sampling units (having 3-5 traps) at an inter-unit distance of 100 m (Fig. 1). Except for field Cz36, the first row of traps was placed 100 m into the field. Traps were filled with 4% formalin and emptied at weekly intervals. In this paper data from the end of March to the end of June were analysed, covering main dispersal phases of spring breeding carabid species. The material caught was stored and later identified to species level.

Trapping data of the dominant spring breeding species *Pterostichus (Poecilus) cupreus* (L.), *Agonum dorsale* (Pont.) and *Bembidion tetracolum* Say were analysed in two ways: a) overall spatial patterns were graphically

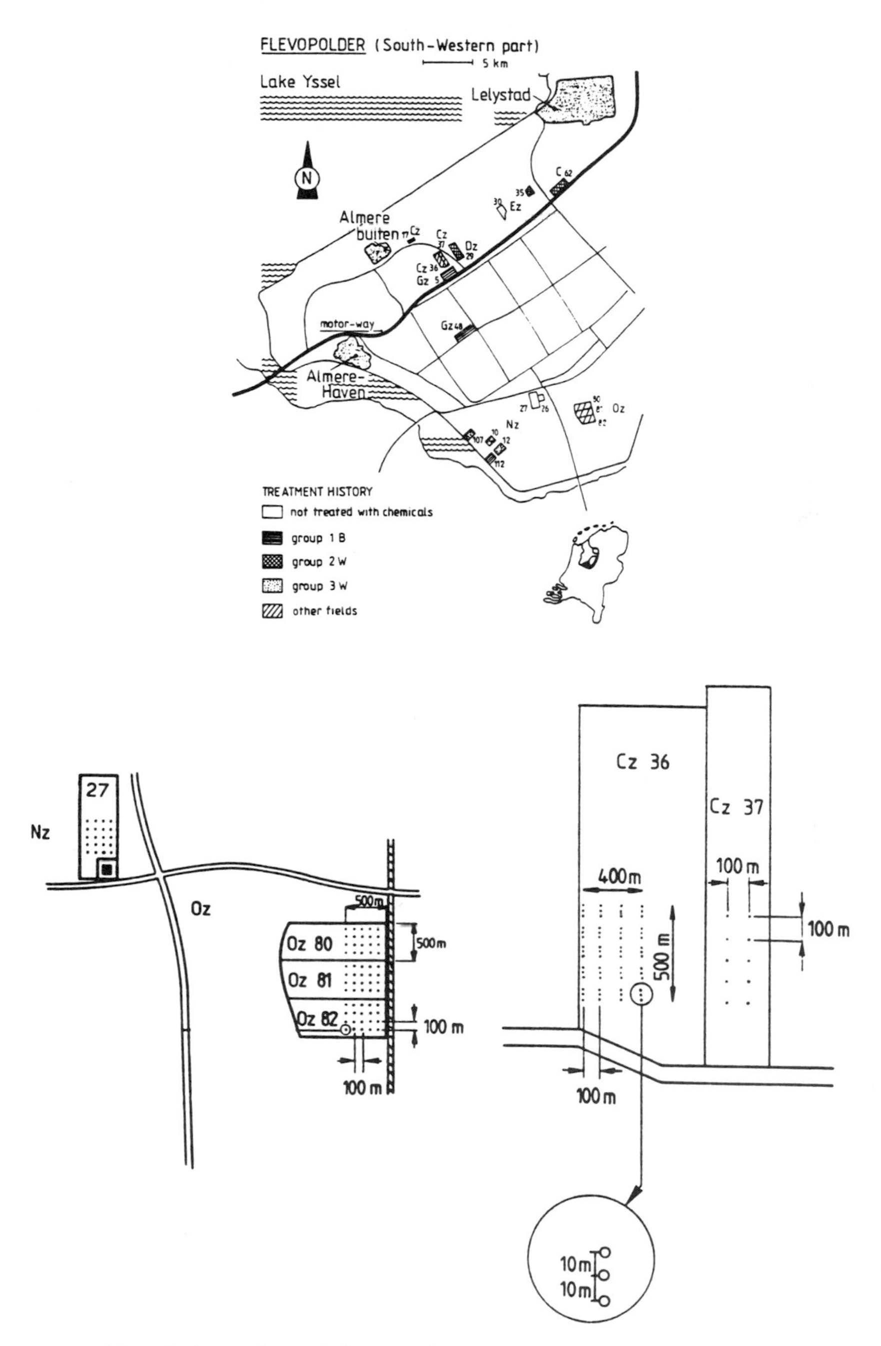

Fig. 1. Location of the study area with sampling design.

studied by plotting activity densities of the total grid for the separate trapping weeks; b) expected invasion effects were analysed by plotting temporal changes in activity densities as a function of distance to the field margins and time. Multiple regressions on log-transformed catch data were used to detect trends.

Results

From the analysis of the grid sample data it appeared that during most of the trapping period activity densities of the species concerned are rather evenly distributed over the grid. No obvious gradients related to nearby field or road margins were found (Fig. 2). In only one field (Cz36), *A. dorsale* was caught in slightly higher numbers near the headland and/or the field margin. The distribution of *B. tetracolum* was correlated to field margins in 3 fields (Table 1). Of the carabid species studied *P. cupreus* reached the highest activity-density levels. Both in early spring and at its maximum during the reproductive period (May-June) no apparent border effects were found. *A. dorsale* was quite abundant in winter wheat, showing its highest activity somewhat later than *P. cupreus*. At the beginning of the activity period more beetles of this species were caught in traps nearest to the borders of field Cz36. The impact of distance from the field margin on activity density, however, was not significant when analysed by multiple linear regression (Table 1). In all fields, beetles were caught in the middle of the field immediately from the beginning of the activity period. A more detailed analysis of the spatial-temporal dynamics of *A. dorsale* in Cz36 showed evidence for invasion both from the field margin (being newly planted shrubwood) and from the headland further away (Fig. 3).

Table 1. Multiple regression analysis of log-activity-density of carabids in several fields, in relation to distance to field margin and time of year. n.s. = not significant; * = $p<0.05$; ** = $p<0.01$; *** = $p<0.001$.

	A. dorsale		*B. tetracolum*		*P. cupreus*	
Field	Distance	Time	Distance	Time	Distance	Time
CZ36	n.s.	***	*** (+)	*	n.s.	***
NZ27	n.s.	***	** (+)	***	n.s.	***
OZ80	n.s.	***	* (-)	**	n.s.	***
OZ81	n.s.	***	-	-	n.s.	***
OZ82	n.s.	n.s.	-	-	n.s.	***

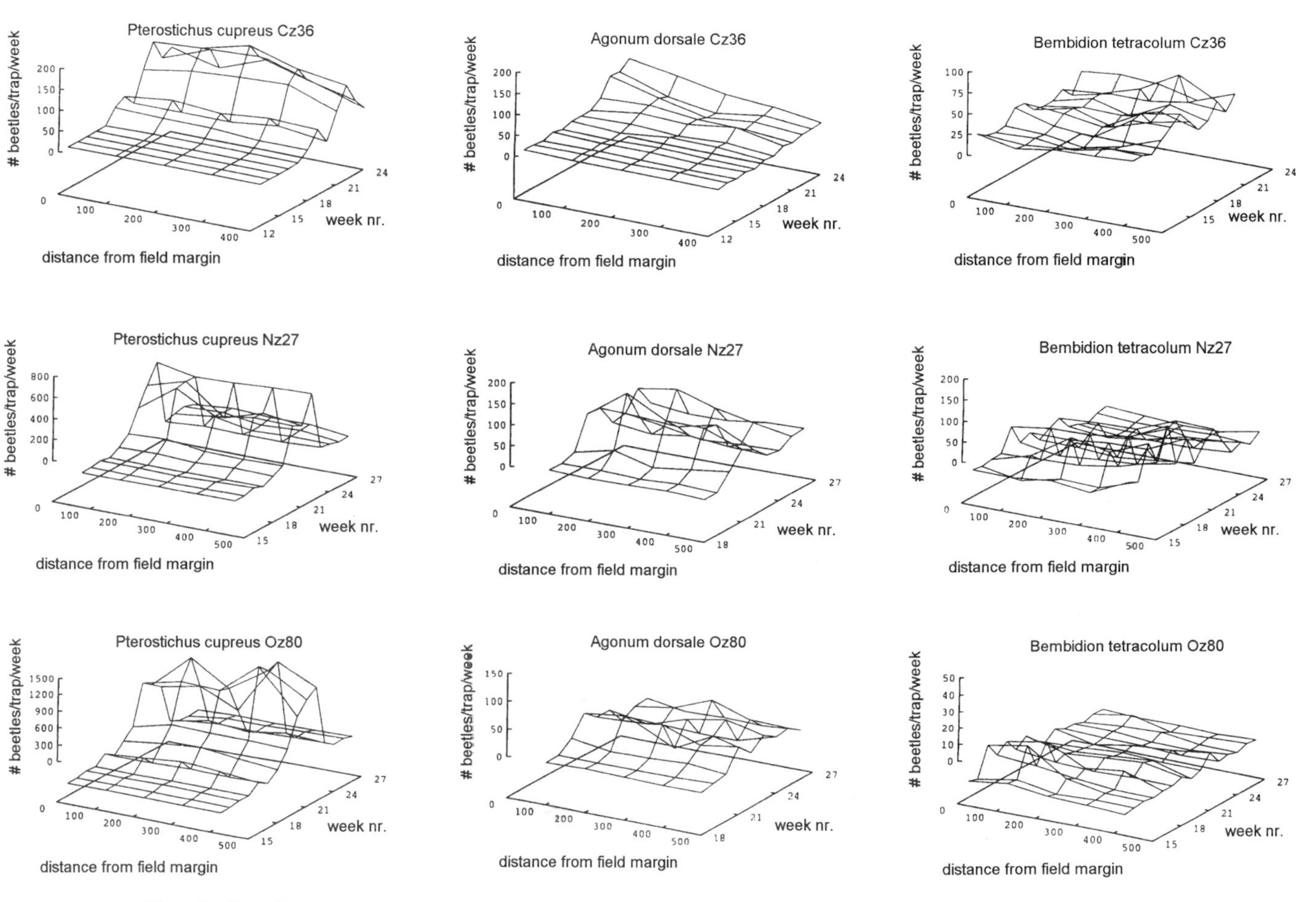

Fig. 2. Spatio-temporal distribution of carabid species in relation to field margins and time.

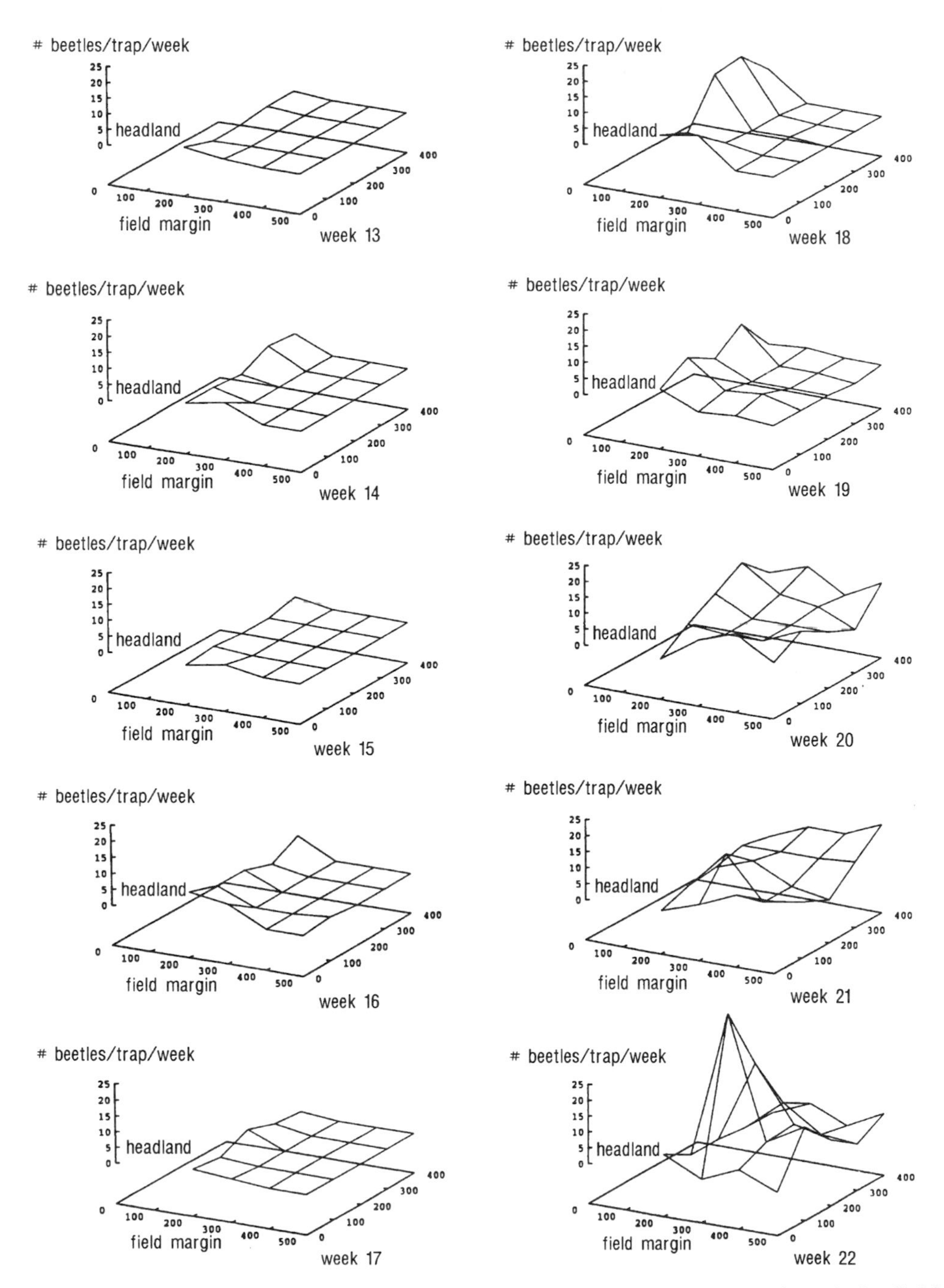

Fig. 3. Seasonal changes in activity density pattern af *Agonum dorsale* in field Cz36, oil seed rape 1986.

For *B. tetracolum,* some gradient in numbers caught was found but its relationship with distance to field margins was inconsistent. In two cases they appeared more numerous in the centre of the field, in only one case more common near the borders. The species was abundant for a long period all over the field, showing aggregation in a rather irregular manner.

Discussion

The within field distribution of beetle activity density found in this study was more even than we expected based on the literature (Sotherton 1984, Coombes & Sotherton 1986, Riedel 1991). The general uniformity of the study area with respect to abiotic and biotic factors may partly explain this lack of obvious patterns. Large scale agricultural fields with a few different crops are only alternated by ditches and roads, and a few newly established woodlands. Field margins in this area are reduced to small strips which can only have a limited function as refuges.

Pterostichus cupreus is found to be a very succesful species in the Dutch polder area, since it is present in high numbers. It is very possible that considerable numbers survive within the field during the winter period, especially in fields with winter crops. The rather even distribution found from the start of the season (see also Wallin 1985) supports this.

Although *Agonum dorsale* is often considered highly dependent on suitable field margins for overwintering (Sotherton 1984, 1985, Coombes & Sotherton 1986) activity density was not obviously related to the proximity of shelter in our study. However, it may be possible that the impact of field margins is only detectable in the area close to the field borders, as we found in field Cz36 where one trap series was closer to the border. In other fields, where the first trap row was 100 m away from the border, no border effect was found. This may explain the discrepancy with the conclusions of Coombes & Sotherton (1986) which were based on gradients found within 100 m from the field border. In their case colonisation was observed as a wave of dispersal from the field margin into the field.

Bembidion tetracolum is a very dispersive species rapidly colonising newly created or disturbed habitats (Haeck 1971). In addition to the resident population, a substantial part of the population may immigrate by flight into suitable patches. The even distribution seen at a large scale in our fields, with irregular aggregation at a smaller scale throughout the season, points to a rapid response by walking and flight, to suitable conditions within the fields. The slight but statistically significant gradients may be related to some

movements in the landscape. *Biol. Conserv.* 54: 209-222.

Riedel, W. 1991. Overwintering and spring dispersal of *Bembidion lampros* (Coleoptera, Carabidae) from established hibernation sites in a winter wheat field in Denmark. In: Polgar, L., Chambers, R.J., Dixon, A.F.G & Hodek, I. (eds.) *Behaviour and Impact of Aphidophaga*, pp. 235-241. SPB Academic Publishing bv, The Hague.

Rossi, R.E., Mulla, D.J., Journel, A.G. & Franz, E.H. 1992. Geostatistical tools for modeling and interpreting ecological spatial dependence. *Ecol. Monogr.* 62: 277-314.

Sotherton, N.W. 1984. The distribution and abundance of predatory arthropods overwintering in farmland. *Ann. Appl. Biol.* 105: 423-429.

Sotherton, N.W. 1985. The distribution and abundance of predatory arthropods overwintering in field boundaries. *Ann. Appl. Biol..* 106: 17-21.

Thomas, G. 1988. The spatial dynamics of linyphiid spider population recovery following exposure to pyrethroid insecticide. *Aspects Appl. Biol.* 12: 269-271.

Thomas, M.B., Wratten, S.D. & Sotherton, N.W. 1991. Creation of habitat "island" habitats in farmland to manipulate populations of beneficial insects. *J. Appl. Ecol.* 28: 907-917.

Wallin, H. 1985. Spatial and temporal distribution of some abundant carabid beetles (Coleoptera,Carabidae) in cereal fields and adjacent habitats. *Pedobiologia* 28: 19-34.

Welling, M. 1990. Dispersal of ground beetles (Col., Carabidae) in arable land. *Med. Fac. Landbouww. Rijksuniv. Gent* 55: 483-491.

The effect of cereal headland treatments on carabid communities

Amanda Hawthorne[1] & *Mark Hassall*

School of Environmental Sciences, University of East Anglia,
Norwich NR4 7TJ, England
[1]Present address: Waltham Centre for Pet Nutrition, Waltham on the Wolds,
Melton Mowbray, Leicestershire, LE14 4RT, England

Abstract
The overall abundance and species composition of carabid beetles were compared in three cereal field headland management regimes in the Breckland area of Eastern England.

Uncropped Wildlife Strips, which were cultivated but not sown, contained more species and a greater overall abundance of carabids than either Sprayed Headlands or "Conservation Headlands", which were sown but received reduced pesticide inputs.

Both carabid abundance and species richness of the community were correlated with percentage cover of dicotyledonous plants and the abundance of other invertebrates. Whilst carabid species richness was also strongly related to total vegetation cover and abundance of aphids and Collembola.

Experimental reduction of vegetation in the Uncropped Wildlife Strip lead to a decrease in abundance of most species of carabid except *Bembidion lampros* which required bare ground.

The number of species found in the Uncropped Wildlife Strip on the same site increased progressively throughout the three year study.

Key words: Carabid beetles, community structure, cereal headlands

Introduction

Since the 1940s increased intensification of agricultural practices has lead to a general decline in the abundance of carabid beetles on arable farmland for a number of reasons: Firstly as a consequence of the use of broad spectrum insecticides killing these, and other, non target animals (Burns 1988); secondly as a result of selective insecticides and herbicides reducing the availability of invertebrate and plant food (Chiverton & Sotherton 1991, Powell et al. 1985); thirdly due to cultivation techniques (Stassart & Gregoire-Wibo 1983) and

***Arthropod natural enemies in arable land* · *I** *Density, spatial heterogeneity and dispersal*
S. Toft & W. Riedel (eds.). *Acta Jutlandica* vol. 70:2 1995, pp. 185-198.

finally as a result of fragmentation of their habitat (Mader 1988, den Boer 1990).

Since the early 1980s changes in agricultural policy have resulted in some reversals of this trend with increased interest in alternative farming practices. Organic farming has increased in importance and has been shown to benefit carabid populations (Dritschilo & Erwin 1982, Booij & Noorlander 1992, Kromp 1989), whilst minimal cultivation techniques have also resulted in greater carabid abundance (Blumberg & Crossley 1983). In addition there has been increasing interest in the creation of overwintering sites to augment predator populations (Thomas, Wratten & Sotherton 1991, Nentwig 1988) and in the selective reduction of pesticide usage through the use of Conservation Headlands in the perimeter of cereal fields (Sotherton, Boatman & Rands 1989, Rands et al. 1985).

Both Conservation Headlands and an alternative headland prescription, the Uncropped Wildlife Strip, have been included in the management prescriptions for Breckland Environmentally Sensitive Area (E.S.A).

Breckland is an area at the heart of East Anglia in the United Kingdom, which is historically associated with arable farming on a temporary basis. A unique community of flora and fauna has developed in association with areas of abandoned and extensive agriculture, the so-called "brecks". The advent of modern agricultural techniques has led to the destruction and fragmentation of these habitats so that many species are now rare (Lambley & Rothera 1990). The concept of an Uncropped Wildlife Strip was created to encourage many of these threatened species of flora and fauna (Nature Conservancy Council 1984).

In this paper we examine the consequences of introducing Uncropped Wildlife Strips for the diversity and abundance of the carabid fauna in winter wheat fields. A randomised block design was used to compare the carabid populations of an Uncropped Wildlife Strip, with those of Conservation and fully Sprayed Headlands. In particular we ask:

- 1. How do the carabid communities subjected to different headland treatments differ in species composition and relative abundance ?
- 2. How are species richness and total carabid abundance related to environmental characteristics of the headlands ?
- 3. How do they vary with time both within and between seasons ?

Study Site

Two strips (120 m x 6 m) of each of the three headland treatments were arranged in a randomized block design along one side of an 18.6 ha winter wheat field, facing southwest. The farm was situated on the southern edge of Breckland at Ixworth Thorpe, Suffolk (Grid reference TL 931 739).

Sprayed Headlands received pesticides and fertiliser applied at the same rates as the rest of the field and followed normal farming practise. The Conservation Headland was treated as above with the exceptions of herbicides and insecticides, in accordance with specific guidelines for choice of compound and timing (Sotherton et al. 1989), thus allowing weeds and associated invertebrates to flourish. The Uncropped Wildlife Strip was rotovated annually in September and left unsown, without pesticides, lime, nor irrigation (Ministry of Agriculture, Fisheries and Food 1993).

Methods

Sampling of carabid community

Beetles were sampled using pitfall traps consisting of a 70 x 60 mm diameter plastic cup set into a plastic sleeve, so that the cup rim was flush with the surface. Traps were filled to approximately 30 mm with one third ethylene glycol, two thirds water with a drop of detergent. Positioned above the trap was a square metal cover 120 x 120 mm, to prevent dilution of the trapping fluid with rainwater. An area of approximately 150 mm immediately around the trap was cleared of vegetation to standardise traps in the different habitats. Whilst species varied in their trapability and activity, these parameters did not differ significantly between treatments (Hawthorne 1995), thus allowing comparisons of catches from different headland treatments. Traps were placed 20 m apart in the middle of the headland treatments, 3 m from, and parallel to, the field boundary. Three traps were used for each headland treatment in 1990 and five in 1991 and 1992. Opposite each of these traps a further trap was established in the crop at 8 m from the field boundary in 1990 and 1992. Catches were emptied every two weeks in 1990 from 8 April until 31 July, weekly from 23 April to 6 August in 1991 and weekly from 25 February until 27 July in 1992.

Measurements of environmental variables

In order to investigate how the composition and structure of the carabid community varied in relation to different headland treatments, the following environmental variables were monitored during 1991 and correlated with both overall carabid abundance and species richness:

- 1. The percentage cover of all vegetation, of dicotyledonous species, monocotyledonous species and bare ground were estimated in five 0.25 m^2 quadrats for each headland treatment at 3 m from the field boundary. Measurements were made on 24 April, 22 May, 17 June and 24 July in 1991.
- 2. Vegetation height diversity, using Simpson's Index of diversity, was calculated from the percentage cover in each 10 cm height interval between ground level (0 cm) and 60 cm for each quadrat. Data were pooled where species were recorded at heights greater than 61 cm.
- 3. Relative humidity was monitored at the soil surface using cobalt thiocyanate paper exposed for 20 minutes, immersed in liquid paraffin and examined in a Lovibond colour comparator in the laboratory. Measurements were made at two-week intervals from the 5 May until 26 July 1991.
- 4. Aphid numbers were monitored on the same days as the vegetation surveys, where aphids were monitored prior to the vegetation being sampled, to prevent disturbance by walking. Cereal aphids were counted on 30 tillers selected at random in each headland treatment.
- 5. All ground-active invertebrates were recorded from pitfall traps and sorted to family level, data presented here included the total number of invertebrates caught and of Collembola in particular, since these are considered to be an important food source for many carabids.

Experimental manipulation of environmental variables

To test hypotheses concerning the influence of vegetation cover and bare earth on the number of species caught and their relative abundance, twelve 4 m x 4 m plots on the most heavily vegetated treatment, the Uncropped Wildlife Strip, were marked out in 1992 in a randomised block design. Four plots were sprayed, using a knapsack sprayer with the herbicide Gramoxone at field strength (0.5 g paraquat per litre) on 14 May to create bare, undisturbed ground. Any emergent vegetation after this period was removed by hand as seedlings. Four other plots were hoed from the last week in April until the end of July, to remove vegetation and create bare, disturbed ground. The remaining plots were left undisturbed as fully vegetated plots.

After two weeks in which free movement of beetles between plots was allowed, 2 m x 2 m areas at the centre of each plot were enclosed on 2 June, this was done using lengths of 4 mm plywood buried 0.1 m deep to leave 0.15 m above ground, with all corners were nailed to prevent beetles escaping. Ground-active carabids were sampled from both control and bare ground plots using covered pitfall traps. These were arranged with a single trap in the centre, one at the centre of each quadrant and one in the middle of and next to each side of the arena. Traps were emptied weekly from 2 June until 27 July.

Results

Community structure

Rank abundance, pooled for the three years, were broadly similar for each headland treament (Table 1). The biggest difference was that the highest ranking species in the Uncropped Wildlife Strip accounted for 28% of the individuals, compared with 16.2% and 18.6% in the Conservation and Sprayed Headlands respectively. In all treatments six species contributed more than 5% to the total catch, out of a total of 35 species for the normal Sprayed Headland; 41 for the Conservation Headland and 43 species in the Uncropped Wildlife Strip (Table 1).

The most abundant species overall was *Bembidion lampros*, which was amongst the six most abundant species in each headland treatment. However it was more than twice as numerous in the Uncropped Wildlife Strip as any other species, and seven times more abundant there than in the Sprayed Headland.

Pterostichus melanarius was the second most abundant species overall and amongst the top four species in all three treatments.

Table 1. Total catch and proportions of the overall carabid total in Conservation Headlands (CH), Uncropped Wildlife Strips (UH) and Sprayed Headlands (SH) for the 10 most highly ranked species, using data pooled from 1990, 1991 and 1992. *A.dors = Agonum dorsale, P.mel = Pterostichus melanarius, B.lamp = Bembidion lampros, B. tetra = B. tetracolum, D.atri = Demetrius atricapilus, T.quad = Trechus quadristriatus, H.ruf = Harpalus rufipes, H.aff = H.affinis, A.bif = Amara bifrons, A.aen = A.aenea, A.sim = A.similata, A.fam = A.familiaris*

	CH			UH			SH		
Rank	Spp.	%	Total	Spp.	%	Total	Spp.	%	Total
1	*A.dors*	16.15	238	*B.lamp*	28.03	697	*A.dors*	18.55	174
2	*P.mel*	14.18	209	*H.aff*	12.95	322	*B.lamp*	12.79	120
3	*A.bif*	13.77	203	*A.bif*	9.81	244	*H.ruf*	11.83	111
4	*T.quad*	8.68	128	*P.mel*	9.81	244	*P.mel*	10.34	97
5	*D.atri*	6.99	103	*H.ruf*	6.35	158	*D.atri*	9.28	87
6	*B.lamp*	6.38	94	*A.sim*	5.43	135	*T.quad*	7.57	71
7	*A.fam*	4.82	71	*A.aen*	4.26	106	*H.aff*	3.20	30
8	*B.tetra*	4.75	70	*B.tetra*	3.70	92	*A.bif*	2.88	27
9	*H.ruf*	4.07	60	*T.quad*	3.38	84	*A.aen*	2.77	26
10	*A.sim*	2.85	42	*A.dors*	2.27	69	*A.sim*	1.92	18
	Total		1474	Total		2487	Total		938

Agonum dorsale was third most abundant overall but less common in the Uncropped Wildlife Strip than in the Sprayed and Conservation Headlands, in both of which it was the most abundant species.

There were thus substantial differences in the relative contributions made by the most abundant species to the different headland communities. This was also the case for the less common species, some of which were restricted to one treatment and others of which completely avoided a particular treatment. Those restricted to the Uncropped Wildlife Strip included *Bembidion femoratum*, *Amara tibialis*, *Acupalpus meridianus*, *Bradycellus harpalinus*, *Bradycellus distinctus* and *Harpalus rubripes*. In contrast *Amara nitida*, *Amara spreta* and *Pterostichus angustatus* were only found in the Conservation Headland. Only one species, *Calathus piceus*, was restricted solely to the Sprayed Headland.

Relationships between environmental variables and carabid community characteristics

Analyses of the relationships between carabid community parameters and environmental variables are summarized in Table 2, which showed a large number of positive relationships. For overall carabid abundance the most significant correlations were those with percentage cover of dicotyledonous plants and abundance of invertebrates. Species richness was most strongly correlated with total vegetation cover, cover by dicotyledonous vegetation and the abundance of Collembola, aphids and general invertebrate abundance. Both carabid abundance and species richness were negatively correlated with relative humidity.

Table 2. Summary of significant correlations between both the number of carabids and the number of species $trap^{-1}$ $week^{-1}$ with environmental measurements. Where each pitfall catch is associated with the corresponding environmemtal variable measured adjacent to the trap. Data were pooled throughout 1991 where n = 120 and ** is $P < 0.01$ and *** is $P < 0.001$ and NS is non significant ($P > 0.001$). Significance was used at the 1 % level to minimise the risk of obtaining a significant correlation at random.

	Carabid abundance	Species richness
No. dicotyledonous species	NS	0.25^{**}
% cover dicotyledonous species	0.302^{***}	0.503^{***}
No. monocotyledonous species	0.287^{**}	0.253^{**}
% cover monocotyledonous species	NS	NS
% cover bare ground	-0.272^{**}	-0.386^{**}
% cover all vegetation	0.247^{**}	0.441^{***}
Height diversity	NS	0.254^{**}
Collembola	0.282^{**}	0.399^{***}
Invertebrates	0.364^{***}	0.458^{***}
Aphids	0.226^{**}	0.371^{***}
Relative humidity	-0.399^{***}	-0.42^{***}

Experimental manipulation of environmental variables

The overall number of carabids caught, fell significantly when the vegetation cover was removed ($F_{(2,9)} = 16.55$, $P = 0.001$) (Table 3). One conspicuous exception to this trend was *B. lampros* of which significantly fewer were caught in the fully vegetated plots than in the herbicide treated or hoed plots ($F_{(2,9)} = 79.71$, $P = 0.001$). This species is characteristic of habitats with open spaces and bare ground (Mitchel 1963) and so was favoured by experimental reduction in vegetation late in the season. This also explained the abundance of this species in the Uncropped Wildlife Strip throughout its peak activity period from

Table 3. Mean densities (numbers m^{-2} ± SE) of species in fully vegetated, herbicide treated and hoed bare ground plots established in the Uncropped Wildlife Strip in 1992.

Species	Fully vegetated (control)	Hoed bare ground	Herbicide treated bare ground
Harpalus affinis (Shrank)	+ 18.31 ± 1.04	5.06 ± 0.60	4.37 ± 0.71
Amara similata (Gyllenhal)	11.75 ± 3.56	0.75 ± 0.36	0.31 ± 0.20
Amara bifrons (Gyllenhal)	9.18 ± 2.25	2.43 ± 0.53	0.87 ± 0.32
Trechus quadristriatus (Schrank)	0.31 ± 0.16	0.06 ± 0.05	0.93 ± 0.42
Amara aenea (De Geer)	+ + 6.43 ± 1.64	0.44 ± 0.26	0.87 ± 0.32
Harpalus rufipes (De Geer)	6.0 ± 1.91	1.94 ± 0.36	2.06 ± 0.56
Calathus fuscipes (Goez)	1.56 ± 0.32	1.06 ± 0.10	0.81 ± 0.14
Harpalus tardus (Panzer)	1.5 ± 0.41	0.69 ± 0.16	0.37 ± 0.14
Harpalus rufibarbis (Fabricius)	1.31 ± 0.52	0.5 ± 0.09	0.69 ± 0.26
Agonum dorsale (Pontoppidan)	0.75 ± 0.29	0.31 ± 0.16	0.37 ± 0.19
Calathus melanocephalus (Linnaeus)	0.56 ± 0.05	0.44 ± 0.10	0.25 ± 0.09
Loricera pilicornis (Fabricius)	0.56 ± 0.05	0.25 ± 0.09	0.12 ± 0.06
Pterostichus melanarius (Illiger)	0.56 ± 0.42	0.12 ± 0.06	0
Agonum muelleri (Herbst)	0.37 ± 0.06	0	0
Amara familiaris (Duftschmid)	0.06 ± 0.05	0	0
Amara ovata (Fabricius)	0.06 ± 0.05	0	0
Metabletus foveatus (Fourcroy)	2.25 ± 1.25	2.06 ± 0.41	2.25 ± 0.45
Amara plebeja (Gyllenhal)	0.06 ± 0.05	0.06 ± 0.05	0.06 ± 0.05
Notiophilus biguttatus (Fabricius)	0.06 ± 0.05	0.06 ± 0.05	0.06 ± 0.05
Synuchus nivalis (Panzer)	0.06 ± 0.05	0	0.06 ± 0.05
Bembidion quadrimaculatum (Linnaeus)	0.25 ± 0.15	2.25 ± 0.15	1.31 ± 0.39
Demetrius atricapillus (Linneaus)	0.18 ± 0.10	0.25 ± 0.09	0.06 ± 0.05
Amara apricaria (Paykull)	0	0.06 ± 0.05	0
Bembidion lampros (Herbst)	+ + + 7.56 ± 1.57	22.44 ± 1.61	37.06 ± 1.04
Bradycellus harpalinus (Serville)	0.12 ± 0.06	0.12 ± 0.06	0.75 ± 0.34
Bembidion tetracolum (Say)	0	0.19 ± 0.10	0.56 ± 0.28
Nebria brevicollis (Fabricius)	0.06 ± 0.05	0.12 ± 0.06	0.19 ± 0.10
Bembidion guttula (Fabricius)	0	0	0.12 ± 0.06
Acupalpus meridianus (Linneaus)	0	0	0.12 ± 0.16
Amara aulica (Panzer)	0	0	0.06 ± 0.05

\+ $F_{(2,9)} = 7.31$, $P < 0.0001$ UWS(control) > Hoed bare ground = Sprayed bare ground
\+ + $F_{(2,9)} = 8.92$, $P = 0.0027$ UWS(control) > Sprayed bare ground = Hoed bare ground
\+ + + $F_{(2,9)} = 79.71$, $P < 0.0001$ Sprayed bare ground > Hoed bare ground >UWS(control)

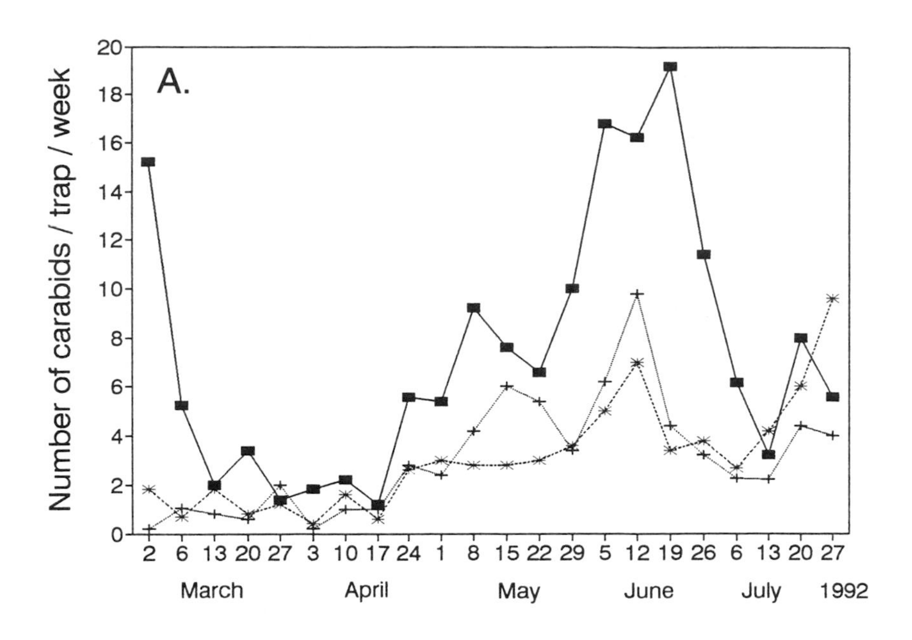

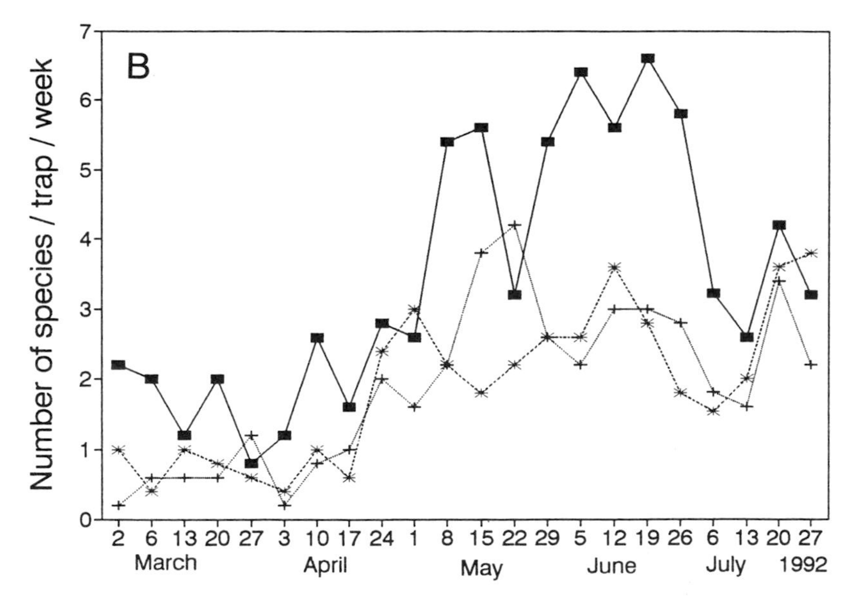

Fig. 1. A. Total numbers of carabids, B. Number of species of carabids caught in pitfall traps during 1992.
Uncropped Wildlife Strip —■—, Conservation Headland ···+···Sprayed Headland – –✳– –

mid-March until mid-June when there are more patches of bare ground in this treatment than in either of the other headland treatments.

Seasonal variation in abundance and species richness

There was a marked difference in both the number of species and the overall abundance of carabids in the Uncropped Wildlife Strip compared with both the other two treatments early in the season (Fig. 1A and 1B). In 1992 this was caused predominantly by activity of *Trechus quadristriatus* at the beginning of March, followed later in the month by *Harpalus affinis*, a largely phytophagous species which also requires patches of bare ground in which to bask (M.L. Luff, pers. comm.). A similar, although larger early season peak was caused by *B. lampros* in the previous year, again utilising the larger areas of bare ground in this treatment, when compared to the others in which winter wheat already dominated the ground cover.

A second pronounced difference between treatments occurs from May through to July when the overall abundance and species richness of carabids are again greater in the Uncropped Wildlife Strip. These are predominantly due to the activity of the largely phytophagous *Amara* species, *A. similata* and *A. bifrons*. Smaller peaks in May and June in the Conservation and Sprayed Headlands were caused by different species, *A. dorsale* and *Demetrius atricapillus*, which actively avoid the patches of bare ground.

Over the whole season the Uncropped Wildlife Strip had significantly higher catches of carabids than the other two headland treatments ($F_{(2,108)} = 57.17$, $P < 0.001$). In addition these catches increased each year when the period between 1 May and 20 July was compared (Fig. 2).

Discussion

The Uncropped Wildlife Strips attract substantially higher payments than the Conservation Headland prescriptions under current E.S.A. agreements (Ministry of Agriculture, Fisheries and Food 1993). Are there commensurate benefits from this option in relation to the carabid fauna of the fields? The results show that both the total abundance of carabids and the number of species present were greatest in the Uncropped Wildlife Strips. This is because of the more complex structure and species diversity of the vegetation which strongly influence the abundance and species richness of carabid faunas (Speight & Lawton 1976; Murdoch et al. 1982, Powell et al. 1985).

In this experiment these parameters were strongly correlated with the percentage cover of all vegetation, but particularly with the cover by dicotyledonous species and with foliage height diversity. There are several

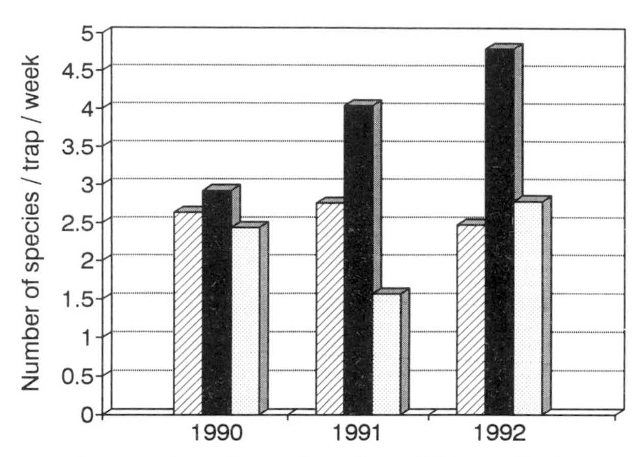

Fig. 2. Changes in the number of species of carabid caught between 1 May and 23 July from 1990 to 1992 in the Uncropped Wildlife Strip (black), Conservation Headland (hatched) and Sprayed Headland (dotted).

reasons for this. Firstly plant tissues can be important directly as a source of food for carabids. Many of the *Amara* and *Harpalus* species that were caught are considered to be largely phytophagous (Hengeveld 1980) while most species of carabids include some vegetation in their diet.

A positive relationship exists between abundance and diversity of many insects and the species and height diversity of vegetation (Sotherton et al. 1985, Murdoch et al. 1972). Thus the availability of potential invertebrate prey for carabids also increases with increasing complexity of the vegetation. The importance of this is confirmed by the strong positive correlations between carabid community characteristics and the density of Collembola and other pitfall trapped invertebrates. High alternative prey densities can be important in encouraging polyphagous predators to persist in a field early in the season, when pest populations are low.

A second indirect effect of increased vegetation complexity is its influence in buffering fluctuations in microclimate (Speight & Lawton 1976). This is considered to be critical in determining the distribution of species such as *Pterostichus melanarius* and *Harpalus rufipes* (Skuhravy, Louda & Sykoro 1971) while a further benefit of structurally diverse vegetation is that it can provide shelter from predators and parasites (Lawton 1978).

The experimental reduction of vegetation from the Uncropped Wildlife Strip provided confirmation that for most of the species the presence of complex vegetation is very important (Table 3). However there are some species of carabid which prefer and actively select bare ground, *B. lampros* is one of these. The abundance of this species increased dramatically in the bare ground plots and it clearly preferred those with a smooth surface to the more disturbed hoed plots. This could account for why, early in the season, when numbers of this species normally peak, it was by far the commonest in the Uncropped Wildlife Strip, where there were no wheat plants and little other vegetation had developed.

Thus there are two peaks in the seasonal patterns of carabid abundance and species richness (Figs. 1A and 1B). The first is early in the season, representing species which prefer bare ground, such as *B. lampros*. The second is later in the season when the vegetation has developed into a more complex and diverse three dimensional matrix, reflecting higher populations of those species which require this dense vegetation and higher humidity (e.g. *A. dorsale* and *P. melanarius*) than are found in the other two headland treatments. The Uncropped Wildlife Strip can thus allow coexistence of potentially mutually exclusive predators (Southwood 1978).

B. lampros, whilst it is one of the smallest carabid species caught in this survey is known to consume more aphids per unit of its body weight than any other species of carabids tested in the laboratory (Sopp & Wratten 1988). Its very high abundance in the Uncropped Wildlife Strips is therefore of considerable importance when evaluating their potential benefits in helping to control pest populations. The strong preference of *P. melanarius* for uncropped strips later in the season may be of similar importance, because whilst less abundant than *B. lampros*, it is a large carabid capable of consuming large numbers of aphids.

If these predators were to remain exclusively within the Uncropped Wildlife Strips they would be unlikely to have a very substantial influence on pests within the crop itself. This is not however the case as carabids sampled at 8 m out into the crop adjacent to the Uncropped Wildlife Strips, during 1992 were significantly ($F_{(2,180)} = 24.69$, $P < 0.0001$) more abundant than they were adjacent to either of the other two headlands.

Corresponding counts of aphid densities showed that these were significantly lower ($F_{(2,348)} = 3.32$, $P = 0.037$) in crops adjacent to the Uncropped Wildlife Strip than adjacent to the other two headlands.

Such evidence of carabids dispersing out from the favourable conditions of the Uncropped Wildlife Strip is of wider significance in relation to the broader status of these beetles within the modern agricultural environment. Cultivation of crops is known to deplete the populations of some species (Blumberg & Crossley 1983). If favourable conditions for larval survival and/or increased adult reproductive rates are provided by the Uncropped Wildlife Strip (as consistent increase in the overall carabid densities and species richness over the three years of this study suggests they do) then these increased populations can provide a reservoir from which depleted populations may be able to be replenished through dispersal.

Populations can survive in an area only if they have sufficient powers of dispersal to "refound" subpopulations that have become extinct, or find new suitable habitats (den Boer 1979). In the modern agricultural landscape suitable habitats are becoming increasingly fragmented (Mader 1988). If appropriately managed headlands provide suitable conditions, albeit temporary, for a species to survive and reproduce before establishing in a more permanent habitat, they may provide an important link between fragments and so help to ensure the survival of the local populations.

The presence of Uncropped Wildlife Strips as a temporary stepping stone to permanent habitats has implications for the survival of many local populations characteristic of the Breckland environment and not just carabids, however this group being highly mobile and easily studied are excellent indicators of habitat suitability for a wide range of invertebrates (Thiele 1977).

Acknowledgements

This work was funded by a CASE studentship between the Ministry of Agriculture, Fisheries and Food and The Game Conservancy Trust. We wish to thank J. J. H. Wilson for the provision of the study sites and P. Pask and the staff of Frederick Hiam Ltd at Ixworth Thorpe for their assistance.

References

Blumberg, A.Y. & Crossley, D.A. 1983. Comparison of soil surface arthropod populations in conventional tillage, no tillage and old field systems. *Agro-ecosystems* 8: 247-253.

Booij, C.J.H. & Noorlander, J. 1992. Farming systems and insect predators. *Agric. Ecosyst. Environ.* 40: 125-135.

Burn, A.J. 1988. Assessment of pesticides on the impact on invertebrate predation in cereal crops. *Asp. Appl. Biol.* 17: 279-288

Chiverton, P.A. & Sotherton, N.W. 1991. The effects of beneficial arthropods on the exclusion of herbicides from cereal crop edges. *J. Appl. Ecol.* 28: 1027-1039.

Boer, P.J. den 1979. Population of carabid beetles and individual behaviour. General Aspects. *Misc. papers. Landbou. Wageningen* 18: 145-150.

Boer, P.J. den 1990. The survival value of dispersal in terrestrial arthropods. *Biol. Conserv.* 54: 175-192.

Dritschilo, W. & Erwin, T.L. 1982. Responses in abundance and diversity of cornfield carabid communities to differences in farming practices. *Ecology* 63: 900-904.

Hawthorne, A. 1995. Validation of the use of pitfall traps to study carabid populations in cereal field headlands. In: Toft, S. & Riedel, W. (eds.) *Arthropod natural enemies in arable land · I*, pp. 61-75. Aarhus.

Hengeveld, R. 1980 Qualitative and quantitative aspects of the food of ground beetles (Coleoptera, Carabidae): review. *Neth. J. Zool.* 30: 555-563.

Kromp, B. 1989. Carabid communities (Carabidae, Coleoptera) in biologically and conventionally farmed agroecosystems. *Agric. Ecosyst. Environ.* 27: 241-251.

Lambley, P.W. & Rothera, S. 1990. Ecological changes in Breckland. Unpublished Report. Nature Conservancy Council, Norwich.

Lawton, J.H. 1978. Host-plant influences on insect diversity: the effects of space and time. In: Mound, L.A. & Waloff, N. (eds.) *Diversity of insect faunas.* Symposia of the Royal Entomological Society 9: 105-125. Blackwell.

Mader, H-J. 1988. Effects of increased spatial heterogeneity on the biocenosis in rural landscapes. *Ecol. Bull.* 39: 169-179.

Ministry of Agriculture, Fisheries and Food (1993). *Breckland E.S.A. Guidelines for farmers.* Reference BD\ESA\2. Her Majestey's Stationery Office, London 12 pp.

Mitchel, B. 1963. Ecology of two carabid beetles *Bembidion lampros* (Herbst) and *Trechus quadristriatus* (Schrank). I. Lifecycles and feeding behaviour. *J. Anim. Ecol.* 32: 289 - 299.

Nature Conservancy Council. 1984. Cherry Hill and Gallops SSSI citation for notification, 27th May 1984. Unpublished report No. 198. Bury St Edmonds.

Nentwig, W. 1989. Augmentation of beneficial arthropods by strip-management. II. Successional strips in a winter wheat field. *Z. Pflanzenkrankh. Pflanzenschutz* 96: 89-99.

Powell, W., Dean, G.J. & Dewar, A. 1985. The influence of weeds on polyphagous arthropod predators in wheat. *Crop Protection* 4: 298-312.

Skuhravy, V., Louda, J. & Sykora, J. 1971. Zur verteilung der Laufkaufer in Feldmonokulturen. *Beiträge zur Entomologie* 21: 539-546.

Speight, M.R. & Lawton, H. 1976. The influence of weed cover on the mortality imposed on artificial prey by predatory ground beetles in cereal fields. *Oecologia (Berl.)* 23: 211-223.

Sopp, P.I. & Wratten, S.D. 1986. Rates of consumption of cereal aphids by some polyphagous predators in the laboratory. *Entomol. Exp. Appl.* 41: 69-73.

Sotherton, N.W., Boatman, N.D. & Rands, M.R.W. 1989. The 'Conservation Headland' experiment in cereal ecosystems. *The Entomologist* 108: 135-143.

Sotherton, N.W., Rands, M.R.W. & Moreby, S.J. 1985. Comparison of herbicide treated and untreated headlands for the survival of game and wildlife. *Proc. Br. Crop. Prot. Conf.* 991-998.

Southwood, T.R.E. 1978. The components of diversity. In: Mound, L.A. & Waloff, N. (eds.) *Diversity of insect faunas.* Symposia of the Royal Entomological Society 9: 19-40. Blackwell.

Stassart, P. & Gregoire-Wibo, C. 1983. Influence of soil tillage on carabid populations in large scale crop fields. Preliminary results. *Med. Fac. Landbouww. Rijksuniv. Gent* 48: 465-474.

Thiele, H.U. 1977. *Carabid Beetles in their Environments.* Spriger-Verlag, Berlin.
Thomas, M.B, Wratten, S.D. & Sotherton, N.W. 1991. Creation of 'island habitats in farmland to manipulate populations of beneficial arthropods: predator densities and emigration. *J. Appl. Ecol.* 28: 906-917.

Carabids in sprayed and unsprayed crop edges of winter wheat, sugar beet and potatoes

G.R. de Snoo, R.J. van der Poll & J. de Leeuw

Centre of Environmental Science, Leiden University
P.O. Box 9518, 2300 RA Leiden, The Netherlands

Abstract
In the Dutch Field Margin Project a management strategy is being developed to promote nature conservation in arable fields and reduce pesticide drift to non-target areas. In 1990 and 1991 carabids were sampled by pitfalls in sprayed and unsprayed crop edges 3 m wide and 100 m long in winter wheat, sugar beet and potatoes. The activity density of carabids in winter wheat (both 1990 and 1991) and in sugar beet (1991) was significantly higher in the unsprayed margins. In these crops the number of species was significantly higher in 1991. In potatoes there was only a significant difference in the number of species (1990). Species of the carabid genera *Amara* and *Harpalus*, both herbivorous, show the most marked increase. The relevance of the results is discussed.

Key words: Carabidae, field margins, pesticides, activity density

Introduction

Carabids are among the most dominant epigeal soil invertebrates of arable fields. Most species are polyphagous predators with a preference for aphids and Collembola. Their main significance for agricultural practice is their role in preventing pest outbreaks (Luff 1982). *Agonum dorsale* is particularly important in this respect because the species also preys very successfully on aphids when these are present in low densities (Sunderland & Vickerman 1980). Adult carabids from the genera *Amara* and *Harpalus* are partly phytophagous, consuming weed seeds. The number of carabids in arable fields is important for the chick survival of some birds, such as pheasants (Hill 1985)

One of the factors determining the presence of carabids in arable fields is pesticide use. Compared with other West European countries, pesticide use is very high in the Netherlands: on arable farmland 19 kg a.i./ha on

Arthropod natural enemies in arable land · I *Density, spatial heterogeneity and dispersal*
S. Toft & W. Riedel (eds.). *Acta Jutlandica* vol. 70:2 1995, pp. 199-211.
 ISBN 87 7288 492

average per year (MJP-G 1991). The use of pesticides on a plot generally has an adverse impact on the number of carabid species and individuals found there, and there is consequently less aphid predation (Vickerman & Sunderland 1977, Edwards et al. 1979, Sotherton et al. 1987, Dritschilo & Wanner 1990, Basedow et al. 1991). At the same time certain species, such as *Trechus quadristriatus* and *Asaphidion flavipes*, appear to be more abundant under conditions of intensive cropping (Dritschilo & Wanner 1990, Basedow et al. 1991). The abundance of some species in arable fields is also influenced by the presence of distinct field boundary elements such as grass strips and hedges. So-called spring breeders such as *A. dorsale*, in particular, overwinter as adults in large numbers in perennial vegetation, from there colonizing fields in spring (Sotherton 1984, Wallin 1986, Coombes & Sotherton 1986, Riedel 1991, Basedow et al. 1991).

Over the past ten years there has been a growing interest in maintaining unsprayed crop edges. In 6 m wide unsprayed cereal crop edges, Hassall et al. (1992) found significantly more carabids (individuals and species) than in sprayed edges. Storck-Weyhermüller & Welling (1991) found more carabids (activity density, species and diversity) and also less aphid predation in 3-5 m wide herbicide-free winter wheat crop edges. However, a study by Felkl (1988) in 2-5 m wide unsprayed cereal crop edges showed no increase in the number of carabid species and only a slight increase in the number of individuals. Using pitfall traps in 6 m wide cereal crop edges where no herbicides has been used, Chiverton & Sotherton (1991) found no increase in numbers of the studied species *Agonum dorsale* and *Pterostichus melanarius*. These studies focused solely on effects in the unsprayed edges of cereal crops. The Dutch Field Margin Project being undertaken in the Haarlemmermeerpolder is also concerned with unsprayed potato and sugar beet crop edges. The aim of this project is to develop a management strategy for promoting nature conservation in arable fields and for reducing pesticide drift to non-target areas. The effects on weeds, invertebrates, vertebrates, yields and pesticide drift into ditches and ditch banks are being investigated.

Methods

Study area

The study was carried out on seven farms in the Haarlemmermeerpolder in 1990 and 1991. In this polder, reclaimed about 150 years ago, most parcels are 1000 m long and 200 m wide and are bordered by ditches. Ditch bank vegetation consists mainly of perennial grasses such as *Elymus repens* and *Festuca rubra*. Two or three times a year ditch banks are mown or chopped, and the swath left lying. The most common rotation on the farms is: winter wheat followed by potatoes and a second winter wheat crop and finally sugar beet. The size of most of the fields investigated in this study was 500 x 100m (average: 5.2 ha). The (clay) soil contains 22.8 ± 9.0% silt (0-16 μm diam.) and 2.5 ± 1.0% organic matter. In 1990 six fields of each crop were investigated, in 1991 8 sugar beet fields, 6 winter wheat fields and 5 potato fields were studied. Along the field edges strips measuring 3 m wide and 100 m long and bordering the ditches (Fig. 1) were not sprayed with herbicides or insecticides. Spraying of fungicides was allowed. These compounds are virtually non-toxic to carabids and no side effects on carabid abundance were found in the field. Organophosphate fungicides such as pyrazophos form an exception in this respect (cf. Sotherton et al. 1987), and this compound was not applied. Conventional fertilizing and tillage regimes were maintained in the fields studied.

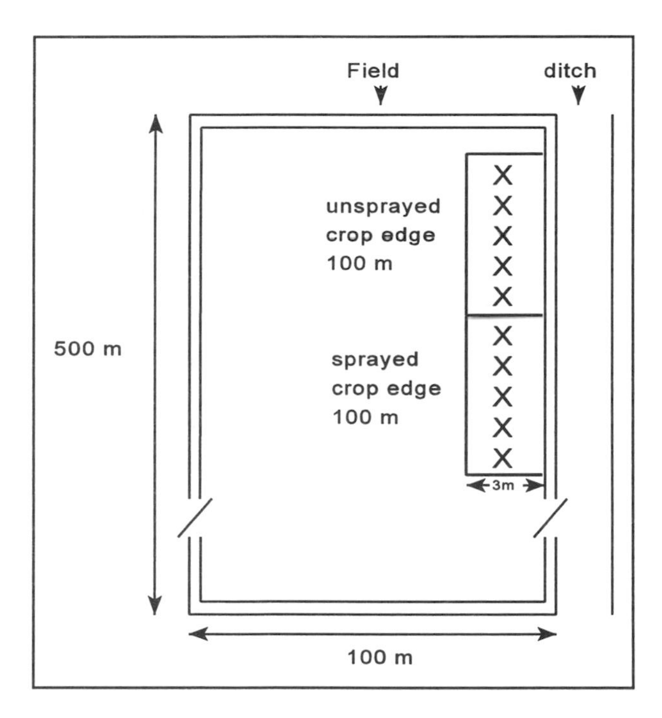

Fig. 1. Sampling of carabids in sprayed and unsprayed crop edges on the fields, x = pitfall trap.

There are differences in the spraying regime on the farms involved, for example active ingredient, spraying time, frequency etc. However, in winter wheat the farmers generally use herbicides twice, once in April and once in May. Insecticides are used once or twice in June against aphids and a mixture of fungicides mostly in June against diseases which cause premature senescence. In sugar beet there are generally three herbicide sprayings, from the end of March until mid-May, and one to three insecticide sprayings from April to July. In sugar beet no fungicides are used. In potatoes finally there is usually one herbicide spraying about mid-May, two insecticide sprayings in May and June, and about ten fungicide sprayings from the end of May until harvest, in September. The pesticides most frequently used on the various crops are summarized in Table 1.

Sampling programme

During the growing seasons carabids were trapped weekly in the sprayed and unsprayed crop edges, using pitfall traps (11.3 cm diam., with a plastic roof) partly filled with 4% formalin in 1990 and 50% ethylene glycol in 1991. In 1990 the sampling campaign lasted 13 weeks, in 1991 12 weeks, from the beginning of May to the end of July. In 1990 sampling in sugar beet and potato fields was continued from the beginning of June to the beginning of September. In 1990 and 1991 four and five pitfalls, respectively, were placed 15 m apart along crop edges, 1.5 m from the field edge (see Fig. 1). To compare the results of the pitfalls filled with formalin with those filled with ethylene glycol, a control experiment was carried out in 1991 in winter wheat and potatoes. At three farms a comparison was made between the two chemicals, with carabids being trapped for a total of three times in June and July: weeks no. 23, 26 and 29 (5 pitfalls per crop edge).

Table 1. The mostly used pesticides in 1990 and 1991 in the sprayed edges of the various crops.

	Herbicides	Insecticides	Fungicides
Winter wheat	MCPA mecoprop bentazone isoproturon	dimethoate phosphamidon	propiconazole maneb prochloraz fenpropimorph
Sugar beet	ethofumesate phenmedipham metamitron	oxydemeton-methyl pirimicarb	-
Potatoes	metribuzin	parathion dimethoate	maneb

The nomenclature of the carabids follows Lindroth (1974).

To assess weed growth on the sprayed and unsprayed edges, weed coverage and species diversity were measured at the end of June (Braun-Blanquet, total sample area 75 m^2). Because of abundant growth on a number of fields, weeds had to be partly removed by hand in some unsprayed edges: in 1990 on 4 crop edges in sugar beet and 3 in winter wheat (in June) and in 1991 on all edges in sugar beet (early July) and 3 in winter wheat (mid-July). In the winter wheat fields only *Matricaria recutita* was removed.

Statistical processing of results

The impact of treatments on the total number of individuals (activity density) and the total number of species was assessed by means of a two-tailed two-way ANOVA, summing the data for each pitfall over the entire sampling period. At the species level the number of carabids were averaged per crop edge and the results were processed using the Wilcoxon matched-pairs ranks test, for species found at half or more of the fields.

Results

Comparison of trap preservatives

The control experiment showed no significant difference in numbers trapped between the pitfalls with formalin and ethylene glycol ($P > 0.05$, two-tailed two-way ANOVA). However, fewer individuals seem to be caught with ethylene glycol in winter wheat. There was also no significant difference in the number of species trapped, but there appeared to be slightly fewer species in the pitfalls with ethylene glycol ($P = 0.08$).

Comparison of crops and treatments

On unsprayed field edges average weed coverage was much higher than on sprayed edges (Table 2). In sugar beet crops weed coverage was particularly

Table 2. Average weed coverage (percentage cover) on sprayed and unsprayed crop edges.

	1990		1991	
	Unsprayed	Sprayed	Unsprayed	Sprayed
Winter wheat	25	1	30	3
Sugar beet	37	7	51	14
Potatoes	17	5	9	2

high. In winter wheat *Matricaria recutita*, *Polygonum aviculare* and *P. convolvulus* predominate, while in sugar beet and potato crops *Chenopodium album* and *P. persica* may also become dominant.

In 1990 a total of 44,233 individuals of 68 carabid species were trapped; in 1991 the corresponding figures were 43,243 and 55, respectively. The lower number of species trapped in 1991, may have been due in part to the preservatives used in the pitfalls. Overall, six species predominated, together accounting for 82% of the total number of individuals: *Pterostichus melanarius* (31%), *Bembidion tetracolum* (16%), *Nebria brevicollis* (10%), *Trechus quadristriatus* (9%), *Harpalus rufipes* (8%) and *Agonum dorsale* (8%).

Because of the different length of the sampling programs in the two years, as well as for the various crops in 1990, due caution should be applied in comparing the two years and the three crops. Table 3 gives the results for the sprayed and unsprayed edges of the various crops. The greatest number of species and individuals were caught in winter wheat and sugar beet crops, with lower numbers in potatoes. In winter wheat in both years a significant increase was found in the number of individuals in unsprayed edges. In one year (1991) also the number of species increased significantly. In sugar beet, a significant increase was found in the number of individuals and the number of species, but only in 1991. In potatoes, finally, a significant effect was found only on the number of species in 1990. In all cases there were significant differences between the locations.

Table 3. Activity density (individuals) and number of species of carabid beetles in the various crops in sprayed and unsprayed crop edges in 1990 and 1991. Mean numbers per pitfall ± standard deviation (per crop edge). Two-tailed two-way ANOVA: ** = $P < 0.01$, *** = $P < 0.001$, ns = not significant, n = number of pitfall traps.

	1990		1991	
	Unsprayed	Sprayed	Unsprayed	Sprayed
Winter wheat	(n = 24)		(n = 30)	
Individuals	333 ± 94	283 ± 98 **	252 ± 67	183 ± 58 ***
Species	22.7 ± 2.5	21.5 ± 2.2 ns	16.5 ± 1.9	14.9 ± 2.9 **
Sugar beet	(n = 24)		(n = 40)	
Individuals	367 ± 214	345 ± 269 ns	318 ± 127	236 ± 115***
Species	16.0 ± 2.5	15.7 ± 3.7 ns	16.1 ± 1.8	14.3 ± 2.7 ***
Potatoes	(n = 24)		(n = 25)	
Individuals	247 ± 104	268 ± 115 ns	164 ± 135	157 ± 121 ns
Species	14.0 ± 2.4	11.8 ± 2.3 **	9.5 ± 1.8	9.4 ± 2.3 ns

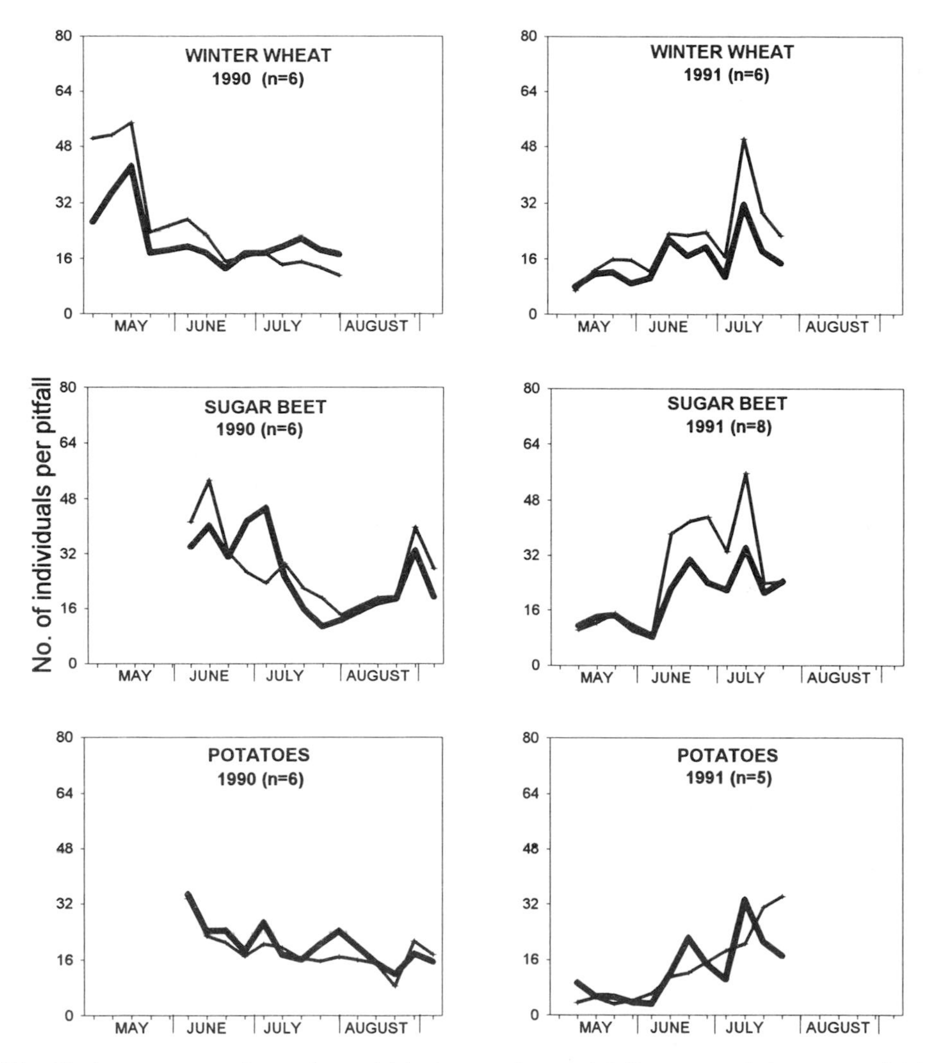

Fig. 2. Average number of carabids trapped per pitfall per week in sprayed (thick lines) and unsprayed (thin lines) edges of winter wheat, sugar beet and potatoes in 1990 and 1991. In 1990: 4 pitfall traps per crop edge, in 1991: 5 pitfall traps per crop edge.

Fig. 2 shows the variation in time in the number of individuals trapped. The pattern of trappings was very different in the two years. In winter wheat in 1990 a large number of carabids was trapped at the beginning of the sampling campaign, in 1991 at the end. The peak in 1990 is due almost entirely to the number of *Nebria brevicollis* in the pitfalls. In 1991 there is similarity among the patterns for the various crops, in each crop there was a peak in the number of carabids trapped around mid-June and in the second week of July. In 1991 there was a difference in carabid activity density on sprayed and unsprayed strips throughout the season in sugar beet and winter wheat.

The impact on the different crops at species level is given in Tables 4, 5 and 6. These tables show the dominant species as well as the species present at at least half the sampling sites and with a significantly higher activity density on unsprayed crop edges. Although there are major differences between the two years, it can still be concluded that some species show a preference for certain crops. In winter wheat relatively large numbers of *Loricera pilicornis*, *Agonum dorsale* and various *Amara* species were found, in potatoes and sugar beet large numbers of *Trechus quadristriatus* and *Pterostichus melanarius*. Compared to the sprayed edge, in winter wheat, a statistically significant increase in activity density was found 1990 for three carabid species and in 1991 for four species (Table 4). The effects were found only for species of the genera *Amara* and *Harpalus*. In sugar beet (Table 5) in 1990 there was a significant increase in five species and in 1991 in eight species. Again most of these species belong to the genera *Amara* and *Harpalus*, but effects were also found on *Bembidion lampros*,

Table 4. Mean number of selected carabid beetles per pitfall trap ± standard deviation (per crop edge) in sprayed and unsprayed winter wheat crop edges in 1990 and 1991. Two-tailed Wilcoxon matched-pairs ranks test: * = $P < 0.05$; ** = $P < 0.01$; ns = not significant; - = not tested: species present on less than 50% of the fields; n = no. of fields. Carabid taxonomy follows Lindroth (1974).

WINTER WHEAT	1990 (n=6)		1991 (n=6)	
	Unsprayed	Sprayed	Unsprayed	Sprayed
Nebria brevicollis	82.0 ± 74.8	37.3± 32.9 ns	35.3 ± 27.1	29.0 ± 26.2 ns
Loricera pilicornis	18.9 ± 15.1	22.8± 22.4 ns	4.9 ± 5.3	3.0 ± 3.7 ns
Trechus quadristriatus	6.2 ± 4.0	8.0± 7.4 ns	6.4 ± 4.0	11.5 ± 7.6 ns
Bembidion lampros	4.5 ± 3.8	6.4± 4.1 ns	2.0 ± 1.6	1.6 ± 1.7 ns
Bembidion tetracolum	56.4 ± 28.1	53.3± 25.3 ns	39.2 ± 31.2	31.1 ± 14.4 ns
Pterostichus melanarius	28.3 ± 20.9	43.7± 46.3 ns	68.1 ± 38.7	51.2 ± 43.9 ns
Pterostichus niger	2.5 ± 1.4	2.5± 1.9 ns	4.5 ± 5.0	5.1 ± 7.0 ns
Agonum dorsale	61.9 ± 30.3	46.3± 33.8 ns	40.5 ± 29.5	24.8 ± 26.0 ns
Amara aenea	10.3 ± 9.2	6.2± 7.4 ns	3.2 ± 5.1	0.4 ± 0.4 *
Amara apricaria	0.6 ± 0.5	0.2± 0.2 ns	0.4 ± 0.4	0.3 ± 0.4 ns
Amara bifrons	0.2 ± 0.2	<0.1± 0.1 *	0.2 ± 0.2	0 *
Amara familiaris	2.5 ± 3.7	0.4± 0.4 *	0	<0.1 ± 0.1 -
Amara similata	15.2 ± 11.7	3.3± 3.9 *	5.8 ± 7.0	0.9 ± 0.8 ns
Harpalus aeneus	11.3 ± 9.0	9.3± 2.9 ns	9.2 ± 10.0	2.6 ± 1.6 *
Harpalus rufipes	9.8 ± 7.8	14.3± 14.9 ns	15.9 ± 6.8	8.5 ± 3.5 *

Table 5. Mean number of selected carabid beetles per pitfall trap ± standard deviation (per crop edge) in sprayed and unsprayed sugar beet crop edges in 1990 and 1991. Two-tailed Wilcoxon matched-pairs ranks test: * = $P < 0.05$; ** = $P < 0.01$; ns = not significant; - = not tested: species present on less than 50% of the fields; n = no. of fields. Carabid taxonomy follows Lindroth (1974).

SUGAR BEET	1990 (n=6)		1991 (n=8)	
	Unsprayed	Sprayed	Unsprayed	Sprayed
Nebria brevicollis	15.6 ± 20.6	11.5 ± 13.4 ns	34.2 ± 50.2	24.3 ± 41.9 ns
Loricera pilicornis	1.9 ± 1.5	1.6 ± 2.1 ns	0.1 ± 0.1	0.3 ± 0.4 ns
Trechus quadristriatus	32.1 ± 29.6	30.1 ± 25.5 ns	36.2 ± 44.2	36.9 ± 43.4 ns
Bembidion lampros	25.8 ± 41.9	15.4 ± 27.3 *	5.1 ± 2.9	4.4 ± 3.6 ns
Bembidion tetracolum	34.8 ± 32.4	84.0 ± 170.5 ns	48.5 ± 29.7	54.9 ± 43.9 ns
Pterostichus melanarius	129.1 ± 151.1	125.0 ± 155.6 ns	75.3 ± 86.1	49.0 ± 47.8 *
Pterostichus niger	11.3 ± 7.4	11.7 ± 8.1 ns	2.3 ± 2.5	1.4 ± 2.2 ns
Agonum dorsale	8.8 ± 8.3	2.8 ± 3.4 *	36.8 ± 33.9	17.6 ± 24.2 **
Amara aenea	0.2 ± 0.2	0.4 ± 0.6 ns	2.6 ± 3.4	0.9 ± 1.3 *
Amara apricaria	5.4 ± 6.1	2.5 ± 3.2 *	0.4 ± 0.2	0.1 ± 0.1 *
Amara bifrons	0.8 ± 0.6	0.7 ± 0.8 ns	0.4 ± 0.4	0.1 ± 0.2 *
Amara familiaris	0	0 -	0.1 ± 0.1	0 -
Amara similata	1.4 ± 1.6	0.8 ± 0.9 ns	9.7 ± 13.7	3.3 ± 4.9 **
Harpalus aeneus	19.4 ± 10.9	11.4 ± 8.1 *	24.1 ± 13.8	14.4 ± 6.7 *
Harpalus rufipes	66.2 ± 40.9	34.2 ± 18.9 *	25.7 ± 17.8	16.3 ± 16.2 *

Table 6. Mean number of selected carabid beetles per pitfall trap ± standard deviation (per crop edge) in sprayed and unsprayed potatoe crop edges in 1990 and 1991. Two-tailed Wilcoxon matched-pairs ranks test: * = $P < 0.05$; ** = $P < 0.01$; ns = not significant; - = not tested: species present on less than 50% of the fields; n = no. of fields. Carabid taxonomy follows Lindroth (1974).

POTATOES	1990 (n=6)		1991 (n=5)	
	Unsprayed	Sprayed	Unsprayed	Sprayed
Nebria brevicollis	11.8 ± 14.9	7.4 ± 8.2 ns	5.0 ± 3.0	4.6 ± 1.5 ns
Loricera pilicornis	0.1 ± 0.1	0.3 ± 0.2 ns	<0.1 ± 0.1	0 -
Trechus quadristriatus	43.5 ± 40.7	39.1 ± 35.6 ns	19.5 ± 22.7	14.0 ± 14.2 ns
Bembidion lampros	1.3 ± 1.3	1.3 ± 1.8 ns	5.0 ± 9.8	3.0 ± 5.8 ns
Bembidion tetracolum	18.8 ± 31.5	15.1 ± 19.5 ns	29.1 ± 32.7	31.4 ± 40.8 ns
Pterostichus melanarius	100.1 ± 60.7	138.5 ± 92.9 ns	91.6 ± 139.6	91.9 ± 122.5 ns
Pterostichus niger	8.8 ± 9.2	12.0 ± 12.8 ns	1.2 ± 1.8	1.0 ± 0.6 ns
Agonum dorsale	2.5 ± 2.3	2.3 ± 4.3 ns	1.0 ± 1.1	0.8 ± 0.7 ns
Amara aenea	0.2 ± 0.2	0.1 ± 0.2 ns	0.1 ± 0.1	0.1 ± 0.1 ns
Amara apricaria	0.9 ± 0.9	<0.1 ± 0.1 *	0.3 ± 0.6	<0.1 ± 0.1 -
Amara bifrons	1.2 ± 1.6	0.6 ± 0.6 ns	0.1 ± 0.1	0.1 ± 0.3 -
Amara familiaris	0	0 -	0	<0.1 ± 0.1 -
Amara similata	0.8 ± 0.6	0.3 ± 0.3 ns	0.2 ± 0.3	0.2 ± 0.2 ns
Harpalus aeneus	9.5 ± 7.3	11.7 ± 11.6 ns	4.6 ± 6.4	4.2 ± 3.9 ns
Harpalus rufipes	38.8 ± 34.2	31.9 ± 38.3 ns	3.9 ± 5.4	3.1 ± 4.9 ns

Agonum dorsale and *Pterostichus melanarius*. In potatoes (Table 6) only in 1990 was a statistically significant increase found in *Amara apricaria*. Some species were trapped much less frequently in the unsprayed margins, such as *Bembidion lampros* and *Trechus quadristriatus* in winter wheat, however, these effects were not statistically significant. Overall, of all ten species showing a significant increase in one or more crops or years, there were seven *Amara* or *Harpalus* species. It is interesting that the species in these genera are herbivorous, and that there seems to be a relation between the number of species showing significant effects and the percentage weed cover in the various crops and years (see also Table 2).

Discussion

The results of this study show that in unsprayed winter wheat crop edges more carabid individuals, and in one year more carabid species were caught than in the sprayed crop edges. Also in sugar beet in 1991 more individuals and species were caught in the unsprayed edges. With respect to individual species, too, the greatest impact was seen in these crops. In potatoes the average number of species was affected in 1990 only. It can be concluded that in all three crops unsprayed edges only 3 metres wide can have a positive effect on carabids.

The changes in the carabid fauna may be due to direct toxic pesticide effects, on the one hand, and indirect (ecological) effects such as the disappearance of prey or changes in habitat, on the other. Insecticides, in particular, may have a toxic impact on carabids. Sensitivity varies from species to species, however, and there is a very wide range of toxicity depending on the compound involved (Edwards & Thompson 1975; Förster 1991). Moreover, the degree of exposure depends on the habits of the species (Dixon & McKinlay 1992). Because the weed vegetation becomes very dense in the unsprayed edges, the microclimate also changes. Felkl (1988) found a lower soil temperature and higher humidity in unsprayed field edges. This can have a major impact on species composition (Wallin 1986; Quinn et al. 1991). Because both direct and indirect effects may be involved, and because different pesticides were used at different times on the various farms, no attempt has been made to establish a direct relationship between spraying regimes and carabid activity density. But it should be remarked that in some crops there was already a difference in activity density before insecticide spraying, due possibly to herbicides applied earlier in the season. It is also noteworthy that it was above all the numbers of *Amara* and

Harpalus species that increased. The abundance of these species correlates with the weed density (Kokta & Niemann 1990; Basedow et al. 1991; Welling et al. 1988). Prevalence of *Agonum dorsale* is likewise positively influenced by the presence of weeds (Coombes & Sotherton 1986). It is possible that the high activity density of phytophagous carabids could play a role in the natural control of weed seeds in the unsprayed edges (cf. Welling et al. 1988; Kokta 1988). The variation in weed cover may possibly also explain the differences in carabid activity density among the various crops or within a given crop (sugar beet, for example). The significance of the higher activity density of carabids in unsprayed crop edges in relation to aphid predation is not entirely clear. Although Storck-Weyhermüller & Welling (1991) found generally less aphid predation in the sprayed cereal edges, as a result of all predators, Chiverton & Sotherton (1991) demonstrated that carabids in the unsprayed cereal edges consumed less aphids, because of greater variation in prey.

Because pitfall traps were used in the study, trapping success was also determined by carabid activity and species trappability (cf. Sunderland et al. 1995). This may in turn be affected by pesticide application, with more carabids being caught post-spraying due to greater activity resulting from food shortages or a more open habitat in sprayed fields (Chiverton 1984; Basedow et al. 1991; Dixon & Mckinlay 1992). It is therefore possible that the use of pitfalls has under-estimated the differences in density between the sprayed and unsprayed edges. Although in the present study a higher activity density was generally measured in the unsprayed crop edges, the results should be interpreted with due care.

Acknowledgment

We would like to thank E. Meelis of the Theoretical Biology Department of Leiden University for his statistical help.

References

Basedow, Th., Braun, C., Lühr, A., Naumann, J., Norgall, T. & Yanes, G. 1991. Abundanz, Biomasse und Artenzahl epigaïscher Raubarthropoden auf unterscheidlich intensiv bewirtschafteten Weizen- und Rübenfeldern: Unterschied und ihre Ursachen. Ergebnisse eines dreistufigen Vergleichs in Hessen, 1985 bis 1988. *Zool. Jb. Syst.* 118: 87-116.

Chiverton, P.A. & Sotherton, N.W. 1991. The effects on beneficial arthropods of exclusion of herbicides from cereal crop edges. *J. Appl. Ecol.* 28: 1027-1039.

Chiverton, P.A. 1984. Pitfall catches of the carabid beetle *Pterostichus melanarius*, in relation to gut content and prey densities, in insecticide treated and untreated spring barley. *Entomol. Exp. Appl.* 36: 23-30.

Coombes, D.S. & Sotherton, N.W. 1986. The dispersal and distribution of polyphagous predatory Coleoptera in cereals. *Ann. Appl. Biol.* 108: 461-474.

Dixon, P.L. & McKinlay, R.G. 1992. Pitfall catches of and aphid predation by *Pterostichus melanarius* and *Pterostichus madidus* in insecticide treated and untreated potatoes. *Entomol. Exp. Appl.* 64: 63-72.

Dritschilo, W. & Wanner, D. 1980. Ground beetle abundance in organic and conventional corn fields. *Environ. Entomol.* 9 (5): 629-631.

Edwards, C.A. & Thompson, A.R. 1975. Some effects of insecticides on predatory beetles. Proc. *Asp. Appl. Biol.* 80: 132-135.

Edwards, C.A., Sunderland, K.D. & George, K.S. 1979. Studies on polyphagous predators of cereal aphids. *J. Appl. Ecol.* 16: 811-823.

Felkl, G. 1988. Erste Untersuchungen über die Abundanz von epigaïschen Raubarthropoden, Getreide Blattläusen und stenophagen Blattlausprädatoren in herbizidfreien Winterweizen-Ackerrandstreifen in Hessen. *Gesunde Pflanzen* 40 (12): 483-491.

Förster, P. 1991. Einfluss von Pflanzenschutzmitteln auf Larven und Adulte von *Platynus dorsalis* (Pont.) (Col., Carabidae) und Adulte von *Tachyporus hypnorum* (L.) (Col., Staphylinidae) in Labor- und Halbfreilandversuchen. *Zeitschrift für Pflanzenkrankheiten und Pflanzenschutz* 98 (5): 457-463.

Haccou, R. & Meelis, E. 1992. *Statistical analysis of behavioural data.* Institute of Theoretical Biology, Leiden University. Oxford University Press.

Hassall, M., Hawthorne, A., Maudsley, M., White, P. & Cardwell, C. 1992. Effects of headland management on invertebrate communities in cereal fields. *Agric. Ecosyst. Environ.* 40: 155-178.

Hill, D.A. 1985. The feeding ecology and survival of pheasant chicks on arable farmland. *J. Appl. Ecol.* 22: 645-654.

Kokta, Ch. 1988. Beziehungen zwischen der Verunkrautung und Phytophagen Laufkäfern der Gattung *Amara*. *Mitt. Biol. Bundesanst. Land- und Forstwirtschaft* 247: 139-145.

Kokta, Ch. & Niemann, P. 1990. Wechselwirkungen zwischen Produktionsintensität und Aktivitätsdichte von Laufkäfern in einer dreiliedrigen Fruchtfolge. DFG-Forschungsbericht "Integrierte Pflanzenproduktion": Weinheim: 126-139.

Lindroth, C.H. 1974. *Handbooks for identification of British insects: Coleoptora, Carabidae.* Vol. IV part 2, Royal Entomological Society of London, London, 148 pp.

Luff, M.L. 1982. Population dynamics of Carabidae. *Ann. Appl. Biol.* 101:164-170.

MJP-G 1991. Dutch Ministry of Agriculture, Nature Management and Fisheries; Meerjarenplan Gewasbescherming. Regeringsbeslissing. Tweede Kamer, vergaderjaar 1990-1991, 21 677, nrs. 3-4. SDU Uitgeverij, Den Haag.

Quinn, M.A., Kepner, R.L., Walgenbach, D.D., Nelson Foster, R., Bohls, R.A., Pooler, P.D., Reuter, K.C. & Swain, J.L. 1991. Effect of habitat characteristics and perturbation from insecticides on the community dynamics of groud beetles (Coleoptera: Carabidae) on mixed-grass rangeland. *Environ. Entomol.* 20 (5): 1285-1294.

Riedel, W. 1991. Overwintering and spring dispersal of *Bembidion lampros* (Coleoptera: Carabidae) from established hibernation sites in a winter wheat field in Denmark. In: Polgár, L., Chambers, R.J., Dixon, A.F.G. & Hodek, I. (eds.) *Behaviour and Impact of Aphidophaga* pp. 235-241. Academic Publishing bv, The Hague, The Netherlands.

Sotherton, N.W. 1984. The distribution and abundance of predatory arthropods overwintering on farmland. *Ann. Appl. Biol.* 105: 423-429.

Sotherton, N.W., Moreby, S.J. & Langley, M.G. 1987. The effects of the foliar fungicide pyrazophos on beneficial arthropods in barley fields. *Ann. Appl. Biol.* 111: 75-87.

Storck-Weyhermüller S. & Welling, M. 1991. Regulationsmöglichkeiten von Schad- und Nutzarthropoden im Winterweizen durch Ackerschonstreifen. *Mitt. Biol. Bundesanst. Land-Forstwirtschaft. Berlin-Dahlem.* Heft 273 Berlin.

Sunderland, K.D. & Vickerman, G.P. 1980. Aphid feeding by some polyphagous predators in relation to aphid density in cereal fields. *J. Appl. Ecol.* 17: 389-396.

Vickerman, G.P. & Sunderland, K.D. 1977. Some effects of dimethoate on arthropods in winter wheat. *J. Appl. Ecol.* 14: 767-777.

Wallin, H. 1986. Habitat choice of some field-inhabiting carabid beetles (Coleoptera: Carabidae) studied by recapture of marked individuals. *Ecol. Entomol.* 11: 457-466.

Welling, M., Pötzl, R.A. & Jürgens, D. 1988. Untersuchungen in Hessen über Auswirkung und Bedeutung von Ackerschonstreifen. 3: Epigäische Raubarthropoden. *Mitt. Biol. Bundesanst. Land- und Forstwirtschaft* 247: 55-63.

Spatial distribution of polyphagous predators in nursery fields

J. K. Holopainen

Ecological Laboratory, Department of Environmental Sciences, University of Kuopio, P.O.B. 1627, SF-70211 Kuopio, Finland.

Abstract
Activity densities of carabids and other polyphagous predators in nurseries were monitored with pitfall traps in the field boundary and in the central area of the field. Spiders in early summer and carabids in early and late summer appeared to be more numerous near field boundaries. An increased microarthropod activity was observed in the marginal area of the fields, where pH of topsoil was lower and the content of organic material in the soil was higher than in the field center. Microarthropods act as a food source for polyphagous predators, so greater availability of food could be one explanation for the higher numbers of polyphagous predators in field margins. A laboratory experiment demonstrated that the carabid beetle *Amara plebeja* preys on the nymphs of the bug *Lygus rugulipennis* that damage pine seedlings. To conserve the natural enemies of nursery pests, the field areas near the edge should not be treated with insecticides

Key words: Pitfall traps, soil properties, Carabidae, Staphylinidae, Collembola, Araneae, Acarina, *Bembidion* sp., *Lygus rugulipennis*

Introduction

Picea and *Pinus* are the dominant tree genera produced for forest regeneration in European nurseries (Nef 1993). In Finland, the annual production of tree seedlings in nurseries is nowadays about 220 million seedlings in an area of about 900 ha (Aarne 1992). Our knowledge of the status of beneficial arthropods in nursery ecosystems is rather limited, while the cultivation technology in forest nurseries is very intensive, and the input of fertilizers and pesticides is high.

Forest nurseries in different parts of the world are facing new pest problems caused by the polyphagous *Lygus* bugs (Heteroptera: Miridae) (South 1991). *Lygus rugulipennis* Popp. attacks the apical meristem and buds of conifer seedlings, stopping growth and development of multiple-leadered seedlings (Holopainen 1986, Lilja 1986, Poteri et al. 1987, Kytö 1992). Increased

Arthropod natural enemies in arable land · I *Density, spatial heterogeneity and dispersal*
S. Toft & W. Riedel (eds.). *Acta Jutlandica* vol. 70:2 1995, pp. 213-220.

Table 2. The total catches (seven trapping periods) of various soil arthropods in the central field area and in the boundary of a 2+1 pine transplant field between July 22 and September 9, 1986.

	Field (n=12) Avg. ± SD	Boundary (n=4) Avg. ± SD	t-value	p
Carabidae	18.6± 11.5	89.8±23.7	8.24	<0.001
Staphylinidae	8.4± 2.6	9.5± 2.1	0.58	0.573
Araneae	31.7± 9.2	20.3±11.9	2.01	0.064
Collembola	552.8±126.3	764.0±98.9	3.03	0.009
Acari: Trombiidae	5.4± 2.8	44.0±35.2	4.05	0.001
Other Acari	5.0± 5.4	69.0±32.2	7.07	<0.001

Spatial distribution of soil arthropod activity in nursery fields

Among predatory animals, the numbers of carabid beetles caught were significantly higher in the boundary than in the central field area in 1986 (Table 2). The most common carabid species in the field were *Bembidion femoratum* Sturm, *B. quadrimaculatum* (L.) and *Calathus erratus* (Sahlb.), while in the field boundary dominating species were *Harpalus affinis* (Schrank) and *A. plebeja*. Numbers of prey animals (springtails and mites) as well as numbers of predatory mites (Trombiidae) were also higher in the field boundary.

In 1987, trapping site affected significantly ($p<0.05$) the total catches of spiders, springtails and mites, but mean total catch per trap of staphylinids and carabids was affected only at the 10% significance level (Table 3). The highest number of carabids was trapped in the field at 3 m from the field margin, while the highest number of spiders was trapped in the hedge. Among the carabid species *B. quadrimaculatum* was not caught in the hedge and it was more evenly distributed in the field (Fig. 2A) than other dominant species. *B. lampros*

Table 3. The total catches of various soil arthropods in the hedge and at various distances from the field margin in a 2+1 pine transplant field from May 18 to June 22, 1987. Values are means of four replicate traps. The values followed by the same letter within a line are not different (at p = 0.05) from each other according to Tukey's mean test. ANOVA and multiple comparisons were performed on log(n+1)-transformed data.

		Distance from the field margin			
	Hedge	3m	18m	63m		
	Avg. ± SD	Avg. ± SD	Avg. ± SD	Avg. ± SD	$F_{3,15}$	p
Carabidae	15.0± 3.4a	55.5± 29.5b	31.0±21.4ab	34.3±13.7ab	3.371	0.055
Staphylinidae	20.3± 11.0a	11.0± 3.4a	10.0± 7.5a	23.8± 8.5a	3.008	0.072
Araneae	33.3± 2.6b	21.0± 6.3b	11.0± 3.4a	11.0± 3.7a	16.411	<0.001
Collembola	268.0±138.2b	214.0± 93.2ab	120.0±55.6ab	85.0±43.6a	5.385	0.014
Acari	246.3±116.3b	4.8± 5.6a	1.0± 2.0a	0 a	33.854	<0.001

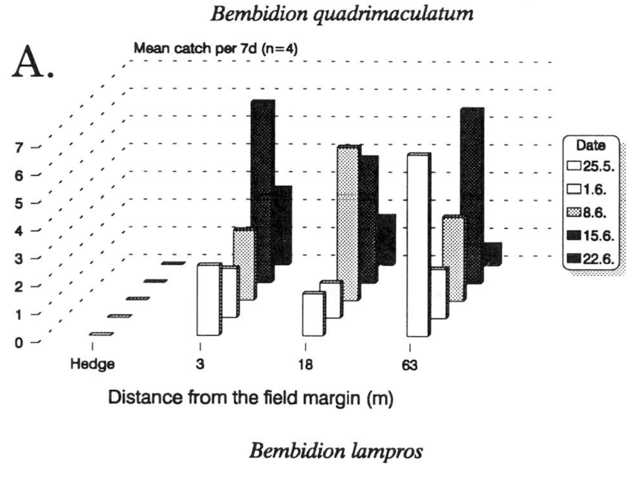

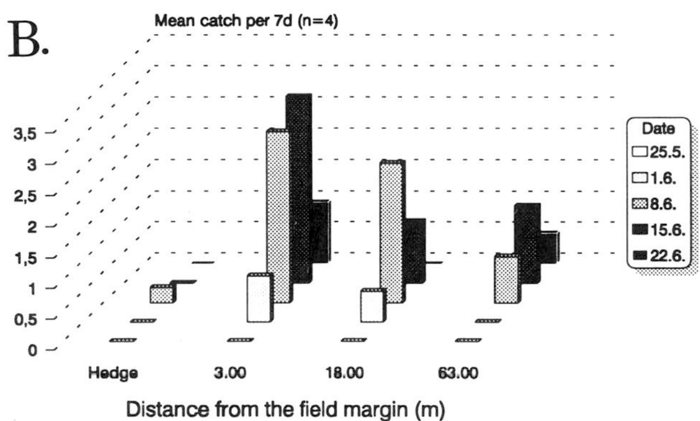

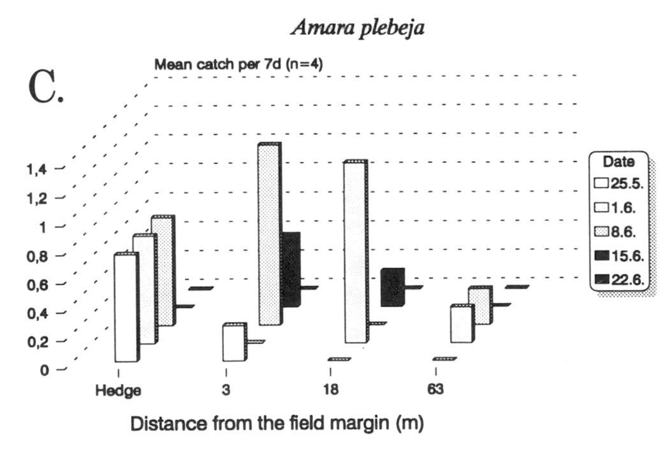

Fig. 2. Mean weekly catches of *Bembidion quadrimaculatum* (A), *Bembidion lampros* (B) and *Amara plebeja* (C) in the hedge and in different parts of a nursery field in 1987.

Holopainen, J.K. & Rikala, R. 1994. Effects of three insecticides on the activity of non-target arthropods in nursery soil. *Acta Zool. Fenn.* (in press).

Krogerus, R. 1960. Ökologische Studien über nordische Moorarthropoden. Artenbestand, ökologische Faktoren, Korrelation der Arten. *Commentat. Biol.* 21: 1-238.

Kytö, M. 1992. *Lygus* bugs cause latent bud disorders in *Pinus sylvestris* L. seedlings. *Scand. J. For. Res.* 7: 121-127.

Lilja, S. 1986. Disease and pest problems on *Pinus sylvestris* nurseries in Finland. *Bull. OEPP/EPPO* 16: 561-564

Luff, M.L. 1975. Some features influencing the efficiency of pitfall traps. *Oecologia (Berl.)* 19: 345-357.

Nef, L. 1993. Damage and control of mites and insect pests, entomological research, forest nursery production in Europe. Report of COST 813, Diseases and disorder in forest nurseries, 24 pp. Comission of the European Communities.

Poteri, M., R. Heikkilä and L. Yuan-li, 1987. Peltoluteen aiheuttaman kasvuhäiriön kehittyminen yksivuotiailla männyntaimilla. (Development of growth disturbance caused by *Lygus rugulipennis* in one-year-old pine seedlings). *Folia For.* 695: 1-14.

South, D.B. 1991. *Lygus* bugs: A worldwide problem in conifer nurseries. In: Sutherland, J.R. & Glover, S.G. (eds.) Proceedings of the first meeting of IUFRO working party S2.07-09, Diseases and Insects in Forest Nurseries. Forestry Canada, Pacific Forest Centre, Information Report BC-X-331: 215-222.

Sotherton, N.W. 1985. The distribution and abundance of predatory Coleoptera overwintering in field boundaries. *Ann. Appl. Biol.* 106: 17-21.

Speight, M.R. & Lawton, J.H. 1976. The influence of weed cover on the mortality imposed on artificial prey by predatory ground beetles in cereal fields. *Oecologia (Berl.)* 23: 211-223.

Sunderland, K. D. & Vickerman, G. P. 1980. Aphid feeding by some polyphagous predators in relation to aphid density in cereal fields. *J. Appl. Ecol.* 17: 389-396.

Thiele, H.-U. 1977. *Carabid beetles in their environments.* Springer-Verlag, Heidelberg.

Wallin, H. 1989. Habitat selection, reproduction and survival of two small carabid species on arable land: a comparison between *Trechus secalis* and *Bembidion lampros*. *Holarct. Ecol.* 12: 193-200.

Spatial distribution of hibernating polyphagous predators within field boundaries

W. Riedel

Department of Zoology, University of Aarhus, Universitetsparken Bldg. 135,
DK-8000 Aarhus C, Denmark, and
Danish Research Centre for Plant Protection, Department of Pest Management,
Lottenborgvej 2, DK-2800 Lyngby, Denmark

Abstract
In the present study the horizontal distribution of hibernating predators in field boundaries is described. A clear preference for the edge of the uncultivated grass vegetation was found for the field inhabiting carabids *Bembidion lampros* and *B. tetracolum* as well as for total carabids, even when the outer edge consisted of a new and rather poor grass cover. In the new grass cover with only a little dead organic material, predators were more abundant than in the original old edge only 50 cm further into the hedge. No significant distribution preference was observed for *Amara familiaris*, *A. plebeja* and the staphylinid *Tachyporus hypnorum*. The two *Amara* species as well as *T. hypnorum* are among the few documented flying predatory beetles inhabiting agricultural land. The data suggests that beetles dispersing by walking hibernate at the outer edge, while flying species extend further into the hedgerow.

Key words: Hedgerows, hibernation, horizontal distribution, Carabidae, Staphylinidae

Introduction

Much attention has been given to the role of ground dwelling predators which are active in the natural control of aphids in cereal crops (e.g. Edwards et al. 1979, Chiverton 1986, Helenius 1990), especially to the spring breeding species of ground beetles (Carabidae), rove beetles (Staphylinidae) and spiders (Araneae). Several of these are known to overwinter as adults in high densities in undisturbed field edge biotopes (Lipkow 1966, Pollard 1968, Desender 1981, 1982, Sotherton 1984, 1985) and occasionally they are found hibernating within fields (Desender 1982, Sotherton 1984, 1985, Wallin 1989, Riedel 1991). Appropriate overwintering sites have been shown to make considerable

Arthropod natural enemies in arable land · I *Density, spatial heterogeneity and dispersal*
S. Toft & W. Riedel (eds.). *Acta Jutlandica* vol. 70:2 1995, pp. 221-226.
 ISBN 87 7288 492

contributions to the field population of the predator complex in spring (Ekbom & Wiktelius 1985, Coombes & Sotherton 1986, Riedel 1992). In that way early aphid predation in the fields may be enhanced (Riedel 1992), which may have an impact on aphid infestations in cereals (Chiverton 1986). However, detailed information about appropriate over-wintering sites and the spatial distribution of the predators during the winter is scarce. The vertical distribution in the soil profile during winter in relation to temperature and soil type has been described for several groups of invertebrates (including Coleoptera) by Dowdy (1944) and some vegetational preferences have been demonstrated for single carabid species as well as for the aphid predator complex (Luff 1966a, 1966b, Pollard 1968, Thomas et al. 1991, Riedel 1992).

To identify and to consider the requirements of appropriate overwintering sites more detailed information about the spatial distribution of predators within the hibernation biotopes is needed. Such information could form a better basis for decisions of how natural overwintering sites should be managed as well as how to create new ones. Furthermore, when estimating densities of overwintering predators, more precise knowledge about their spatial distribution will lead to more reliable results.

Materials and methods

In order to describe the spatial distribution of hibernating beneficial arthropods in grassy field boundaries, a total of 66 soil samples (20 × 20 × 20 cm) were taken with a frame of stainless steel in December 1991 in two experimental setups (A and B).

A: 60 soil samples were taken as paired samples with one being taken in the outermost 0-20 cm of the field boundaries and the other one just behind the first one, at a distance of 25-45 cm from the edge. The samples were taken on two different farms: 12 pairs in the field boundary of a 65 ha winter wheat field on a commercial farm on Zealand, Denmark, with a humus rich soil type; and 18 pairs in a south facing hedge bottom next to a 6 m wide bare boundary strip on the experimental farm Foulum, Jutland, Denmark, with a coarse loamy soil. All samples were taken in grass vegetation.

B: In the hedgerow at Foulum two distinct 50 m sections of the grassy hedge bottom extended half a metre out into the uncultivated boundary strip. This was caused by a deliberately displaced ploughing in the spring which resulted in new hedge bottom margins in the two sections with only a sparse cover of grass (mainly of *Agropyron repens*). In each of these two sections three triplets of soil samples were taken. Each triplet consisted of one soil sample taken in the outermost 0-20 cm, one at 25-45 cm and the third at 50-70 cm from the new edge. The latter corresponded to the old edge of the hedge bottom and thus had a long, dense grass cover.

The 12 samples in the new vegetation are also a part of the experimental setup marked A above.

Soil samples were kept in plastic bags and stored at 5°C before the beetles were extracted by floating in a specially designed container with water entering at the base and an outflow at the top (Riedel 1992). Subsequently the beetles were collected directly on a fine meshed sieve and transferred to 70% alcohol for later identification.

Results

The distribution of the dominating carabids and staphylinids in the grassy field borders is shown in Table 1. Only species known to inhabit cultivated land during spring/summer are included.

A: Results from the 30 paired samples shows a significant preference for the outer edge for *Bembidion lampros* (73.3% in the outermost samples), *B. tetracolum* (85.5%), "Other Carabids" (85.8%) and "Total Carabids" (71.6%). No distribution preference was seen for the carabids *Agonum dorsale*, *Amara familiaris*, *A. plebeja* and for the staphylinid *Tachyporus hypnorum*. No preference for the internal of the grassy field boundaries was seen for any species or groups.

B: The results show the distribution of the predators in the hedge bottom with a 50 cm edge of sparse grass cover. Again no species or group showed significant preference for the most central part of the hedgerow bottom despite the old and dense grass vegetation present, and lower numbers per soil sample in the old vegetation seems to be the rule except for *A. plebeja*. A significant preference for the new hedge bottom edge was established for *B. lampros*, *B. tetracolum*, "Other Carabids" and "Total Carabids".

The same consistent result was not seen for *A. dorsale* which this time showed a marked preference for the outer edge; a significant negative relation between overwintering density and distance from the outer edge was observed ($P < 0.05$).

Table 1. The mean number of hibernating polyphagous predators per soil sample (20 × 20 × 20 cm) in different distances from the outer edge of grassy field boundaries. For the paired samples a t-test for paired observations was performed. For the triplet samples an ANOVA-regression analysis was performed. The statistical tests were carried out on square root transformed data. However, all figures presented in the table are non-transformed means. n.s. = not significant (p>0.05).

	Paired samples (n=30)			Triplet samples (n=6)			
	Distance (cm)			Distance (cm)			
	0-20	25-45	p	0-20	25-45	50-70	p
A. dorsale	1.07	0.97	n.s.	2.17	0.5	0.33	<.05
B. lampros	11.00	4.00	<.001	23.00	12.50	1.83	<.01
B. tetracolum	1.00	0.17	<.05	3.00	0.50	0.00	<.05
A. familiaris	0.67	0.33	n.s.	0.50	0.83	0.00	n.s.
A. plebeja	0.93	0.87	n.s.	0.33	1.17	1.50	n.s.
Other carabids	2.23	0.37	<.01	3.33	1.00	0.83	<.05
Total carabids	16.90	6.70	<.001	32.33	16.50	4.50	<.001
T. hypnorum	3.17	2.97	n.s.	4.67	1.17	2.17	n.s.

Discussion

A clear preference for the outer edge of field boundaries was shown for *B. lampros*, *B. tetracolum*, "Other Field Inhabiting Carabids" and "Total Field Inhabiting Carabids". When the outer edge consisted of a new and rather poor grass cover, this spatial distribution was repeated in the bottom of a hedgerow for the same taxa. Furthermore, in a poor grass cover with only a little dead organic material in the outer edge, predators were more abundant than in the original old edge only 50 cm further into the hedge.

High densities of hibernating polyphagous predators have frequently been found in dense vegetation (e.g. Sotherton 1985, Thomas et al. 1991, Riedel 1992). Furthermore, Desender (1982) showed a positive relation between the mean sod layer in grass and the densities of hibernating carabids. This might not be in discrepancy with the present results. However, higher densities of field inhabiting carabids in the outer edge, whether with poor or dense grass cover, indicates a strong edge preference irrespective of the vegetation. This edge preference could be explained by the fact that many field inhabiting carabids disperse by walking. Consequently, in their search for overwintering sites, the grassy ecotone of a field border might be the first appropriate site they come across which then apparently gives no further stimuli for immigration into the interior.

No edge preference was observed for *A. familiaris*, *A. plebeja* and *T. hypnorum*. For *T. hypnorum* this is in accordance with Kennedy et al. (1986) who found overwintering specimens in a grassy wasteland as far as 50 metres from the nearest cultivated field. The two *Amara* species as well as *T. hypnorum* are among the few predatory beetles inhabiting agricultural land for which frequent flying is documented (den Boer 1971, Lipkow 1966). This character might be the reason why they are more evenly distributed in the grassy field borders.

A clear distribution pattern of several polyphagous predators during hibernation was observed, with high densities of total overwintering predators in the outer edge of the field boundaries. Special attention should therefore be made to this ecotone, e.g when soil treatment like ploughing in the autumn is carried out.

References

Boer, P.J. den 1971. On the dispersal power of carabid beetles and its possible significance. In: *Dispersal and dispersal power of carabids beetles.* Misc. Papers Landbouwhogsch. Wageningen 8: 119-138.

Chiverton, P.A. 1986. Predatory density manipulation and its effect on populations of *Rhopalosiphum padi* (Hom.: Aphididae) in spring barley. *Ann. Appl. Biol.* 109: 49-60.

Coombes, S.D. and Sotherton, N.W. 1986. The dispersal and distribution of polyphagous predatory Coleoptera in cereals. *Ann. Appl. Biol.* 108: 461-474.

Desender, K. 1981. Ecological and faunal studies on Coleoptera in agricultural land. I. Seasonal occurrence of Carabidae in the grassy edge of a pasture. *Pedobiologia* 22: 379-384.

Desender, K. 1982. Ecological and faunal studies on Coleoptera in agricultural land. II. Hibernation of Carabidae in agroecosystems. *Pedobiologia* 23: 295-303.

Dowdy, W.W. 1944. The influence of temperature on vertical migration of invertebrates inhabiting different soil types. *Ecology* 25: 449-460.

Edwards, C.A., Sunderland, K.D. & George, K.S. 1979. Studies on polyphagous predators of cereal aphids. *J. Appl. Ecol.* 16: 811-823.

Ekbom, B.S. & Wiktelius, S. 1985. Polyphagous arthropod predators in cereal crops in central Sweden, 1979-1982. *Z. ang. Ent.* 99: 433-442.

Helenius, J. 1990. Effect of epigeal predators on infestation by the aphid *Rhopalosiphum padi* and on grain yield of oats in monocrops and mixed intercrops. *Entomol. Exp. Appl.* 54: 225-236.

Kennedy, T.F., Evans, G.O. & Feeney, A.M. 1986. Studies on the biology of *Tachyporus hypnorum* F. (Col: Staphylinidae), associated with cereal fields in Ireland. *Ir. J. agric. Res.* 25: 81-95.

Lipkow, E. von 1966. Biologisch-ökologische Untersuchungen über *Tachyporus*-Arten und *Tachinus rufipes* (Col.: Staphyl.). *Pedobiologia* 6: 140-177.

Luff, M. 1966a. The abundance and diversity of the beetle fauna of grass tussocks. *J. Appl. Ecol.* 35: 189-208.

Luff, M. 1966b. Cold hardiness of some beetles living in grass tussocks. *Entomol. Exp. Appl.* 9: 191-199.

Pollard, E. 1968. Hedges III. The effect of removal of the bottom flora of a hawthorn hedgerow on the Carabidae of the hedge bottom. *J. Appl. Ecol.* 5: 125-139.

Riedel, W. 1991. Overwintering and spring dispersal of *Bembidion lampros* Herbst (Col.: Carabidae) from established hibernation sites in a winter wheat field in Denmark. In: Polgár, L., Chambers, R.J., Dixon, A.F.G. & Hodek, I. (eds.) *Behaviour and Impact of Aphidophaga*, pp. 235-241. SPB Academic Publishing bv.

Riedel, W. 1992. Hibernation and spring dispersal of polyphagous predators in arable land. Unpublished Ph.D.-thesis, Aarhus University.

Sotherton, N.W. 1984. The distribution and abundance of predatory arthropods overwintering on farmland. *Ann. Appl. Biol.* 105: 423-429.

Sotherton, N.W. 1985. The distribution of predatory Coleoptera overwintering in field boundaries. *Ann. Appl. Biol.* 106: 17-21.

Thomas, M.B., Wratten, S.D. & Sotherton, N.W. 1991. Creation of "island" habitats in farmland to manipulate populations of beneficial arthropods; densities and emigration. *J. Appl. Ecol.* 28: 906-917.

Wallin, H. 1989. Habitat selection, reproduction and survival of two small carabid species on arable land: A comparison between *Trechus secalis* and *Bembidion lampros*. *Holarct. Ecol.* 12: 193-200.

Spatial variation of *Sitobion avenae* (F.) (Hom.: Aphididae) and its primary parasitoids (Hym.: Aphidiidae, Aphelinidae)

P. Ruggle & N. Holst

Zoological Institute, Department of Population Biology, University of Copenhagen,
Universitetsparken 15, DK-2100 Copenhagen, Denmark

(Poster presentation)

Abstract
We analyzed the spatial variation of the grain aphid, *Sitobion avenae* (F.), and its primary parasitoids (Hymenoptera: Aphidiidae, Aphelinidae) with geostatistical methods. Aphids and their endoparasitoids were sampled on grid points five times during the 1993 winter wheat season. The mapping of aphid infestations indicated that aphids tended to concentrate in the center of the field, when population levels were high, but they were more uniformly distributed after the start of heavy precipitation. The distributions of aphids and parasitoids were similar and indicated a constant parasitization rate over most of the area. Average aphid and parasitoid densities were similar to the results of the spatial analysis. However, wrong density estimates are expected, if samples are taken exclusively from the margins of the field. We generally recommend spatially referenced sampling and spatial data analysis as a reliable and efficient method to estimate actual population densities.

Key words: Geostatistics, sampling, spatial distribution

Introduction

Is the use of average densities adequate to study insect population dynamics in the field, or is it necessary to also include their spatial variation? To answer this question, we compared conventional mean densities of the grain aphid, *S. avenae* (F.), and its parasitoids (Hym.: Aphidiidae, Aphelinidae) to spatially referenced data. The geostatistical methods employed here have only rarely been used in agricultural entomology (Liebhold et al. 1993), despite their potential utility.

Arthropod natural enemies in arable land* · *I Density, spatial heterogeneity and dispersal
S. Toft & W. Riedel (eds.). *Acta Jutlandica* vol. 70:2 1995, pp. 227-233.
 ISBN 87 7288 492

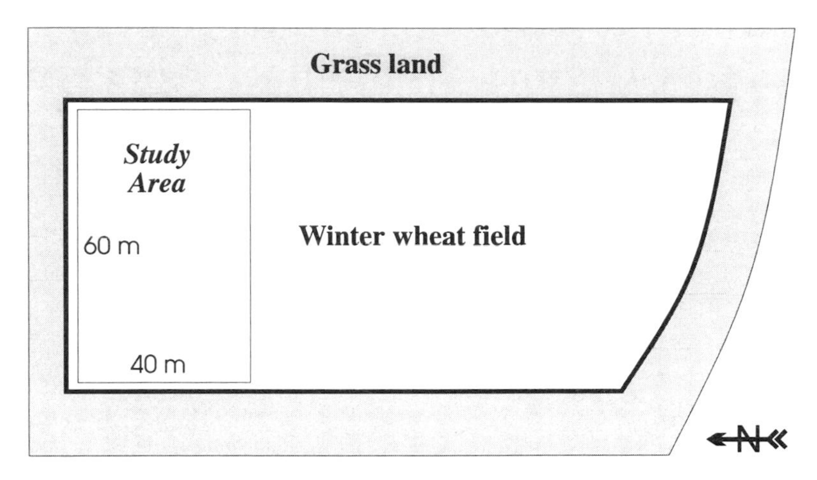

Fig. 1. Map of the study area

Materials and methods

S. avenae were collected in a winter wheat field (Fig. 1) near Høje Tåstrup, Denmark, on five occasions during the 1993 season. We used a variable sampling plan with 25-30 grid points and a mesh size of 10-20 m. At each point we collected 10-12 tillers. The aphids were extracted in the laboratory. We dissected third and fourth instar nymphs and kept the adults for several days at 20°C to allow potential parasitoid to form mummies. To analyze the data and interpolate between the points we used geostatistical methods, which are described by Burrough (1987). The maps and semivariograms were produced with the GS® geostatistical analysis system (Gamma Design Software, Plainwell, MI 49080) and with the GRIDZO™ griding and contouring software (RockWare, Wheat Ridge, CO 80033). Terms commonly used in geostatistics are explained in the appendix section.

Results

The mapping of aphid infestations indicated a general density gradient from the edges of the field to the center when the average population density was high (Fig. 2A; July 6, 20, and 27). Aphids tended to concentrate in the center of the field but were more uniformly distributed after the onset of heavy precipitation (Figs. 2 and 3; July 13). In general, aphid densities (Fig. 2A) varied more than the densities of the parasitoids (Fig. 2B), but low aphid and parasitoid densities did not allow for conclusive results.

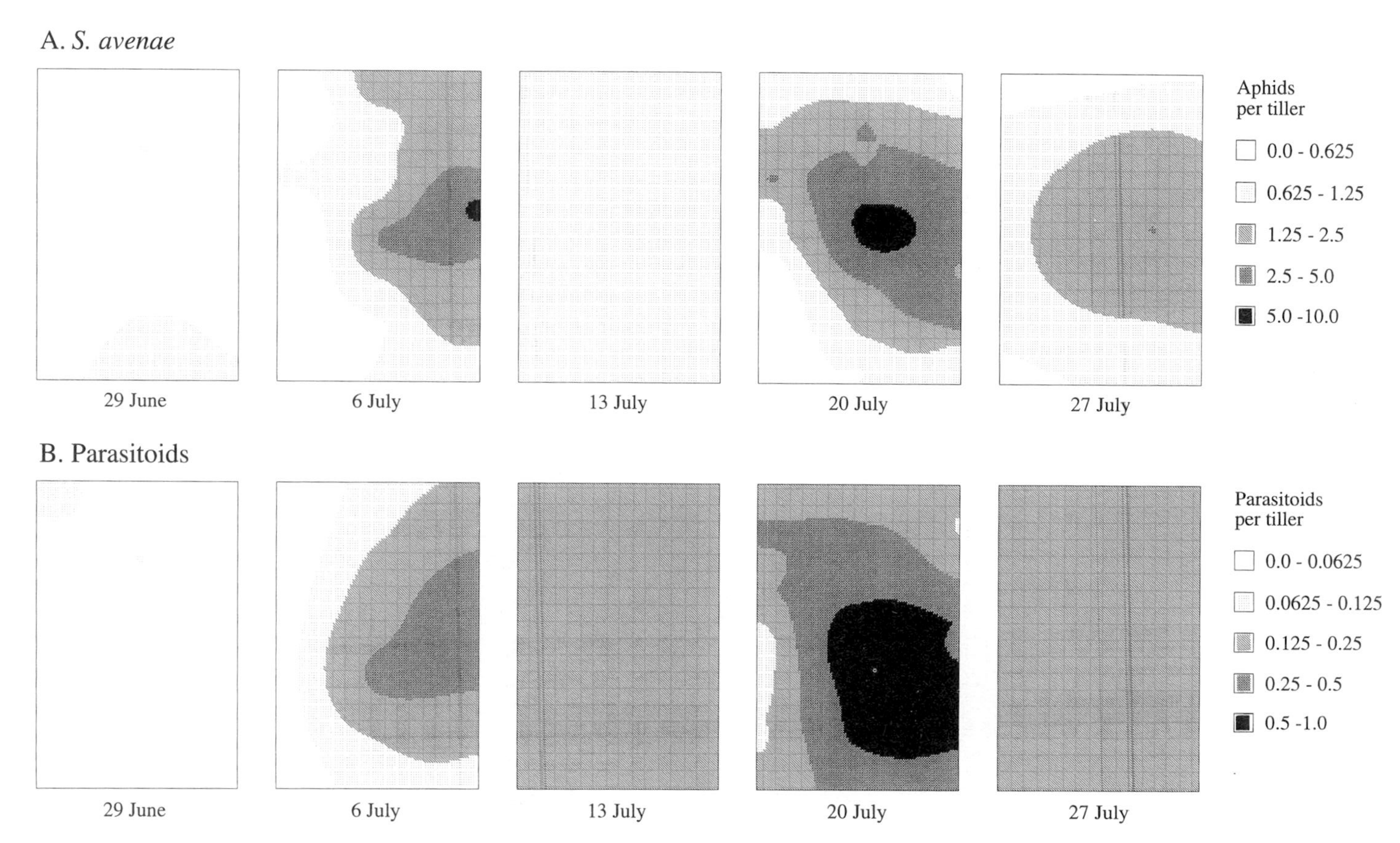

Fig. 2. Distribution maps of *S. avenae* (A) and its parasitoids (B) in the study area.

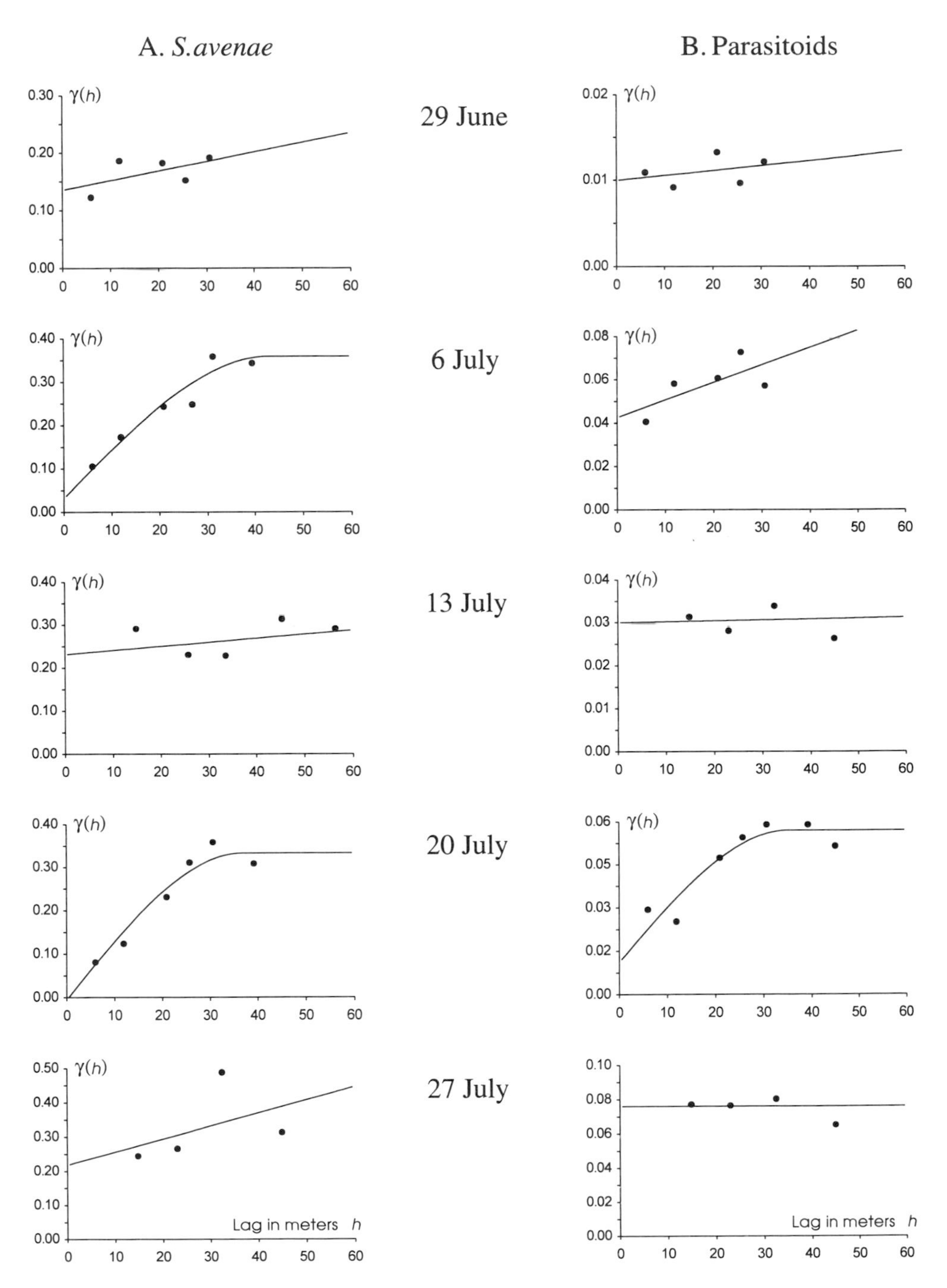

Fig. 3. Semivariograms of *S. avenae* (A) and its parasitoids (B). The components of a semivariogram are explained in the appendix section.

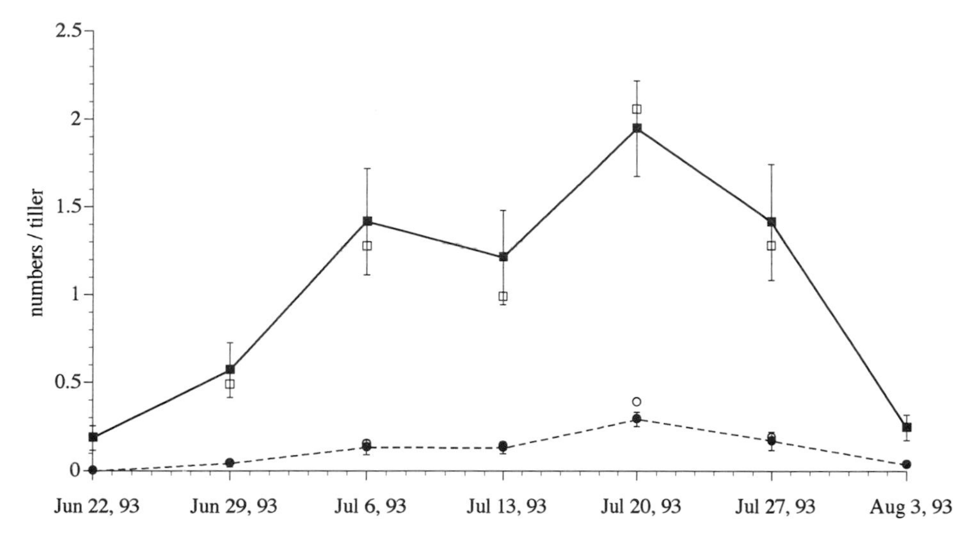

Fig. 4. Phenology based on average (mean ± s.e.) *S. avenae* (squares) and parasitoid (dots) densities. The open squares and circles respectively represent the average of interpolated aphid and parasitoid densities (Fig. 2).

The distributions of aphids and parasitoids were very similar (Fig. 2), because parasitoids were only recovered from collected aphids, and parasitization rates appeared to be proportional to aphid densities over most of the area.

According to the range (30-40 m) of the semivariograms (Fig. 3), the grid sizes used in this study seemed adequate for assessing the spatial distribution of *S. avenae* (Fig. 3A; July 6 and 20) and its parasitoids (Fig. 3B; July 20), when densities were increasing. At low densities, however, the grid size seemed inadequate (Fig. 3; June 29; July 13 and 27).

Average aphid and parasitoid densities compared favorably to the more detailed results of the spatial analysis, as the averages of actual and spatially interpolated densities were similar (Fig. 4). Wrong density estimates would have resulted from samples exclusively taken from within a 10 m margin of the field.

Conclusions

The distributions of *S. avenae* and parasitoids were aggregated. Elaborate sampling plans are required, even if the data lack spatial reference (Feng et al. 1993). We suggest that average cereal aphid densities should be used, if the samples are representative of the insect populations over the whole space. This could be checked with the methods used here. A more formal test, however, is considered time-consuming (Simard et al. 1991). We recommend spatially referenced data sampling for field studies in cereals. The data collection effort is not substantially increased, yet spatial data analysis is made possible. If necessary, sampling plans could be optimized for spatial variation (Simard et al. 1991). Other agroecosystems with similar distributions of key species may also profit from the techniques of spatial data analysis.

Acknowledgments

We thank Per Kølster (Royal Veterinary and Agricultural University, Copenhagen, Denmark) for letting us sample a wheat field in his Systems Research Area. Trine S. Nielsen assisted in collecting and processing the samples. P. Ruggle and N. Holst were supported by the Centre for Agricultural Biodiversity of the Danish Environmental Research Programme.

References

Burrough, P.A. 1987. Spatial aspects of ecological data. In: Jongman, R.H.G., Braak, C.J.F. ter and Tongeren, O.F.R. van (eds.) *Data analysis in community and landscape ecology.* Pudoc, Wageningen. p. 303-327.

Feng, M. G., Nowierski, R.M. & Zeng, Z. 1993. Populations of *Sitobion avenae* and *Aphidius ervi* on spring wheat in the southwestern United States. Spatial distribution and sequential sampling plans based on numerical and binomial counts. *Entomol. Exp. Appl.* 67: 109-117.

Liebhold, A. M., Rossi, R.E. & Kemp, W.P. 1993. Geostatistics and geographic information systems in applied insect ecology. *Ann. Rev. Ent.* 38: 303-327.

Simard, Y., Legendre, P., Lavoi, G. & Marcotte, D. 1991. Mapping, estimating biomass, and optimizing sampling programs for spatially autocorrelated data: Case study of the Northern Shrimp (*Pandus borealis*). *Canadian Journal of Fisheries and Aquatic Sciences* 49: 32-45.

Appendix: glossary and explanations
(cf. Fig. 5)

Kriging: Method used to create a distribution map based on the estimated semivariogram function.

Lag (h): Distance between points.

Nugget: Estimate of variance due to sampling error.

Range: Distance at which the sill is reached. It marks the longest distance at which samples show a level of similarity. The absence of a range may indicate that the sampling points were too far apart.

Semivariance (γ): Measure of dissimilarity between points, as estimated in equation (1):

$$\gamma(h)=\frac{1}{2N(h)}\sum_{i=1}^{N(h)}[Z(x_i+h)-Z(x_i)]^2, \tag{1}$$

where h is the distance (lag) between points, N is the number of pairs, and Z is the observed value at points x_i and x_i+h.

Semivariogram: Standard graph of the relationship between semivariance (γ) and lag (h). It is used to estimate functional relation between the two variables.

Sill: Asymptote of semivariogram function. It is equal to the sample variance.

Spatially referenced data: Data associated with spatial coordinates.

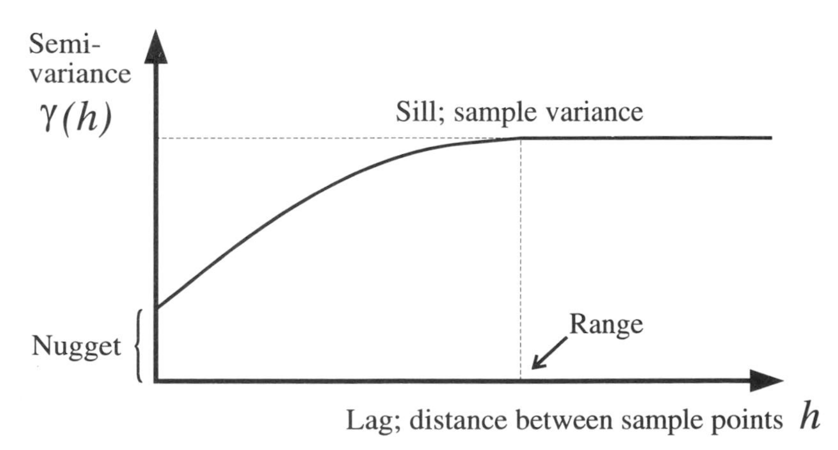

Fig. 5. Generic form of a semivariogram.

Within-field spatial heterogeneity of arthropod predators and parasitoids

W. Powell[1], A. Hawthorne[2], J.-L. Hemptinne[3], J.K. Holopainen[4], L.J.F.M. den Nijs[5], W. Riedel[6] & P. Ruggle[7]

[1]Entomology & Nematology Department, IACR-Rothamsted, Harpenden, Herts., AL5 2JQ, England
[2]Waltham Centre for Pet Nutrition, Waltham on the Wolds, Melton Mowbray, Leicestershire, LE14 4RT, England
[3]Faculté des Sciences Agronomique UER Zoologie Générale et Appliquée, 2 Passage des Déportés, 5030 Gembloux, Belgium
[4]Ecological Laboratory, Department of Environmental Sciences, University of Kuopio, P.O.Box 1627, 70211 Kuopio, Finland
[5]Research Institute for Plant Protection, P.O.Box 9060, 6700 GW Wageningen, The Netherlands
[6]Institute of Biological Sciences, Department of Zoology, University of Aarhus, Universitetsparken Building 135, DK-8000 Århus C, Denmark
[7]Department of Population Biology, University of Copenhagen, Universitetsparken 15, DK-2100 Copenhagen Ø, Denmark

(2nd to 7th author in alphabetical order)

Introduction

This paper is based upon discussions held during an EC-Workshop entitled "Estimating population densities and dispersal rates of beneficial predators and parasitoids in agroecosystems" which took place at the University of Aarhus, Denmark in October 1993. The main aim of this discussion session was to consider ways of investigating the within-field spatial heterogeneity of arthropod predators and parasitoids.

What do we mean by within-field spatial heterogeneity?

The results of predator or parasitoid sampling programmes usually indicate some degree of heterogeneity in the spatial distribution of natural enemy populations within individual fields. The most likely explanation for this is a

***Arthropod natural enemies in arable land · I** Density, spatial heterogeneity and dispersal*
S. Toft & W. Riedel (eds.). *Acta Jutlandica* vol. 70:2 1995, pp. 235-242.

response by natural enemies to spatial heterogeneity in one or more attributes of their environment. These attributes may be physical factors such as soil surface roughness, botanical factors such as vegetation type or density, microclimatic factors or spatial variability in prey density. Obviously, some of these attributes are linked; for example heterogeneity in vegetation density is usually accompanied by heterogeneity in microclimate at the soil surface. Spatial heterogeneity within a crop field can, of course, exist in the vertical plane as well as in the horizontal (e.g. microclimate, vegetation structure, prey densities).

In pitfall trap studies of carabid beetles, for example, considerable inter-trap variability in catches of individual species can occur at small spatial scales (Speight & Lawton 1976, Gruttke & Weigmann 1990, Niemelä 1990). This has been attributed to spatial heterogeneity in weed density (Speight & Lawton 1976), to variation in small-scale habitat structure around traps (Niemelä 1990), to interactions with other arthropods such as ants (Niemelä 1990) and to variation in soil type (Thiele 1977) or the pH of the topsoil (Gruttke & Weigmann 1990). In addition, proximity to the field margin can have a significant effect on polyphagous predator densities (Coombes & Sotherton 1986, Holopainen 1995), although these effects often do not extend very far into the field (Booij et al. 1995). Edge effects are probably much less important for highly mobile species such as active fliers (Riedel 1995).

When considering small-scale spatial heterogeneity in natural enemy distributions associated with microhabitat factors, it is important to distinguish between heterogeneity which involves distinct boundaries between habitat patches (e.g. discrete patches of weeds within a crop) and heterogeneity in the form of gradual changes in some microhabitat factor across a large area. In the latter case distinct habitat patches may be impossible to define. It is also important to consider that the perception of spatial heterogeneity by arthropods may be different from our own perceptions, and important factors which are not apparent to us may be influencing natural enemy distributions. Furthermore spatial heterogeneity often changes with time. Equally, the behavioural responses of individual predators or parasitoids to spatial heterogeneity may change with changes in reproductive state.

Is within-field spatial heterogeneity important?

Honek (1979, 1981 & 1982) has shown that the composition of field communities of adult aphidophagous Coccinellidae is related to several factors such as plant density and distance from field margins. However, in order to assess the potential importance of spatial heterogeneity at the within-field scale, we need a lot more information about the influence of environmental heterogeneity on both the behaviour of individual predators and parasitoids and on the population dynamics of species. For example, the foraging behaviour of individuals can be affected by spatial heterogeneity in habitat features, in prey distribution and by the presence of conspecifics (Hemptinne et al. 1992). Vegetation density at ground level can affect the locomotory activity of carabid beetles thereby influencing rates of encounters with prey (Thiele 1977, Hawthorne 1995). Therefore, it could be argued that sampling methods which measure activity-density (e.g. pitfall traps) may give better estimates of predatory effects than methods which measure absolute densities. Theoretical models of parasitoid foraging behaviour have highlighted the importance of spatial heterogeneity in stabilizing host-parasitoid interactions (Holt & Hassell 1993). However, some field experiments have contradicted model predictions. Patchiness in goldenrod stands favoured outbreaks of the aphid *Uroleucon nigrotuberculatum* because it interfered with the nonrandom searching behaviour of ladybird predators (Kareiva 1987). Therefore, further field data are urgently needed to test these ideas, particularly at different spatial scales and for specific communities of prey and natural enemy species.

We know very little about the influence of small-scale habitat heterogeneity, as opposed to large scale habitat fragmentation, on the local population dynamics of predators. For example, although we have some knowledge about habitat requirements for optimal overwintering survival of many of the adult carabid and staphylinid beetles which inhabit arable crop fields (Sotherton 1985), we know very little about the microhabitat factors which determine the choice of oviposition sites or which affect larval survival. Theoretically, factors such as soil structure, soil moisture, vegetation cover and microclimate could all have an important impact on population recruitment and survival rates, but the significance of their within-field heterogeneity in the long-term population dynamics of natural enemy species needs clarification. Population studies of carabid beetles are notoriously difficult because of problems with the sampling of pre-adult stages. However, a better understanding of microhabitat factors influencing the spatial distribution of these stages may allow the design of more efficient sampling programmes.

Knowledge of the effects of within-field spatial heterogeneity is important for the development of integrated pest management strategies which involve the conservation and enhancement of natural enemies. Strategies involving the deliberate creation of habitat heterogeneity (unsprayed headlands, set-aside field margins, beetle conservation banks, reservoirs for non-pest hosts of parasitoids) are being actively explored in an effort to maximise the impact of natural enemies on pest populations (Powell 1986, Wratten & Powell 1990).

Methodologies for investigating within-field spatial heterogeneity

Two approaches can be taken when investigating within-field spatial heterogeneity:

1) Quantify and describe any observed heterogeneity in the spatial distribution of the predator or parasitoid species being studied, and then investigate the causes of this heterogeneity in terms of microhabitat variability or prey distribution.

2) Quantify and describe any observed spatial heterogeneity in microhabitat or prey density and investigate the effect of this on natural enemy distribution.

In the first approach it is essential to make sure that observed heterogeneity in the distribution and abundance of a species is a real phenomenon and not simply an artefact arising from the sampling method being used. For example, the efficiency of a pitfall trap partly depends upon how well it has been installed and its siting within the field. Cracks in the soil around the lip of the trap, the slope of the soil immediately around the top of the trap and strands of vegetation overhanging the trap can all affect the probability of an arthropod falling into the trap or escaping from it. It is very easy when setting up a large network of traps to introduce this kind of inter-trap variability, particularly if more than one person is involved in the work. In the case of pitfall trapping studies, it would be instructive to record details of the microhabitat immediately surrounding each trap in order to relate inter-trap catch variability to habitat heterogeneity. Similarly, heterogeneity in vegetation density can affect the efficiency of "D-vac" type samplers, thereby creating apparent heterogeneity in the spatial distribution of sampled species in situations where such heterogeneity may not actually exist.

There are a number of ways in which the relationship between microhabitat or prey spatial heterogeneity and within-field natural enemy distribution can be investigated. Firstly, if sampling data suggest that certain habitat factors are determining the spatial distribution of a specific predator or natural enemy group, manipulative field and laboratory experiments should be devised to test the relationship. Powell et al. (1985, 1986) manipulated weed density in a cereal crop in a replicated plot trial in order to investigate the influence of weed cover on the distribution of a range of natural enemies.

Field sampling programmes can be devised specifically to investigate spatial heterogeneity in a descriptive study. By setting up a grid system of sampling points geostatistical mapping methods can be employed to reveal spatial distribution patterns. If habitat factors are recorded and quantified together with predator or parasitoid abundance data this methodology will aid the detection of spatial associations. Ruggle & Holst (1995) used geostatistical methods to map the spatial distribution of cereal aphids and their parasitoids within individual fields.

Ultimately the spatial distribution of predators and parasitoids within a field is the result of behavioural responses to environmental factors by individuals in the population. Aggregation within a defined habitat patch may result from active or passive responses; that is individuals may actively move to the patch in response to a visual or olfactory cue, or they may be retained within a patch, which contains favourable microclimatic conditions or provides a resource such as food or shelter, after encountering the patch during random movement (Smith 1971). There is evidence that insect parasitoids and predators are attracted towards patches containing hosts by volatile semiochemicals and also are often retained in microhabitats containing hosts by contact semiochemical cues (Vet & Dicke 1992, Budenberg et al. 1992, Carter & Dixon 1984).

Therefore, in order to understand relationships between microhabitat spatial heterogeneity and natural enemy distributions it is essential to study the behavioural responses of individuals. First of all, it is important to determine which life stage is the most sensitive to heterogeneity. In the case of ladybirds, there is much information on larvae but the foraging behaviour of adults has been neglected, although the latter are much more mobile and appear to react strongly to environmental factors (Kindlmann & Dixon 1993, Hemptinne et al. 1992). Behavioural studies are most easily done in the laboratory where different factors can be controlled and manipulated, but there is also a need for good field observations. There is some evidence that ground-dwelling predators detect and respond to distinct habitat boundaries, often changing their direction of movement (Mader et al. 1990). Such information is essential for predicting aggregation and dispersal patterns following habitat manipulation within fields. The harmonic radar tracking system devised for use with large carabid beetles (Wallin & Ekbom 1988) could be a valuable tool for studying locomotory activity in spatially heterogeneous field arenas. Laboratory data on responses to microclimatic gradients and to prey distribution patterns will also help in the interpretation of field observations. Video recorders have become valuable research tools for studying insect behaviour and, although they are mainly used in laboratory situations there is considerable potential for their use in field studies (Wratten 1993).

Conclusions

Small-scale spatial heterogeneity within individual fields may have important effects on the spatial distribution of natural enemy species. Understanding these effects is essential for the development of habitat manipulation strategies aimed at conserving and enhancing natural enemy populations. Therefore, there is an urgent need for both laboratory and field studies to elucidate the importance of these effects in the population dynamics and dispersal behaviour of predator and parasitoid species. Useful methodologies include manipulative field studies involving the creation of spatial heterogeneity, tracking of individuals using harmonic radar or tagging techniques, video recording of behavioural responses to changes in microhabitat factors such as microclimate, vegetation structure and prey density, and geostatistical mapping techniques applied to field data collected on a structured grid system.

References

Booij, C.J.H., Nijs, L.J.M.F. den & Noorlander, J. 1995. Spatio-temporal patterns in activity density of some carabid species in large scale arable fields. In: Toft, S. & Riedel, W. (eds.) *Arthropod natural enemies in arable land · I*, pp. 175-184. Aarhus.

Budenberg, W.J., Powell, W. & Clark, S.J. 1992. The influence of aphids and honeydew on the leaving rate of searching aphid parasitoids from wheat plants. *Entomol. Exp. Appl.* 63: 259-264.

Carter, N.C. & Dixon, A.F.G. 1984. Honeydew: an arrestant stimulus for coccinellids. *Ecol. Ent.* 9: 383-387.

Coombes, D.S. & Sotherton, N.W. 1986. The dispersal and distribution of predatory Coleoptera in cereals. *Ann. Appl. Biol.* 108: 461-474.

Gruttke, H. & Weigmann, G. 1990. Ecological studies on the carabid fauna (Coleoptera) of a ruderal ecosystem in Berlin. In: N.E. Stork (ed.) *The Role of Ground Beetles in Ecological and Environmental Studies,* pp. 181-189. Andover

Hawthorne, A. 1995. Validation of the use of pitfall traps to the study of carabid populations in cereal field headlands. In: Toft, S. & Riedel, W. (eds.) *Arthropod natural enemies in arable land · I*, pp. 61-75. Aarhus.

Hemptinne, J-L., Dixon, A.F.G. & Coffin, J. 1992. Attack strategy of ladybird beetles (Coccinellidae): factors shaping their numerical response. *Oecologia (Berl.)* 90: 238-245.

Holopainen, J.K. 1995. Spatial distribution of polyphagous predators in nursery fields. In: Toft, S. & Riedel, W. (eds.) *Arthropod natural enemies in arable land · I*, pp. 213-220. Aarhus.

Holt, R.D. & Hassell, M.P. 1993. Environmental heterogeneity and the stability of host-parasitoid interactions. *J. Anim. Ecol.* 62: 89-100.

Honek, A. 1979. Plant density and occurence of *Coccinella septempunctata* and *Propylea quatuordecimpunctata* (Coleoptera, Coccinellidae) in cereals. *Acta entomologica bohemoslovaca* 76: 308-312.

Honek, A. 1981. Aphidophagous Coccinellidae (Coleoptera) and Chrysopidae (Neuroptera) on three weeds: factors determining the composition of populations. *Acta entomologica bohemoslovaca* 78: 303-310.

Honek, A. 1982. Factors which determine the composition of field communities of adult aphidophagous Coccinellidae (Coleoptera). *Z. Ang. Ent.* 94: 157-168.

Kareiva, P. 1987. Habitat fragmentation and the stability of predator-prey interactions. *Nature* 326: 388-390.

Kindlmann, P. & Dixon, A.F.G. 1993. Optimal foraging in ladybird beetles and its consequences for their use in biological control. *European Journal of Entomology* 90: 443-450.

Mader, H.J., Schell, C. & Kornacker, P. 1990. Linear barriers to arthropod movements in the landscape. *Biol. Conserv.* 54: 209-222.

Niemelä, J. 1990. Spatial distribution of carabid beetles in the southern Finnish taiga: the question of scale. In: N.E. Stork (ed.) *The Role of Ground Beetles in Ecological and Environmental Studies*, pp. 143-155. Andover.

Powell, W. 1986. Enhancing parasitoid activity in crops. In: J. Waage & D. Greathead (eds.), *Insect Parasitoids.* 13th Symposium of the Royal Entomological Society of London, pp. 319-340. London.

Powell, W., Dean, G.J. & Dewar, A. 1985. The influence of weeds on polyphagous arthropod predators in winter wheat. *Crop Protection* 4: 298-312.

Powell, W., Dean, G.J. & Wilding, N. 1986. The influence of weeds on aphid-specific natural enemies in winter wheat. *Crop protection* 5: 182-189.

Riedel, W. 1995. Distribution of hibernating polyphagous predators in field boundaries. In: Toft, S. & Riedel, W. (eds.) *Arthropod natural enemies in arable land · I*, pp. 221-226. Aarhus.

Ruggle, P. & Holst, N. 1995. Spatial variation of *Sitobion avenae* (F) (Hom.: Aphididae) and its primary parasitoids (Hym.: Aphidiidae, Aphelinidae). In: Toft, S. & Riedel, W. (eds.) *Arthropod natural enemies in arable land · I*, pp. 227-233. Aarhus.

Smith, B.C. 1971. Effects of various factors on the local distribution and density of coccinellid adults on corn (Coleoptera: Coccinellidae). *Can. Ent.* 103: 1119-1120.

Sotherton, N.W. 1985. The distribution and abundance of predatory Coleoptera overwintering in field boundaries. *Ann. Appl. Biol.* 106: 17-21.

Speight, M.R. & Lawton, J.H. 1976. The influence of weed cover on the mortality imposed on artificial prey by predatory ground beetles in cereal fields. *Oecologia (Berl.)* 23: 211-223.

Thiele, H.-U. 1977. *Carabid Beetles in their Environments.* Berlin.

Vet, L.E.M. & Dicke, M. 1992. Ecology of infochemical use by natural enemies in a tritrophic context. *Ann. Rev. Entomol.* 37: 141-172.

Wallin, H. & Ekbom, B.S. 1988. Movements of carabid beetles (Coleoptera: Carabidae) inhabiting cereal fields: a field tracing study. *Oecologia (Berl.)* 77: 39-43.

Wratten, S.D. (ed.) 1993. *Video Techniques in Animal Ecology and Behaviour.* London.

Wratten, S.D. & Powell, W. 1990. Cereal aphids and their natural enemies. In: Firbank, L.G., Carter, N., Darbyshire, J.F. & Potts, G.R. (eds.) *The Ecology of Temperate Cereal Fields.* 32nd Symposium of the British Ecological Society, pp. 233-257. Oxford.

DISPERSAL

Methods for monitoring aerial dispersal by spiders

C.J. Topping[1] *& K.D. Sunderland*

Horticulture Research International, Worthing Road, Littlehampton, West Sussex BN17 6LP, England
[1]Present address: Scottish Agricultural College, Land Resources Dept., 581 King Street, Aberdeen, AB9 1UD, Scotland

Abstract
A suction trap, a rotary trap and water filled tray traps (deposition traps) were used to monitor aerial dispersal in a population of spiders in and around a field of winter wheat between April and November 1991. All three methods proved adequate for describing trends in aerial activity during this period. Suction trapping and rotary sampling were less labour intensive than using deposition traps and provided the facility to record daily fluctuation in activity. Deposition traps had the advantage that they could be left unattended for longer periods of time. The main factor affecting the efficiency of the rotor trap was considered to be the vertical velocities of ballooning spiders. Calibration of the rotor trap was attempted but the difficulties of accurately assessing the number of spiders ballooning from an area of ground rendered this task problematical. A method of assessing the effects of aerial dispersal on the density of spiders on the ground is presented. This method relied on inclusion caging to prevent dispersal and was only partially successful due to statistical problems resulting from the aggregated dispersion of spiders in the field. There was no indication from this study of any large net immigration or emigration of spiders from the field.

Key words: Araneae, Linyphiidae, ballooning, dispersal rate

Introduction

For many agriculturally important species of spider, aerial dispersal is a significant event in the life-cycle. As such this phenomenon has received much attention in the literature (see Weyman 1993 for a review), and consequently a number of techniques are used to measure the number of spiders in the air. However, many of these methods can be criticised for their sensitivity to both weather conditions and size of the animals being trapped (e.g. Duffey 1956, Taylor 1962), and are therefore imperfect for estimating the number of spiders involved in aerial dispersal. Other problems arise from the methods used by

Arthropod natural enemies in arable land · I *Density, spatial heterogeneity and dispersal*
S. Toft & W. Riedel (eds.). *Acta Jutlandica* vol. 70:2 1995, pp. 245-256.
 ISBN 87 7288 492

spiders to take to the air. Ballooning is achieved by releasing a long silk line into a body of upwardly moving air. Once the drag on the line exceeds the gravitational pull of the spiders mass, it releases its hold on the substrate and becomes airborne. Interception traps, such as sticky traps, require the spider body to be impacted onto the trap surface to provide an accurate measure of spider numbers in a volume of air. Unfortunately the silk line used by spiders can be several metres in length and can easily become stuck on the trap even when the spider body is not in the immediate vicinity. The spiders will then often climb up this silk and become caught, causing potentially large inaccuracies in catch.

This paper describes methods developed during an attempt to take measurements of the numbers of spiders involved in ballooning and the quantification of the net effect of this dispersal on population density. Also presented is an alternative approach to quantifying aerial dispersal. This method is based on measuring the density of spiders on the ground before and after dispersal has occurred. This approach cannot accurately quantify the numbers of animals engaged in dispersal but could be used to assess the overall result of the processes of immigration and emigration, i.e. net migration.

Methods

Sampling of spiders was performed in a commercial field of winter wheat (c.v. Riband) in West Sussex, UK (Grid Ref. TQ 045 035) between April and September 1991. Three methods of capturing ballooning spiders were utilised, as well as an inclusion caging method for assessing the change in density as a result of spider dispersal.

The rotor trap

The rotor trap (see Topping et al. 1992 for technical details) was constructed to provide an accurate estimate of aerial density. The sampling head design follows Taylor (1962) and at the rotation speeds used in this study, provided an iso-kinetic air-flow into the trap, thus reducing turbulence across the net entrance. A collecting bottle containing a solution of ethylene glycol and detergent was attached to the rear of the net. The rotation speed was set at 6 rpm which was adequate for the very slow horizontal vectors encountered when studying airborne spiders (they do not take to the air in wind speeds above 3 m sec^{-1} (Richter 1970, Wingerden & Vugts 1974, Greenstone 1990)). At this speed the re-sampling of any volume of moving air was prevented at horizontal wind speeds above 5.55 x 10^{-2} m sec^{-1}; there was also sufficient time (10 sec) for the silk, which will probably precede the spider (see Humphrey 1987), to enter the sample space without interference. A boom length of 10 m was required in order to keep the revolution rate low whilst maintaining a high head speed. The height of the sampling head could be adjusted by altering the angle of the boom. Power was provided by an electric motor and the drive transmitted via a modified car differential; a more efficient method than the belt drive used in previous rotary trap designs (e.g. Nicholls 1960), which, due to the size of this trap, would have allowed slippage as the boom turned into the wind.

The trap was designed to be much less affected by the wind speed and behaviour of the spider than sticky or suction traps. The sampling head travelled at 6.28 m sec^{-1} and therefore would always travel much faster than the ballooning spider. Thus, any changes in wind speed would be compensated for by the rotation, i.e. any reduction in numbers of spiders due to some being blown away from the net at one point will be balanced by the same number being blown into the net five seconds (180°) later. There would, however, be no compensation for changes in the vertical component of wind speed. As a consequence, if the mean rate of descent differed markedly from the mean rate of ascent, the estimates would be biased. The problems of over sampling due to catching the spiders silk are also overcome by this design because a spider will only be sampled if its body is scooped out of the air by the sampling head. During the sampling period the sampling head was regularly adjusted to sample 0.5 m above the crop surface. The rotor trap was operated 30 m from the field boundary and sampled dawn to dusk by means of a solar clock. Before ploughing, in September, the rotary trap was moved to an area of mown grass near the edge of the field, until November. The trap was emptied each morning.

Three attempts were made to calibrate the rotor trap. On 12 & 19 October 1990 calibration was attempted by operating the trap whilst observing spider take-off on two small (0.36 m^2) areas of ground. Sticky trays, placed on the ground, were also employed to record spiders landing. Spiders were recorded as taking-off only if actually observed taking to the air. The number of spiders caught by the rotor trap during this period was recorded. A third rotor calibration was attempted in a similar way on 14 October 1991, but using much larger areas for observation (9 m^2 observed for take-off and 6.75 m^2 of sticky mats for landing), over 210 minutes. This larger observation area was made possible by recording those spiders which showed typical pre-take-off behaviour (raising the abdomen and releasing silk into the air). These spiders were removed before take-off significantly reducing the amount of time required to record an area of ground. Two experimenters each recorded three 1.5 m^2 areas continuously.

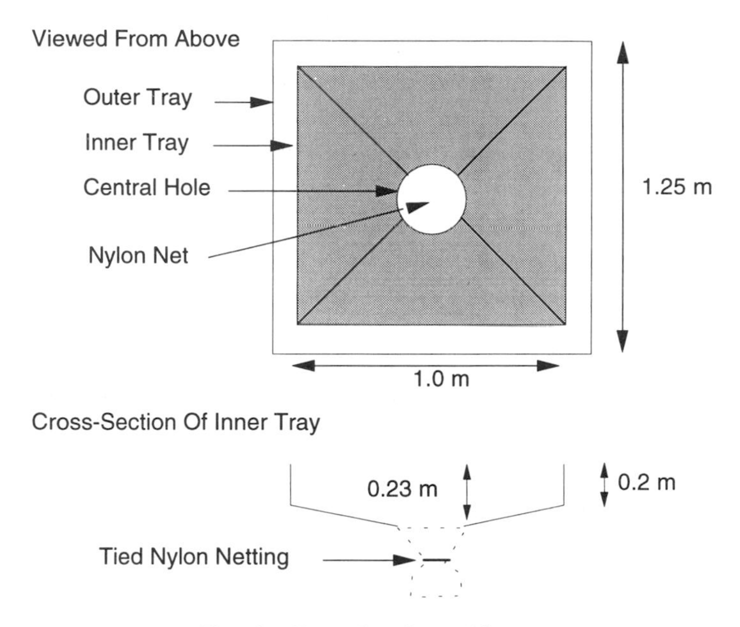

Fig. 1. Plan of a deposition trap

Suction trap

A 46 cm Enclosed Cone Propeller Suction Trap (Taylor 1955), with an air through-put of 75 m^3min^{-1} and a sampling height of 142 cm above the ground surface was located at a distance of 30 m from the rotary trap, on grass at the edge of the crop. The suction trap was run continuously and emptied each morning at the same time as the rotor trap.

Deposition traps

These traps were constructed to provide a measure of aerial activity related to an area of ground. Each deposition trap consisted of a 10 cm deep, 1 m^2, fibreglass tray, filled with water and ethylene glycol (20:1) plus 1% detergent, fitting inside a 1.6 m^2 metal tray containing the same fluid (Fig. 1). The inner tray sloped towards a central hole from which a muslin tube was attached and tied. The outer tray acted as a barrier to prevent spiders walking from the crop into the inner tray, which therefore received only aerial immigrants. Six traps were used during March - September 1991 in the wheat crop, adjacent to the rotor trap. During September to November 1991 the traps were operated on the grass area adjacent to the rotor trap. As the crop grew, the deposition traps were progressively raised on wooden supports to maintain the level of the fluid surface constantly at c. 5 cm above the top of the crop canopy. The traps were emptied at 1-2 weekly intervals by raising the inner tray out of the fluid in the metal tray and placing it onto a portable metal frame containing a plastic collecting bottle. The end of the muslin tube was untied and the tube placed into the collecting bottle. The contents of the trap were then washed out of the tray and into the bottle using a gentle water jet.

The deposition trap catch was compared to another measure of aerial activity per unit area by the following method: Seven deposition traps were placed on a mown lawn. Two 1.0 m^2 plastic sheets were placed adjacent to each trap. A central area of 0.75m^2 was marked out in the centre of each sheet, the whole sheet was coated with an eco-tak trapping grease and was anchored to the ground using wire stakes. The traps and sticky sheets were operated between 8:30 November 15 and 10:00 November 18 1991, during a period of active spider ballooning. All spiders caught stuck to the central area of the sticky sheets or caught in the deposition traps were counted.

Inclusion caging

Density estimation entailed a combination of suction sampling and ground searching on an area of 0.5 m^2 (Topping & Sunderland 1993). Sets of fifteen density samples were taken at intervals of one to two weeks between March and September 1991. All samples were taken at least 60 m from the field boundary with the placement of each sample following a pre-determined plan to avoid sampling the same area twice.

Ten circular 0.5 m^2 x 1 m tall inclusion cages were constructed out of steel rings supported by reinforcing mesh. These were covered with mesh with 3 x 2.5 perforations mm^{-2} (too small to allow the passage of first instar linyphiid spiderlings). The cages were randomly placed over undisturbed areas of crop. In order to seal the cages any vegetation protruding from under the cages was removed and the bases sealed with sufficient compacted soil to prevent entry or exit of spiders. Spider density inside the cages was assessed by density estimation immediately upon removing the cages, on the same day that the field spider density was next measured. The net migration occurring between samples was measured by comparing the mean cage density to the mean field density. Any differences were attributed to migration since mortality and recruitment were assumed to be unaffected by such a short period of caging. Net migration can therefore be given as field density minus cage density.

Results

Rotor trap calibration

Over a period of 105 minutes, five spiders were observed to take-off from two 0.36 m^2 quadrats on the 12/10/90 and no spiders were recorded landing. On 19/10/90 11 spiders were observed to take off and three spiders landed in two quadrats over a period of 52 minutes. The rotor catch on 12/10/90 was five spiders over 240 minutes and on 19/10/90 was 52 spiders in 170 minutes. The area swept over by the rotor trap was 33.61 m^2. Assuming that the rotor trap will catch all spiders passing through this area (but see discussion), the number of spiders caught by the rotor trap can therefore be given as 0.62×10^{-3} m^{-2} min^{-1} on 12/10 and 9.10×10^{-3} $m^{-2}min^{-1}$ on 19/10. The combined observed take-off and landing for these dates was 66.1×10^{-3} m^2 min^{-1} and 373.9×10^{-3} m^{-2} min^{-1}, respectively. Hence, based on this sample, the rotor caught 1/107th and 1/41th of the spiders taking off or landing within the swept area on these two dates.

On 14 October 1991 the rotor trap caught 80 spiders over 33.61 m^2, sticky mats captured 201 spiders on an area of 6.75 m^{-2} and 421 spiders were recorded as about to take-off from 9 m^2. Thus over 210 minutes, the rotor caught 11.3×10^{-3} spiders m^{-2} min^{-1} and 364.6×10^{-3} spiders m^{-2} min^{-1} were recorded as engaging in aerial activity, a ratio of 1:32.

Deposition traps vs. sticky sheets

The sticky traps caught a total of 630 spiders compared to a deposition catch total of 1075. The trapping area of the sticky traps was 7.8 m^2 compared to 7.0

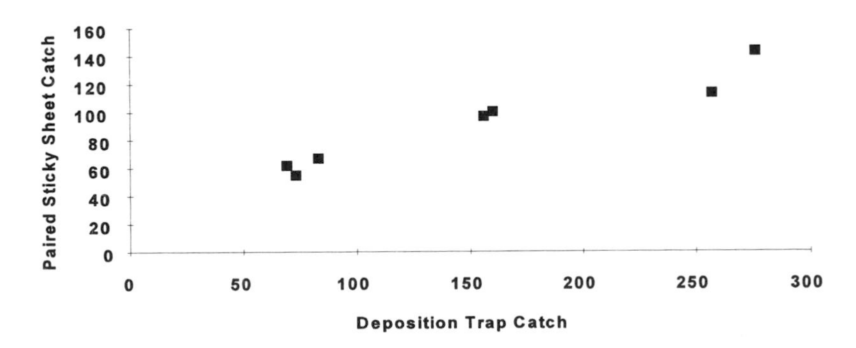

Fig. 2. Individual deposition trap total spider catch vs. total spider catch of paired sticky sheets.

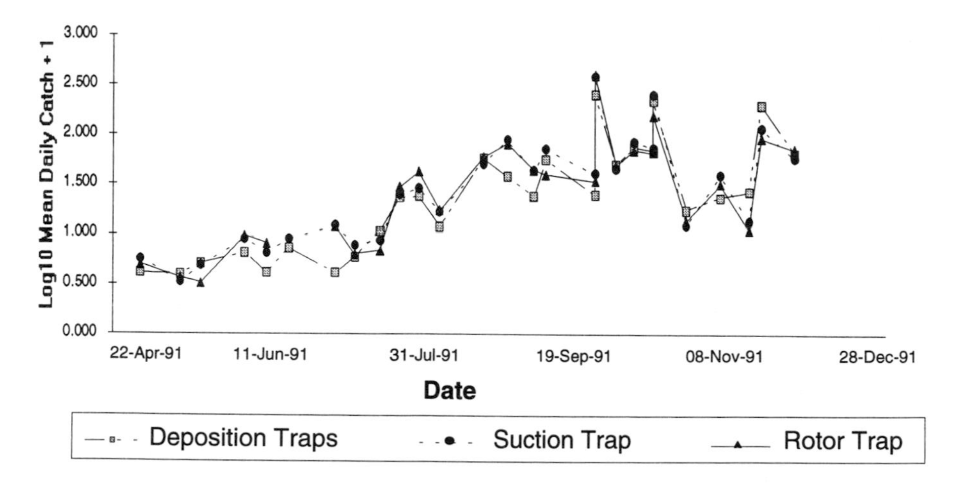

Fig. 3. Mean daily catch of spiders from the deposition traps, suction and rotor traps.

m^2 for the deposition trap catching area. Thus, after adjusting for trapping area, the ratio of deposition trap catch to sticky mat catch was 1.90:1. Fig. 2 shows the relationship between the individual deposition trap catch and the sum of the catches of the two adjacent sticky mats. The graph suggests that there may be a bias towards always obtaining a positive catch from sticky mats towards zero numbers in the deposition traps. However, the sample size is too small to draw any firm conclusions.

Trends in catch numbers

Regression analysis of the relationship between rotary and suction trap spider catches was carried out using square-root transformations. Pre-harvest, post-harvest and September to November samples were positively correlated with the suction trap catch of total spiders ($P < 0.001$, df 114, 32, 71 respectively). The slope of the lines did not differ significantly from each other (slopes with 95% confidence limits: 0.90 ± 0.07, 0.96 ± 0.09, 0.86 ± 0.08 respectively). Hence the two traps were trapping spiders at roughly equivalent efficiencies and the distance between the traps did not affect the catch. Fig. 3 shows the relationship between the catches from the deposition traps, rotor trap and suction trap between May and November (rotor and suction trap daily catches were summed over the periods corresponding to the deposition trap catches). All three types of trap show very similar trends, although the suction trap and rotor trap catch trends are more closely similar to each other than to the deposition catch. This is surprising since the deposition traps were always

Table 1. Relative proportions of the total rotor, suction and deposition trap catches of the most numerous spider taxa between 1 May 1991 and 29 November 1991.

Taxa	Rotary Trap	Suction Trap	Deposition Traps
Erigone atra (Blackwall)	0.146	0.140	0.187
Erigone dentipalpis (Wider)	0.040	0.039	0.041
Lepthyphantes tenuis (Blackwall)	0.132	0.137	0.099
Linyphiinae immatures	0.107	0.130	0.102
Erigoninae immatures	0.304	0.306	0.305
Others	0.271	0.248	0.266
Totals caught	6725	6986	7456

nearer to the rotor than was the suction trap. Rotor trap and suction trap catches had almost identical catches of spiders per unit volume of air despite the difference in catching height. This was not the case with other invertebrates which were caught in greater numbers by the rotor trap (Topping et al. 1992). A comparison between the most common groups of spiders caught (Table 1), shows that very similar proportions of all the main taxa were caught by rotor, suction and deposition traps.

Measurement of net migration

No evidence was found to indicate that spiders could escape or enter the cage once it was installed. The composition of the density samples was very similar (Table 2) with only *Porrhomma microphthalmum* (O.P.-Cambridge) occurring in disproportionate numbers in the cage sample. However, the attempt to use the caging method to assess migration was only partially successful. This was due to the very large spatial variation in density of spiders in the field. Confidence limits for the estimate are given by: $1.96\sqrt{SE^2_{density} + SE^2_{cage}}$.

Table 2. The total catch of the ten most abundant spiders from field and caged density samples.

Species	Field Density Catch m^{-2}	Cage Density Catch m^{-2}
Lepthyphantes tenuis (Bl.)	60.53	67.20
Erigone atra (Bl.)	24.67	20.20
Oedothorax fuscus (Bl.)	22.27	15.00
Bathyphantes gracilis (Bl.)	16.67	14.60
Oedothorax apicatus (Bl.)	13.20	12.20
Erigone promiscua (O.P.-C.)	11.33	9.00
Oedothorax retusus (Westr.)	8.40	5.60
Meioneta rurestris (C.L.K.)	6.93	3.00
Erigone dentipalpis (Wider)	6.13	3.40
Porrhomma microphthalmum (O.P.-C.)	1.73	6.80

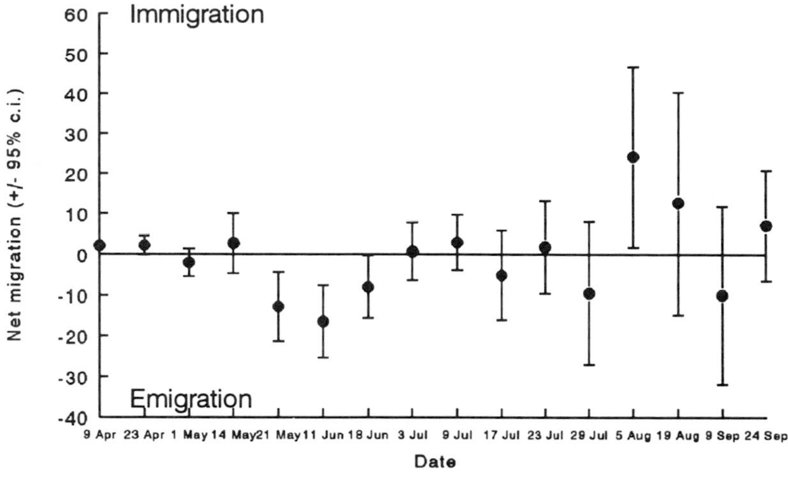

Fig. 4. Plot showing net migration m^{-2} as measured by a caging technique, together with 95% confidence limits.

These values were particularly high between 17 July and 24 September when the spiders were highly aggregated (Topping & Sunderland 1993). Before this, net migration was apparently limited until 21 May; some net emigration probably then occurred between 21 May and 18 June (Fig. 4).

Discussion

All of the techniques used to monitor aerial spider activity were found to be adequate with possibly the deposition traps being least reliable. The variability in the deposition catch may have been due to errors of sorting spiders from the very large insect catch resulting from the long trapping periods used. Alternatively, the trap may have been suffering from a bias. Problems due to spiders climbing from the vegetation into the catching area of the trap were overcome by the moat effect of the outer tray but this trap still appeared to catch a greater number of spiders than a simple sticky sheet of the same area. This was probably due to the fact that the deposition trap also had a physical height as well as a catching area. This could result in capture of spiders which snagged their silk lines on the edge of trap and then climbed along them over the "moat" to become trapped. In fact these traps were regularly observed to be festooned with silk during peak periods of ballooning. This feature would also allow some spiders to escape if they were to land on the silk lines of others, but this event is probably comparatively rare. Deposition traps were also the most labour intensive to operate and thus rotor and suction traps, which can more easily be emptied daily, are probably preferable if a power supply is available. However, the deposition traps do have the advantage that they can be left unattended for longer periods. The only factors limiting efficiency over long trapping periods is the speed of decomposition of the catch, which is accelerated if rainfall dilutes the trap fluid, and increased possibility of sorting errors as the catch volume increases.

In order to assess how many spiders are involved in ballooning within an area of habitat, calibration of the sampling methods would be necessary. However, suction traps cannot be calibrated because catch cannot be related to an area of ground due to their fixed sampling height and variability in the size and shape of the air plume from which spiders are drawn. Calibration of the deposition traps suggested an over-recording of spiders landing by a factor of 1.90, if it is assumed that the sticky sheets were themselves producing an unbiased sample. For the rotor trap, the factors affecting the capture of spiders leaving or landing on an area of habitat are the mechanical efficiency and the effective area (A) that is sampled continuously. The two attempts at calibration were designed to determine the combination of these two factors. There were two main problems with these estimates. Firstly, the area over which an experimenter can monitor spider take-off has to be very small since continuous observation is required to be certain of the number of spiders taking to the air. Secondly, calibration can only be achieved on days when ballooning activity is high and at least a few spiders can be observed taking off. This could be unrepresentative if conditions on good ballooning days differ from other days and influence the catch. Alternatively "A" could be deduced by an indirect method. "A" is controlled by the speed of spider descent or ascent. The proportion of spiders caught (p) can be determined by the formula: $p = (H/V)/T$ where $H = 0.25$ m, the trap height, V = vertical velocity of spiders and $T = 10$ sec, time for one rotation. "A" can be given as 33.61p. "A" is therefore very sensitive to changes in V. Thus if vertical velocity varied between 1.0 m^{-1} sec^{-1} and 0.05 m^{-1} sec^{-1} the "A" would vary between 0.84 m^2 and 16.80 m^2. This type of variation may not be unrealistic since greater than ten-fold variation in V has been observed within a single ballooning period (Topping & Sunderland unpubl.). As a result "A" cannot be considered to be fixed and so calibrations or measures of V would be required for a wide range of weather conditions. These problems underline the difficulties in obtaining a reliable measure of the numbers of spiders taking off or landing from any area of ground, even over a short period.

Although the caging technique used did not achieve precise estimates of net migration due to problems of spider aggregation in the field (Topping &

Sunderland 1993), caging suggested that, despite there often being large numbers of spiders ballooning there was no dramatic immigration or emigration. One possible explanation is that the spider ballooning observed in this study was synchronised over a large area resulting in a general mixing of spiders with little net flow from one habitat to another. This would result in some movement from high to low density areas but this "net migration" would be far less dramatic than the number of spiders actually ballooning would imply. Over such a short sampling period (1-2 weeks), this net migration could easily be too small to be recorded by these sampling techniques. The caging design was deemed successful as a method of containing spiders in the field for short periods and could therefore be used as a method for measuring net migration in future studies, if enough cages could be used to reduce the sampling error, or if spider aggregation was not a problem. Generally caging was thought not to influence behaviour. However, caged samples regularly revealed a number of *Porrhomma microphthalmum* specimens; a species commonly encountered ballooning across the site but which was rare in the field density samples. Since spiders in this genus are often partly subterranean in habit, the cage may have prevented these spiders ballooning, leaving them on the surface to be sampled. Alternatively, the lower light levels in the cage may have tempted these spiders out of the ground. As a result behavioural changes induced by caging cannot be ruled out.

Conclusion

Assessing the number of spiders engaged in aerial dispersal is problematical. Very large numbers of spiders may balloon with little resulting change in ground density. Suction, rotor and deposition traps can be used to determine trends of activity. The rotor trap is perhaps the most useful technique if samples are required from the air-space immediately above the habitat, although relating the catch to the area swept out by the rotor is not possible due to uncertainties about vertical velocities of ballooning spiders altering rotor efficiency. Caging may be a useful technique for assessing the effects of aerial dispersal on spider density but the dispersion of spiders on the ground must be accounted for in the sample size used.

Acknowledgements

This work was supported by the Ministry of Agriculture Fisheries and Food and by a Natural Environment Research Council (Joint Agriculture and the Environment Programme) grant. We would also like to thank Mr. C. Passmore and Mr. D. Bowerman for allowing us to sample on their land. We thank HRI engineering section for building the rotor trap.

References

Duffey, E. 1956. Aerial dispersal in a known spider population. *J. Anim. Ecol.* 25: 85-111.

Humphrey, J.A.C. 1987. Fluid mechanic constraints on spider ballooning. *Oecologia (Berl.)* 73: 469-477.

Greenstone, M.H. 1990. Meteorological determinants of spider ballooning: the roles of thermals verus the vertical windspeed gradient in becoming airborne. *Oecologia (Berl.)* 84: 164-168.

Nicholls, C.F. 1960. A portable mechanical insect trap. *Can. Entomol.* 92: 48-51.

Richter, C.J.J. 1970. Aerial dispersal in relation to habitat in eight wolf-spider species. *Oecologia (Berl.)* 5: 200-214.

Taylor, L.R. 1955. The standardisation of air-flow in insect suction traps. *Ann. Appl. Biol.* 43: 390-408.

Taylor, L.R. 1962. The absolute efficiency of insect suction traps. *Ann. Appl. Biol.* 50: 405-421.

Topping, C.J., Sunderland, K.D. & Bewsey, J. 1992. A large improved rotary trap for sampling aerial invertebrates. *Ann. Appl. Biol.* 121:707-714.

Topping, C.J. & Sunderland, K.D. 1993. Methods for quantifying spider density and migration in cereal crops. *Bull. Brit. arachnol. Soc.* 9: 209-213.

Weyman, G.S. 1993. A review of the possible causative factors and significance of ballooning in spiders. *Ethology, Ecology and Evolution* 5: 279-291.

Wingerden, W.K.R.E. Van & Vugts, H.F. 1974. Factors influencing the aeronautic behaviour of spiders. *Bull. Brit. arachnol. Soc.* 3: 6-10.

Two functions of gossamer dispersal in spiders?

Søren Toft

Department of Zoology, University of Aarhus, Building 135,
DK-8000 Århus C, Denmark.

Abstract
Gossamer dispersing spiders have been collected in Denmark using two methods: 1) A Rothamsted suction trap (height 12.2 m) operated through seven years, 2) Hand collecting at sheep-wire fences (height 1 m) for two years. The seasonal patterns of spider activity revealed by the two methods were very different. The suction trap catches overall had a peak at the end of the summer and low catches in early spring and autumn. The wire catches in the field layer showed two maxima, one in early spring and one in autumn. These findings are interpreted to mean that gossamer dispersing spiders tend to reach high in the air in the warm season, and disperse close to the ground in spring and autumn. Summer aeronautic activity seems to have potential for long-range migration between breeding habitats, whereas autumn and spring movements may have evolved as short-distance migrations in which the same individuals first move from breeding to hibernation habitat, then from hibernation to new breeding habitat.

Key words: Araneae, aeronautic dispersal, ballooning, *Erigone atra*

Introduction

Following Bristowe (1939) aeronautic behaviour, also known as ballooning, has usually been seen as an efficient mechanism for arriving at newly created areas of suitable habitat, and/or to escape from the present living area if it becomes crowded or otherwise inhospitable. In both situations dispersal behaviour is believed to be adaptive, especially for species inhabiting disturbed habitats (Southwood 1962, den Boer 1990). In agreement with this, aerial dispersal is particularly prevalent in species of agricultural fields and tidal areas.

Gossamer is the silken thread used by the spiders when ballooning (gossamer ~ goose-summer: when masses of spiders disperse at the same time,

Arthropod natural enemies in arable land · I *Density, spatial heterogeneity and dispersal*
S. Toft & W. Riedel (eds.). *Acta Jutlandica* vol. 70:2 1995, pp. 257-268.
 ISBN 87 7288 492

bundles of coalescing silk may be formed by turbulent air, resembling goose down). Such threads are also used for horizontal movements between vegetation or over a field. In the latter case only a slight breeze is needed to carry the thread to an anchoring point, and the so-called bridge thread is used as a walking line. In order to become airborne, buoyant air near the ground is needed to lift the spiders against gravity (Humphrey 1987). The weather conditions creating buoyancy are prevalent during the summer, much less so during the cooler seasons. Gossamer activity is therefore expected to result in aeronautic, long-range dispersal during the summer months, and in short-range movements along bridge threads in the cool months of the year.

Several groups of animals inhabiting agricultural fields, e.g. spring breeding carabid beetles (Thiele 1977), are known to perform seasonal migrations between different habitats, for example a reproduction habitat and a hibernation habitat. The biological function of such movements differ from that of migrations in which animals leave one reproduction habitat in order to get to a new and better one, as is generally thought to be the case for aeronautic spiders. As regards spiders, seasonal locomotory movements between habitats have been demonstrated for a few species (Nørgaard 1945, Edgar 1971, Vangsgaard et al. 1990). In these cases females were found to change habitat (from forest to open habitat), presumably in order to optimize environmental conditions for egg development. Kajak (1959) seems to be the only one so far to suggest that autumn balloning of spiders might serve to bring the animals to suitable hibernation habitats. The purpose of this paper is to present arguments in favour of Kajak's idea as regards gossamer activity in early spring and late autumn.

Methods

Sampling

Gossamer dispersing spiders have been obtained from two sources:

1. A Rothamsted suction trap was operated by Dr. Holger Philipsen, Royal Danish Veterinary and Agricultural University, at Højbakkegård Experimental Station by Tåstrup near Copenhagen during the years 1971-1977. The trap is 12.2 m high and the catch is considered independent of populations in the immediate surroundings of the trap. In 1971 trapping was started by 1 July, in the remaining years early, mid or late February, depending on the severity of the winter. Trapping stopped every year at the end of November or

beginning of December. The trap was emptied daily during most of the warm season, and weekly early and late in the year. Spider numbers were consistently very low at the marginal seasons. In Denmark dispersal activity seems to be extremely low during the winter months, contrary to Southern England (Duffey 1956). Detailed data are presented here only for 1972, which was the year of highest activity.

2. During two years (1992-93) spiders were regularly collected from a sheep-wire fence (height c. 1 m) surrounding a horse grazed pasture at Stjær near Århus. A 340 m long stretch of fence was walked a single time during hours 10.00 to 15.00, preferably on days where some activity was presumed to occur (i.e. low wind speeds). Spiders observed on the upper strand of the fence plus those on an electric wire running parallel above the fence, were collected. Data are presented here only for 1992.

Identification

All adult spiders have been identified and the same is the case with juveniles of some species. Several species of small linyphiids are among the most active aeronauts. Such species are notoriously difficult to identify as juveniles. However, it was possible to assign a large fraction of the linyphiid juveniles to a few species also being the most common in the same collections as adults. This was done using the standard characters for subfamily and generic level (i.e. tibial spines, presence and position of metatarsal trichobothria etc., Locket & Millidge 1953) as well as colour patterns and direct habitus comparison with adult specimens. Still, juveniles of *Erigone atra* (Bl.) and *E. dentipalpis* (Wider) are indistinguishable; the same is generally true for *Oedothorax* spp.

Results

Fig. 1 shows that the main seasonal trend for aerial dispersal as revealed by the suction trap, was a gradual increase from a low level in spring to a peak in early autumn (September), then a steep fall-off to zero level by December. Fig. 2 indicates that the clear peak in Fig. 1 was due to extreme catches of a single year (1972), but otherwise confirm the trend. There is no indication of an October peak though a comparatively high level is maintained into October. The pattern is roughly what is expected if flight activity reflects the population build-up through the season in agricultural habitats (cf. Toft et al. 1995).

In the fence catches (Fig. 3) there were two periods with days of high

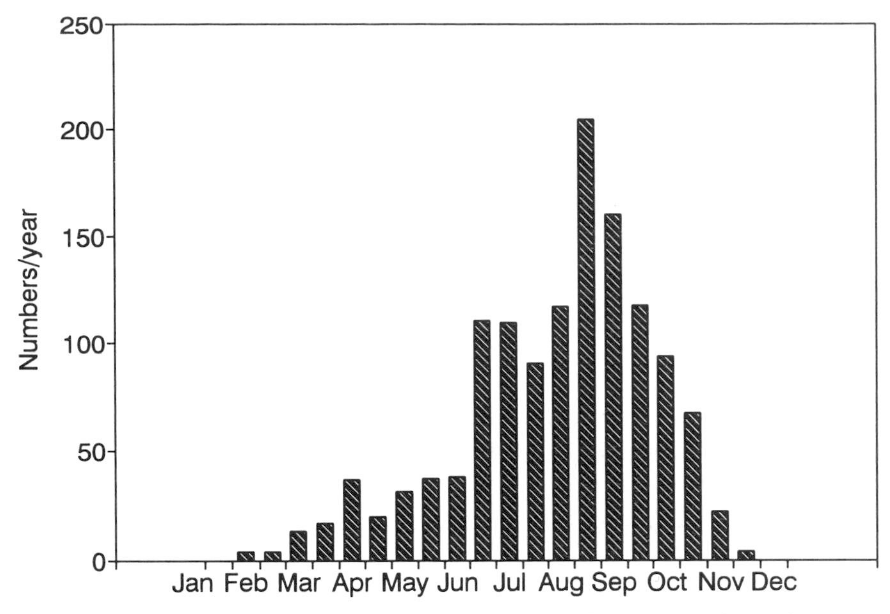

Fig. 1. Seasonal variation in total spider catches in a Rothamsted suction trap near Copenhagen, Denmark, from the years 1972-1977. Bars show numbers per half-month period per year.

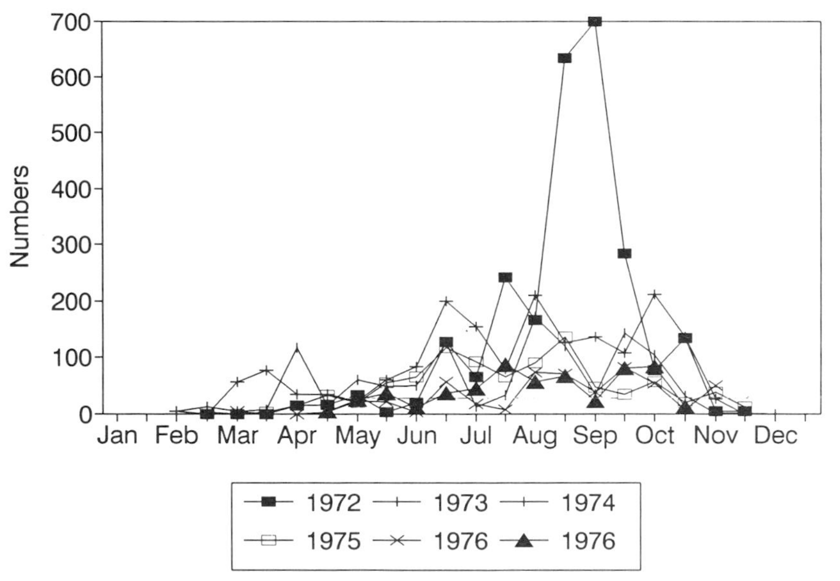

Fig. 2. Seasonal variation in total spider catches in a Rothamsted suction trap near Copenhagen, Denmark, during each of the years 1972 to 1977. Points show numbers per half-month period.

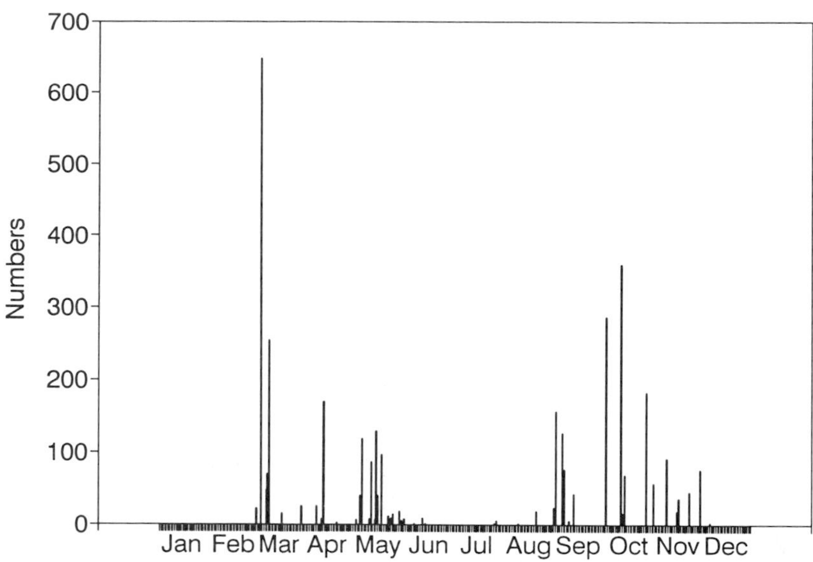

Fig. 3. Daily catches of spiders (all species) from a sheep-fence Denmark 1992

activity, viz. spring and autumn, while during the summer months hardly any activity has been recorded.

Obviously, lack of spiders on the fence is no indication of lack of aeronautic activity. I interpret these patterns to mean that during summer, gossamer activity brings the animals up into the air (vertical lifting predominant), while in spring and autumn they remain close to the ground (horizontal movement predominant). This is in agreement with Nielsen's (1932) statement about a mass dispersal event in October, "... that such floating masses of web originate — not from the flight of spiders through the air — but from the passage of a multitude of spiders across the ground by means of bridge-threads ...". According to wire catches by Thomas (1992), peak activity during the summer is early in the morning. I interpret this in a similar way as above: early in the day before heating has created sufficient buoyancy for vertical lifting, horizontal displacement is prevailing, with resulting increased chances of getting caught by a wire.

Erigone atra

E. atra is a dominant species of agricultural fields as well as in samples of aeronauts. The species has two generations a year (DeKeer & Maelfait 1988; for Denmark: J. Axelsen & S. Toft, unpublished). Hibernating adults breed in spring, giving rise to a new generation of adults in July. Descendants of these mature in September-October.

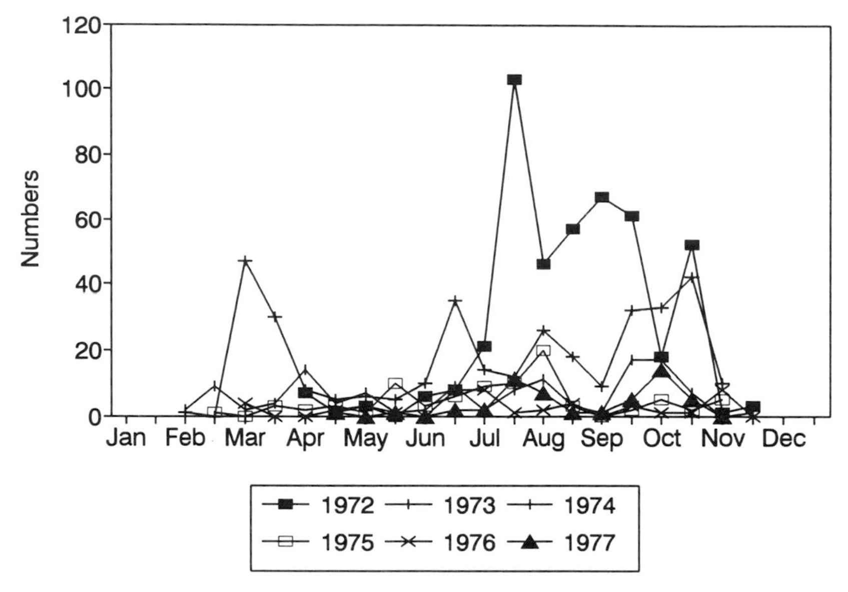

Fig. 4. Seasonal variation in catches of *Erigone atra* in a Rothamsted suction trap at Copenhagen, Denmark, during the years 1972-1997. Graphs show numbers per half-month period.

The suction trap catches of adults give variable results for the years (Fig. 4). One year (1973) had peaks corresponding to the periods of adult occurrence. The year 1972 had high activity in summer and early autumn, but not in spring. Remaining years had low activity with no clear seasonal peaks. Identification of developmental stages have been made only for 1972 (Fig. 5). It is seen that the juveniles occur nearly exclusively in late summer and early autumn. Thus, except for one year there was no early spring activity; juveniles of the spring generation were very poorly represented; and summer adults were active in only two years. In early autumn offspring of the summer generation may show high activity, but only in two years did their activity extend well into October (adults by then).

In the fence catches very few juveniles have been caught at all, mostly in September. The adults (Fig. 6) showed some extreme peak days very early, i.e. on the first good flying days of spring, and again in autumn with peak numbers definitely later than in the suction trap. There was no trace of the generation becoming adult in July. Suction trap samples were variable between years, though in no case similar to the fence catches.

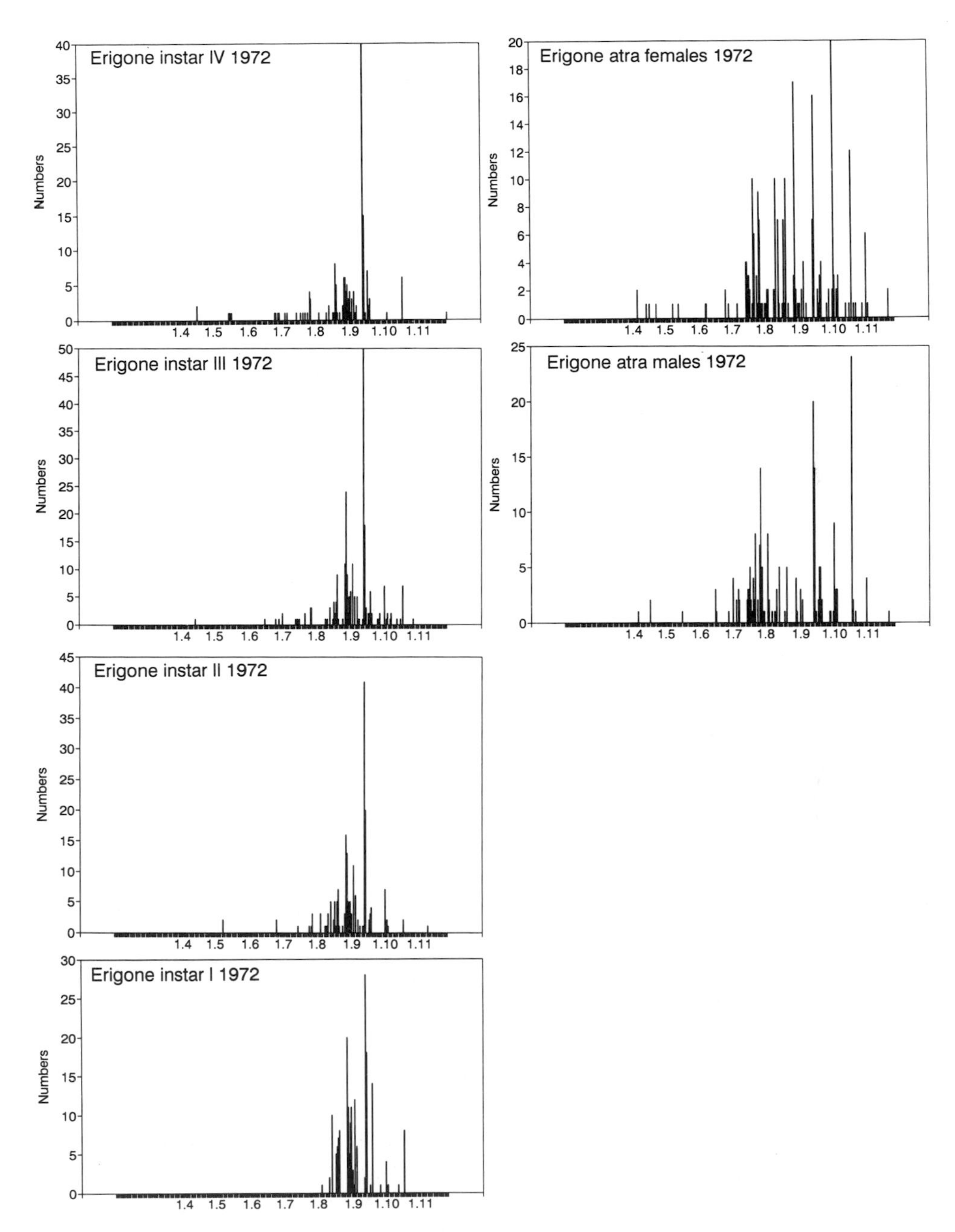

Fig. 5. Daily catches of *Erigone atra* adults and juvenile instars in a Rothamsted suction trap at Copenhagen, Denmark, through 1972.

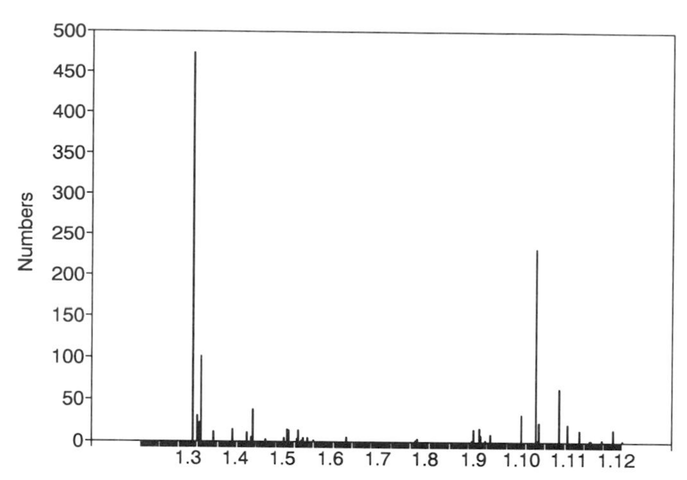

Fig. 6. Daily catches of *Erigone atra* adults from a sheep-fence Denmark 1992. Inclusion of the very few juveniles caught would change the picture very little.

Discussion

The results suggest the following scenario: During the warm season (May to September) the conditions are generally good for take-off: spiders that attempt to fly are relatively successful and ascend vertically from low positions. Aeronautic activity is going on continuously at intensities that vary with population densities, weather and habitat conditions (crowding, food availability, microclimate etc.). This is "true" aerial dispersal activity, ascribed to finding another habitat spot (including a newly created habitat) with better fitness prospects than the one presently inhabited.

In early spring (March-April) and autumn (October-November), conditions are generally unsuitable for dispersal. Even on the few good days, climate generally does not allow the spiders to ascend high up in the air; therefore dispersal takes place close to the ground, the spiders fly only short distances at each try, or else climb along bridge-threads. This activity leads only to local displacements, since total migration distance supposedly is short.

This interpretation implies that gossamer dispersing spiders reach very different distances depending on the season they disperse. Thomas (1992) modelled dispersal distances based on data mainly from July and concluded, that whole-day dispersal displacements most likely would fall between one and six kilometres. Similar calculations for late autumn conditions certainly would produce much smaller dispersal distances.

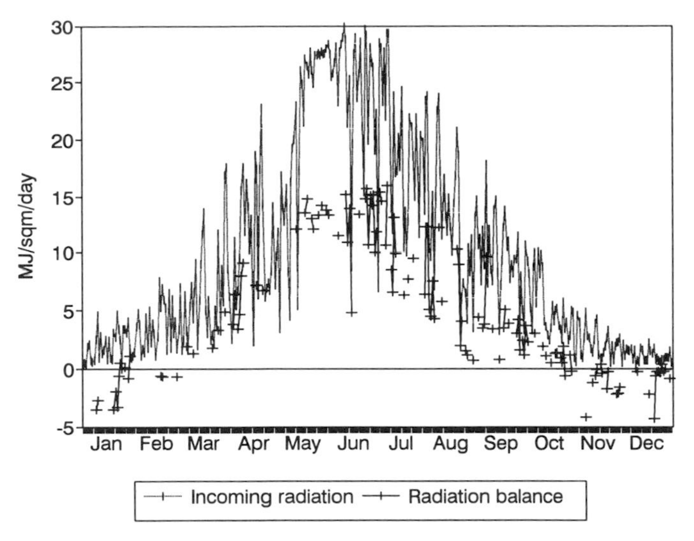

Fig. 7. Daily variation in solar radiation in Denmark 1992, data from Foulum, Central Jutland. For radiation balance only days with av. windspeeds below 4 m/sec are shown.

The proposal that gossamer flight leads to aerial dispersal in summer but merely to walking along bridge-threads in colder months, is easily explained by the general meteorological conditions prevailing at the two seasons. Upward currents created by the sun's heating of the ground is considered the main force by which spiders become airborne. For the wind speeds at which spiders fly (<3 m/sec, Vugts & van Wingerden 1976) buoyancy is strongly correlated with the radiation balance. In Denmark 1992 the radiation balance was very low for the first three months of the year and changed from positive to negative at the beginning of November (Fig. 7). If actual aeronautic activity shows any relation at all to optimal weather conditions for flight, it should be much lower in late autumn and winter than in the summer, which is what appears to be the case with the Rothamsted suction trap samples but not with the fence catches. Also, reports from Britain (Bristowe 1939, Duffey 1956) state mass dispersal events to take place in "the cold season of the year" (i.e. October and the winter months) (Duffey 1956), and in Denmark all such reports are from October (Nielsen 1932; S. Toft, own observations) when the suction trap catches were reduced drastically. I propose that the differing results are due mainly to the different sampling heights for the two data sets. The Rothamsted suction trap supposedly did not record the cold season activity because the spiders did not reach the height at which it collects.

In the autumn good days for dispersal are always sunny calm days following clear nights (Duffey 1956), thus with strong outgoing radiation and consequent cooling of the surface. A stable air layering is created. Incoming radiation during the day will heat up the air from below, creating a narrow instable air layer just above the ground. The spiders may try to take advantage of this, but only slightly higher they meet the stable air and can get no higher. Thus, the spiders are carried mainly by the weak horizontal wind, and are likely to hit vegetation or the ground again very soon.

If they make several take-off attempts over a homogenous surface (cereal or fallow field), this repeated bridge-thread production of numerous individuals may create a dense carpet of silk, which is sometimes (and only in October-November) observed over vast areas of agricultural land. I suggest that these events reported as so called mass dispersals are partly artefacts of bad weather conditions for lifting, resulting in many spiders making several take-off attempts but staying near the ground and not moving very far: lots of silk is produced in a narrow zone at the ground, easy to see for the human eye. If the same number of spiders dispersed on a summer day, each individual would succeed in becoming airborne in fewer attempts; i.e. there would be no accumulation of silk on the ground and a human observer would discover very little.

It is interesting that Duffey (1956) using sticky canes positioned close to the ground found peak activity in the cold months, whereas Sunderland (1991) using the Rothamsted suction trap found a peak in July. These conflicting results, both obtained in Southern England, compare with those of the present study, and can likewise be attributed to the difference in sampling height.

If gossamer activity is still high late in the season, when the climate is generally unsuitable for long-range dispersal, the function of this behaviour may be different. I suggest that the low horizontal movements in autumn can be compared with the out-of-field migrations of other arthropods, i.e. a habitat shift as proposed by Kajak (1959), with possible back migration in spring. I am not implying a return migration, only a further migration to a summer habitat locality. The agricultural spider fauna, as also carabids that are well known to perform such migrations (Thiele 1977), are thought to originate from littoral habitats (Tischler 1980) that may be flooded during the winter season. Under such circumstances a short-range migration to a higher winter habitat would be adaptive. In the present day agricultural situation the behaviour may be purposeless, as the spiders concerned are not known to migrate to uncultivated habitats for the winter, as do the carabids. Perhaps they just move from one field to another.

Acknowledgements

I am indebted to Holger Philipsen, Royal Veterinary and Agricultural University, Copenhagen, for handing me the large and invaluable spider material from his suction trap catches. Also to Harald Mikkelsen and Jørgen E. Olesen, Research Station Foulum, for the radiation data and meteorological discussions. C.F.G. Thomas and C. Topping gave constructive criticism to an earlier draft.

References

Boer, P.J. den. 1990. The survival value of dispersal in terrestrial arthropods. *Biol. Conserv.* 54: 175-192.

Bristowe, W.S. 1939. *The comity of spiders,* Vol. 1. Ray Society, London.

DeKeer, R. & Maelfait, J.-P. 1988. Observations on the life cycle of *Erigone atra* (Araneae, Erigoninae) in a heavily grazed pasture. *Pedobiologia* 32: 201-212.

Duffey, E. 1956. Aerial dispersal in a known spider population. *J. Anim. Ecol.* 25: 85-111.

Edgar, W.D. 1971. The life-cycle, abundance and seasonal movement of the wolf spider *Lycosa (Pardosa) lugubris*, in Central Scotland. *J. Anim. Ecol.* 40: 303-322.

Humphrey, J.A.C. 1987. Fluid mechanic constraints on spider ballooning. *Oecologia (Berl.)* 73: 469-477.

Kajak, A. 1959. Remarks on autumn dispersal of spiders. *Ekol. Polska Seria B* 5: 331-336.

Locket, G.H. & Millidge, A.F. 1953. *British Spiders,* Vol. II. Ray Society, London.

Nielsen, E. 1932. *The biology of spiders,* Vol. 1. Levin & Munksgaard, Copenhagen.

Nørgaard, E. 1945. Økologiske undersøgelser over nogle danske jagtedderkopper. *Flora og Fauna* 51: 1-37.

Southwood, T.R.E. 1962. Migration of terrestrial arthropods in relation to habitat. *Biol. Rev. Cambridge Phil. Soc.* 37: 171-214.

Sunderland, K.D. 1987. Spiders and cereal aphids in Europe. *Bull. SROP/WPRS* 10: 82-102.

Sunderland, K. D. 1991. The ecology of spiders in cereals. *Proc. 6th Int. Symp. Pest & Diseases of Small Grain Cereals and Maize, Halle/Saale, Germany* 1: 269-280.

Thiele, H.-U. 1977. *Carabid beetles in their environment.* Springer, Berlin.

Thomas, C.F.G. 1992. Spatial dynamics of spiders in farmland. Ph.D. thesis, University of Southhampton.

Tischler, W. 1980. *Biologie der Kulturlandschaft.* Gustav Fischer Verlag, Stuttgart.

Toft, S. 1989. Apects of the ground-living spider fauna of two barley fields in Denmark: species richness and phenological synchronization. *Ent. Meddr.* 57: 157-168.

Toft, S., Vangsgaard, C. & Goldschmidt, H. 1994. Distance methods used to estimate densities of web spiders in cereal fields. In: Toft, S. & Riedel, W. (eds.) *Arthropod natural enemies in arable land · I*, pp. 33-45. Aarhus.

Vangsgaard, C., Gravesen, E. & Toft, S. 1990. The spider fauna of a marginal agricultural field. *Ent. Meddr.* 58: 47-54.

Vugts, H.F. & Wingerden, W.K.R.E. van. 1976. Meteorological aspects of aeronautic behaviour of spiders. *Oikos* 27: 433-444.

Dispersal of ground beetles in a rye field in Vienna, Eastern Austria

B. Kromp & M. Nitzlader

L. Boltzmann-Institute for Biological Agriculture and Applied Ecology
A-1110 Vienna, Rinnböckstraße 15, Austria

Abstract

In 1991, species composition, spatial distribution and dispersal movements of carabids were investigated by directional pitfall trapping and a mark-recapture method in a rye field adjacent to a hedge and a woodland strip in Obere Lobau, Vienna, Austria.

From 61370 live-trapped individuals of 86 species, 8035 (except small carabids, e.g. *Bembidion sp.*) were marked with elytral ink-marks and released near their trapping locations (nine trapping stations of two directional traps, located 5, 65 and 125 m from a hedge and a woodland strip, respectively; sampling period: April 3 - October 10).

Concerning spatial distribution of total catches in the field, "edge species" like *Asaphidion flavipes* and *Syntomus pallipes* had the highest catches near the field border, whereas "field species" like *Poecilus cupreus*, *Brachinus explodens* and *Harpalus albanicus* concentrated in the central traps.

By directional trapping, spring dispersal into the field was detected only for *Bembidion lampros* in early April, autumnal emigrations into adjacent habitats for *B. lampros* and *Asaphidion flavipes*.

Mark-recapture results showed an average recapture rate of 4.9%. Of dominant species, the mobile *Brachinus explodens* (recapture rate 8.5%) showed seasonal changes in dispersal patterns. Recapture data of *Platynus dorsalis* are in close agreement with the literature.

Key words: Carabidae, organic rye field, adjacent habitats, directional pitfall trapping, mark-recapture, dispersal

Introduction

Since pesticides are to be avoided in organic agriculture (Maurer 1989), plant protection in this cultivation system is highly dependent on self-regulatory processes. Carabid beetles are mostly polyphagous predators and therefore important antagonists of pests (Luff 1987).

Yearly renewal of carabid populations in fields (Basedow 1990) is based

Arthropod natural enemies in arable land · I *Density, spatial heterogeneity and dispersal*
S. Toft & W. Riedel (eds.). *Acta Jutlandica* vol. 70:2 1995, pp. 269-277.
 ISBN 87 7288 492

on the reproduction both of populations overwintering directly in the field and, for certain species, overwintering in adjacent habitats. The latter immigrate into fields in early spring (e.g. Jensen et al. 1989, Welling 1990, Dennis & Fry 1992). Calculations based on the few available experimental data on overwintering carabids (Desender et al. 1989, Riedel 1991) indicate that the contribution of immigrating carabids to total population densities of carabids in fields may have been overestimated.

In order to estimate the spatial distribution and dispersal of carabids in a field in relation to adjacent habitats, live-trapping and mark-recapture methods were applied in a rye field in Vienna, Eastern Austria.

Material and methods

Study site: an organically cultivated field of winter-rye (4 ha) adjacent to a hedge in the north and a woodland strip in the east (riverside nature-reserve "Lobau", Vienna; 48°10'N and 16°30'E; 152 m above sea level; annual average temperature 9.6°C, annual average total precipitation 510 mm; greyish alluvial soil).

Sampling methods: nine stations, each with two directional pitfall traps set at right angles to each other (plastic boxes measuring 100 x 15 x 15 cm, divided longitudinally by a PVC partition; transparent plastic roof), positioned 5, 65 and 125 m from the hedge and the woodland, respectively (Fig. 1). Traps were emptied twice a week; living individuals were identified to species level and marked using coloured ink-pens (except small carabids, e.g. *Bembidion*) in a pattern reflecting the time and place of release; after marking they were released near their trapping location the same day.

Sampling period: April 3 - October 10, 1991 (42 sampling dates, 1272 samples). During periods of harvest, a reduced number of traps were used; during autumnal soil cultivation, six directional traps were placed directly on the field border.

Results and discussion

Species composition and within-field spatial distribution

A total of 61,370 carabid individuals, representing 86 species, were live-trapped in the field. Twelve species comprised over 95% of total individuals (Fig. 2). Due to the warm and dry microclimate in the organic rye crop, spring-breeding species were predominant. Among autumn breeders, only *Harpalus rufipes*

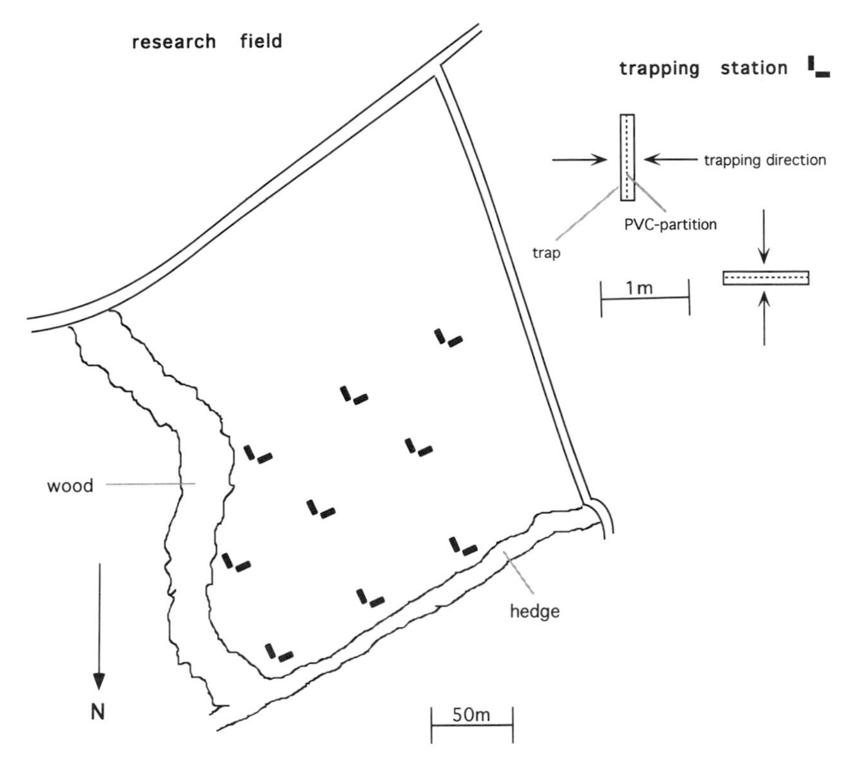

Fig. 1. Research field and trapping stations.

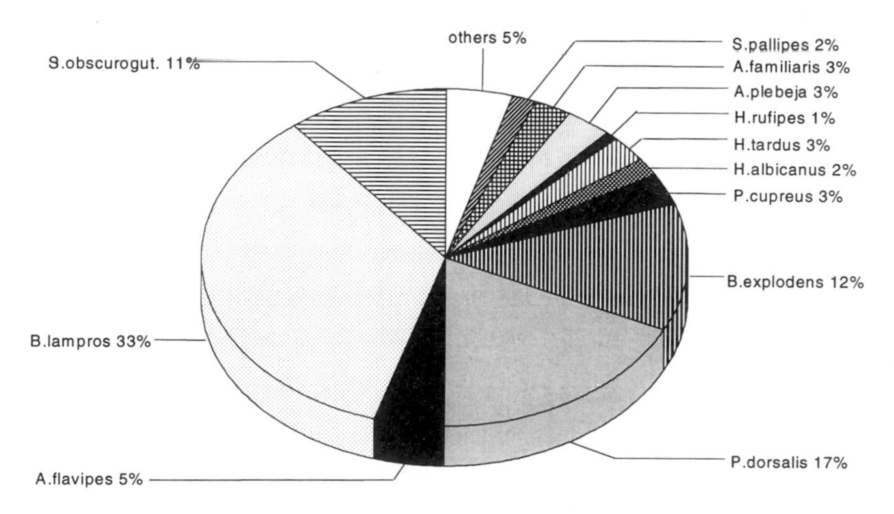

Fig. 2. Carabid dominance (> 1 %; n=49309; April 4 - Sep. 16, 1991).

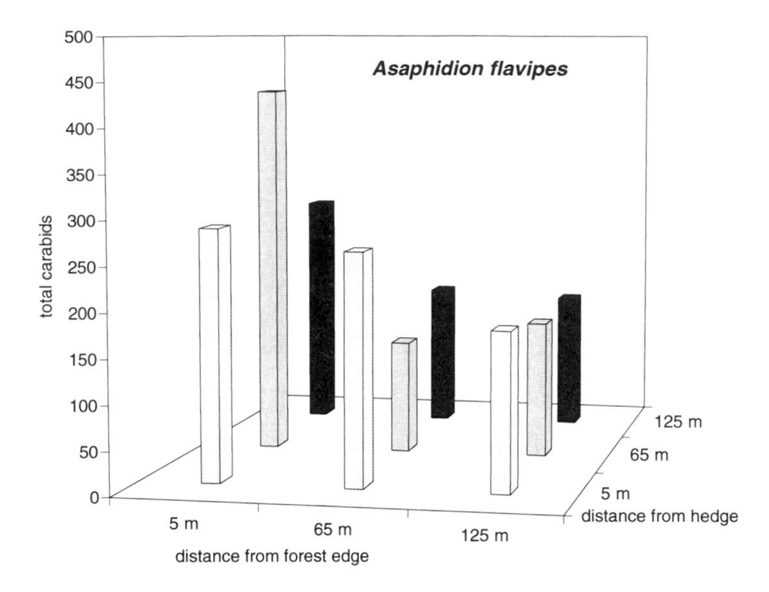

Fig. 3. Spatial distribution of *Asaphidion flavipes* catches in a rye field.

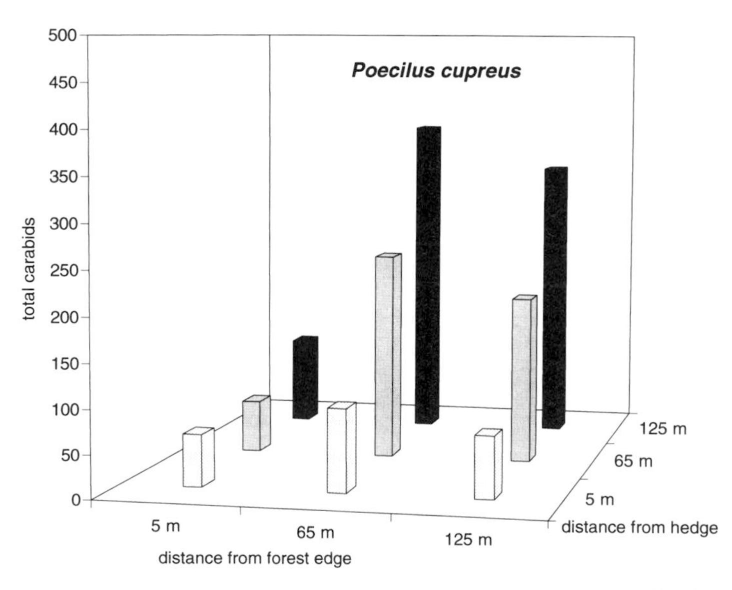

Fig. 4. Spatial distribution of *Poecilus cupreus* catches in a rye field.

(Degeer) and *Amara plebeja* Gyllenhal were abundant. Concerning spatial distribution in the field, *Asaphidion flavipes* (Linnaeus) (4.5% dominance) showed highest total catches in the traps near the field border (Fig. 3, "edge species"), which was similar to *Syntomus pallipes* Dejean (1.5%). *Poecilus cupreus* (Linnaeus) (2.9%), on the other hand, dominated in the trapping stations in the central parts of the field (Fig. 4, "field species"), as did *Brachinus explodens* Duftschmid (12.3%) and the xerothermic *Harpalus albanicus* Reitter (1.7%). *Bembidion lampros* (Herbst) (34.4%) catches were distributed rather evenly over the field.

Directional trapping

In spring, directional trapping data revealed a certain tendency for directed dispersal only for *B. lampros*. It migrated from the border area into the central parts of the field in early April, as the resultant vectors of the catches from the four trapping directions in Fig. 5A indicate. Only three weeks later this directed dispersal apparently ceased, the resultant vectors being small or pointing in different directions (Fig. 5B). At the same time, the total catches indicated by the areas of the circles in Fig. 5 were quite similar in all nine trapping stations. Jensen et al. (1989) found that *Platynus dorsalis* (Pontoppidan) dispersed away from the winter quarters in the field boundaries within a period of two weeks.

For the other abundant species, such as *P. dorsalis*, *B. explodens*, *A. flavipes* and *P. cupreus*, directional trapping provided no evidence for spring dispersal from the field borders into the field. As soon as activity started in early spring, catches were distributed rather equally over all nine trapping stations.

In directional traps placed directly along the field border in autumn, *B. lampros* and *A. flavipes* outward catches clearly were higher than inward catches, indicating emigration into adjacent habitats for overwintering.

Mark-recapture

Of 8035 marked and released beetles, representing 24 species, 396 individuals of 16 species were recaptured (average recapture rate: 4.9%). *Poecilus punctulatus* (Schaller) (marked ind.: n=31) showed the highest recapture rate at 32%, followed by *B. explodens* (n=1961; 8.5%), *P. dorsalis* (n=3773; 3.9 %), *Harpalus tardus* (Panzer) (n=411; 3.4%) and *P. cupreus* (n=722; 3.3%). Based on the code of elytral ink marks, individual movements could be reconstructed. *B. explodens*, a highly mobile species, for instance, showed seasonal changes in dispersal patterns. In early May, rather long distances were

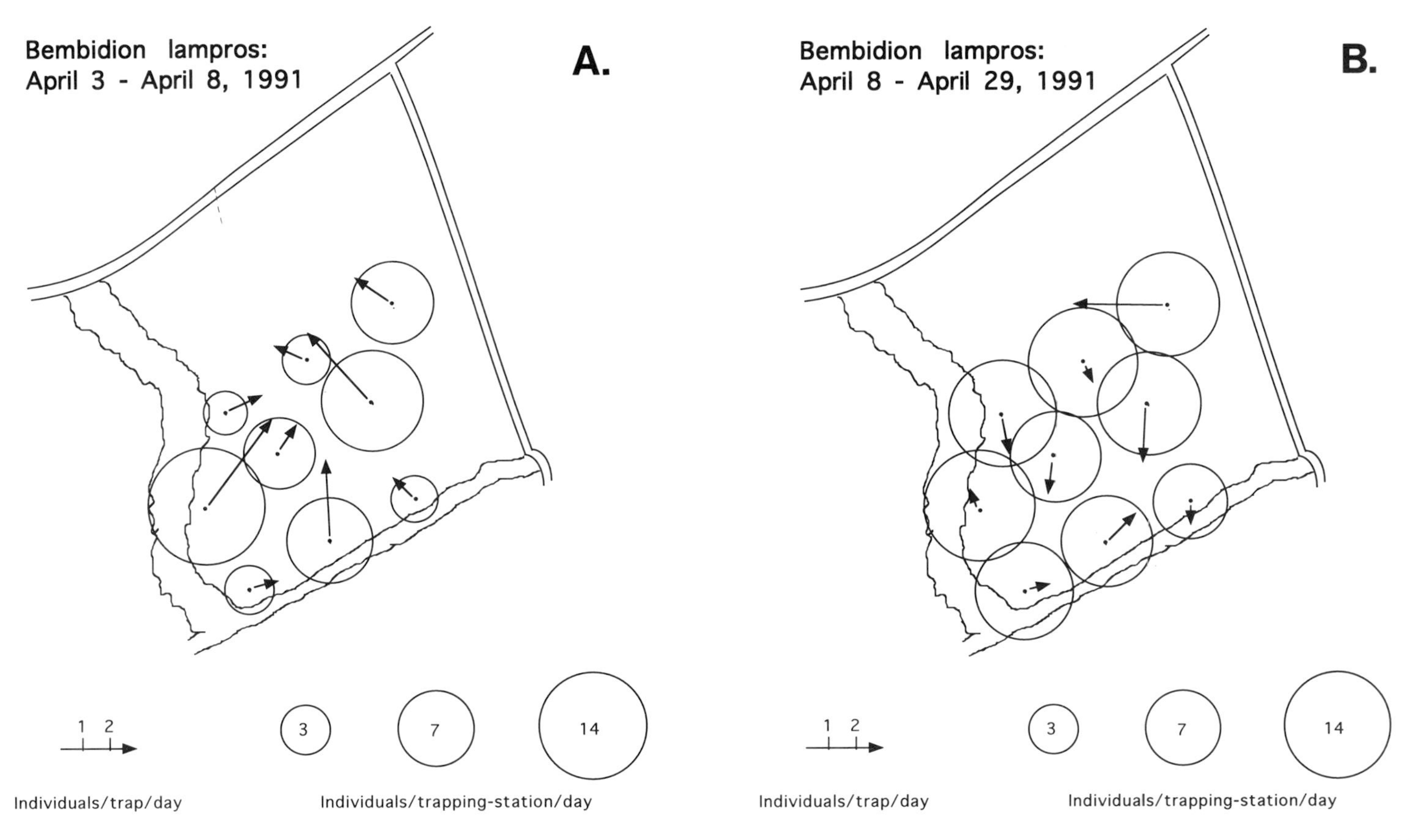

Fig. 5. *Bembidion lampros*: daily catches obtained by directional trapping (A: April 3 - April 8; B: April 8 - April 29, 1991). The arrows are the resultant vectors of the daily catches from the four trapping directions; the areas of the circles correspond to the total daily catches of each trapping station.

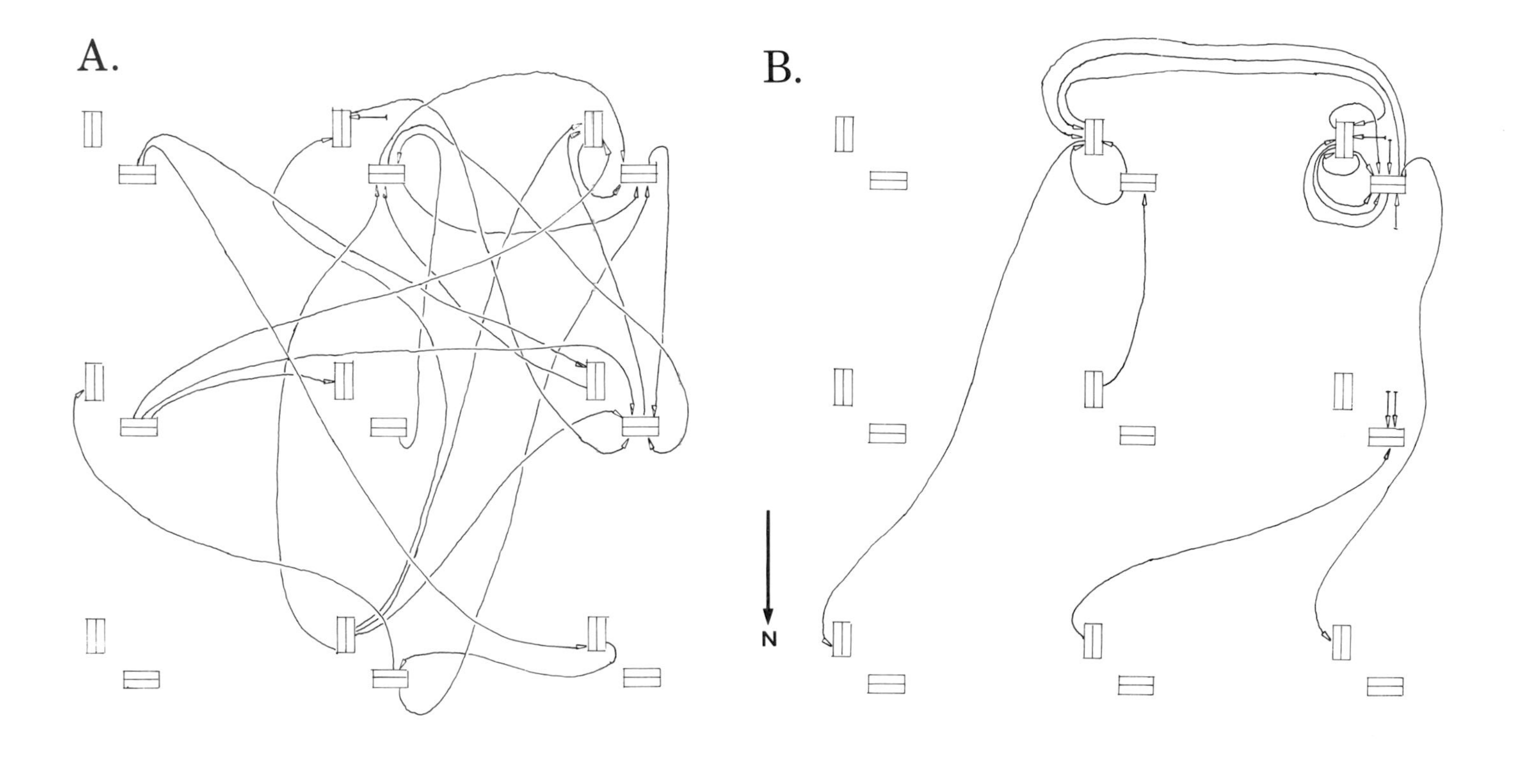

Fig. 6. Individual migrations of marked and released *Brachinus explodens* (trapping stations enlarged ca. 9 fold): A: Trapping period May 5 - May 9; B: Trapping period June 13 - Sep. 9, 1991.

covered across the field (Fig. 6A); in summer, on the other hand, short movements in the central parts of the field predominated (Fig. 6B), possibly due to aggregational behaviour (Wautier 1971).

One individual of *P. dorsalis*, another mobile species, showed a maximum dispersal distance of 46 m/day. The calculated average daily dispersal velocities of recaptured carabids, however, were much lower (1.9 - 5.5 m/day), assuming that the beetles move linearly between point of release and point of recapture. With regard to *P. dorsalis*, the mark-recapture results of this study are in close agreement with those obtained by Jensen et al. (1989): they found a recapture rate of 5.1% (Vienna: 3.9%), an average dispersal velocity of 3.7 m/day (Vienna: 4.1 m/day) and a maximum dispersal distance of 42.3 m/day (Vienna: 46 m/day). According to Wallin (1985) and Luff (1987), the dispersal movements of the dominant field carabids found in this study (Fig. 3) were most likely achieved by walking, with the exception of *Syntomus obscuroguttatus* (Duftschmid); the latter species, together with freshly emerged *Trechus quadristriatus* (Schrank) and *Demetrias atricapillus* Linnaeus, were the only carabids found in the head traps of ground photoeclectors deployed in the Vienna research fields (Kromp et al., unpublished).

Conclusions

Evidence was found for migrations between adjacent habitats and the field for only a few carabid species. Within-field habitat quality for overwintering and reproduction is therefore assumed to be a crucial factor for carabid population densities in fields (see also Kromp & Steinberger 1992).

Acknowledgements

This study is part of a research study supported by the Austrian Ministry for Science and Research. Thanks are due to A. Velimirov and A. Russegger for doing the diagrams.

References

Basedow, T. 1990. Jährliche Vermehrungsraten von Carabiden und Staphyliniden bei unterschiedlicher Intensität des Ackerbaus. *Zool. Beitr.* N.F. 33: 459-377.

Dennis, P. & Fry, G.L.A. 1992. Field margins: can they enhance natural enemy population densities and general arthropod diversity on farmland? *Agric. Ecosyst. Environ.* 40: 95-115.

Desender, K., Alderweireldt, M. & Pollet, M. 1989. Field edges and their importance for polyphagous predatory arthropods. *Med. Fac. Landbouww. Rijksuniv. Gent* 54(3a): 823-833.

Jensen, T.S., Dyring, L., Kristensen, B., Nielsen, B.O. & Rasmussen E.R. 1989. Spring dispersal and summer habitat distribution of *Agonum dorsale* (Coleoptera, Carabidae). *Pedobiologia* 33:155-165.

Kromp, B. & Steinberger, K.H. 1992. Grassy field margins and arthropod diversity: a case study on ground beetles and spiders in eastern Austria (Coleoptera: Carabidae; Arachnida: Aranei, Opiliones). *Agric. Ecosyst. Environ.* 40:71-93.

Luff, M.L. 1987. Biology of polyphagus ground beetles in agriculture. *Agric. Zool. Rev.* 2: 237-278.

Maurer, L.J. 1989. An ecological approach to agriculture. The Austrian example. *Agric. Ecosyst. Environ.* 27: 573-578.

Riedel, W. 1991. Overwintering and spring dispersal of *Bembidion lampros* (Coleoptera: Carabidae) from established hibernation sites in a winter wheat field in Denmark. In: Polgár, L., Chambers, R.J., Dixon, A.F.G. & Hodek,I. (eds.) *Behaviour and Impact of Aphidophaga*, pp. 235-241. SPB Academic Publishing bv., The Hague.

Wallin, H. 1985. Spatial and temporal distribution of some abundant carabid beetles (Coleoptera: Carabidae) in cereal fields and adjacent habitats. *Pedobiologia* 28: 19-34

Wautier, V. 1971. Un phenomène social chez les Coléoptères: le grégarisme des *Brachinus* (Caraboidea Brachinidae). *Insectes Soc.* 18: 1-84.

Welling, M. 1990. Dispersal of ground beetles (Col., Carabidae) in arable land. *Med. Fac. Landbouww. Rijksuniv. Gent* 55(2b): 483-491.

Movement patterns and foraging behaviour of carabid beetles on arable land

Henrik Wallin[1] & *Barbara Ekbom*

Swedish University of Agricultural Sciences, Department of Entomology
Box 7044, S-750 07 Uppsala, Sweden
[1] Present address: Anticimex, P.O. Box 47025, S-100 74 Stockholm, Sweden

Results of over 10 years of work in Sweden, on the importance of predators for aphid populations, has shown that predators can be effective biological control agents of the bird-cherry oat aphid *Rhopalosiphum padi* L. in spring cereals (Ekbom et al. 1992). The aim of this study has been to give a concentrated picture of the intrinsic (physiological) and extrinsic (environmental) factors which affect predation behaviour of generalist predators. We have tried to answer several questions: 1) What characteristics are important for a generalist predator to be an effective natural enemy of aphids in cereals? 2) Can this knowledge be used to enhance predator numbers, or better, the predator's chances of survival in an agricultural landscape? Three carabid species have been used as model predators because we have good information on occurrence, life cycles (Wallin 1989), and prey consumption (Wallin et al. 1992). The three species also vary in size, diurnal activity and have specific needs as to microclimate and environment. The three species are the day active *Pterostichus cupreus* L. (10 mm), and the night active *P. melanarius* Illiger (15 mm) and *P. niger* Schaller (20 mm). The behaviour of these three species was studied in areas of low and high prey (aphid) density. Earlier radar-tracking studies on another carabid beetle, *Calosoma affine* Chaudoir, showed strong behavioural differences related to variations in prey density and availability (Wallin 1991).

A modified version of the portable harmonic radar was used to investigate fine-scale movements of carabid beetles in cereal fields near Uppsala, Sweden in 1989 and 1990. An outline of the general principle of the tracing technique (that was originally developed for locating avalance victims), and diode-tagging of beetles are presented elsewhere (Mascanzoni & Wallin 1986, Wallin & Ekbom 1988). Several version of the harmonic radar have been commercially available since 1986 of which the equipment consisting of a circular polarity

Arthropod natural enemies in arable land · I *Density, spatial heterogeneity and dispersal*
S. Toft & W. Riedel (eds.). *Acta Jutlandica* vol. 70:2 1995, pp. 279-281.
 ISBN 87 7288 492

Helix antenna has been found to operate best when tracking insects. Moreover, the portability of the radar system has been improved; the transmitter and receiver units have been made smaller and have actually been built into the antenna. Loctite Tak Pak glue was used in the present investigation to tag the diode to each beetle's elytra. In addition, one diode leg was attached to conductive silver paint which covered the beetle's elytra. This was found to be particularly useful when tagging individuals of the small *P. cupreus*, since the length of the antenna (cf. Mascanzoni & Wallin 1986) could be decreased. The antenna soldered to the diode increases the detection distance but may, if it is much longer than the beetle, also act as an obstacle. The present and modified version of the portable harmonic radar offers considerably longer detection ranges, and the technique is still being improved. A number of new types of commercially available Schottky diodes, and even "beam-lead" and extremely small diodes will make it possible to use a passive reflector (i.e. diode) without any antenna and to tag beetles even smaller than 10 mm. Each beetle was individually marked on its thorax (Wallin & Ekbom 1988).

At 5-min intervals the positions of traced beetles were recorded using an X, Y coordinate system. The short time interval between consecutive moves (a maximum of 11 moves) was chosen to obtain a better resolution of displacements over a relatively limited area, which differs from previous tracking experiments (cf. Wallin & Ekbom 1988, Wallin 1991). All beetles were released under a wooden plate (20 x 20 cm) in order to avoid immediate activity due to handling. Using the X, Y coordinates, turning angles (Θ) and move lengths (D) were calculated and used in a correlated random walk model (Kareiva & Shigesada 1983). This model provides a framework for making predictions about an animal's rate of dispersion by incorporating the distributions of turning angles and move lengths (McCulloch & Cain 1989, Wallin 1991). We tested the null hypothesis that the beetles follow a predicted (theoretical) movement pattern that corresponds to a correlated random walk (cf. Wallin 1991). Any statistically significant deviation, on a move-by-move comparison, of observed data from predicted data can therefore be interpreted as directed movements. A method based on a z-test (5% level) allows a statistical comparison of observed movement data with predicted data (cf. McCulloch & Cain 1989, Wallin 1991). All beetles exhibited an equal probability of turning left or right (χ^2, $P>0.5$), and a simplified version of the correlated random walk model was therefore used (cf. Wallin & Ekbom 1988). The two basic patterns were directed movement and a correlated random walk (i.e. random movements). Both species-specific differences in movement and behavioural patterns, as well as sex-related variations in sprint speed affected the ability and motivation of the three carabid species to search for and consume cereal aphids when this type of prey was abundant. Results of these studies have been published elsewhere (Wallin & Ekbom 1994).

References

Ekbom, B.S., Wiktelius, S. & Chiverton, P.A. 1992. Can polyphagous predators control the bird cherry-oat aphid in spring cereals?. *Entomol. Exp. Appl.* 65: 215-223.

Karieva, P. & Shigesada, N. 1983. Analyzing insect movements as a correlated random walk. *Oecologia (Berl.)* 56: 234-238.

Mascanzoni, D. & Wallin, H. 1986. The harmonic radar: a new method of tracing insects in the field. *Ecol. Entomol.* 11: 387-390.

McCulloch, C.E. & Cain, M.L. 1989. Analyzing discrete movement data as a correlated random walk. *Ecology* 70: 383-388.

Wallin, H. 1989. The influence of different age classes on the seasonal activity and reproduction of four medium-sized carabid species inhabiting cereal fields. *Holarct. Ecol.* 12: 210-212.

Wallin, H. 1991. Movement patterns and foraging tactics of catepillar hunter inhabiting alfalfa fields. *Funct. Ecol.* 5: 740-749.

Wallin, H., Chiverton, P.A., Ekbom, B.S. & Borg, A. 1992. Diet, fecundity and egg size in some polyphagous predatory carabid beetles. *Entomol. Exp. Appl.* 65: 129-140.

Wallin, H. & Ekbom, B. 1988. Movements of carabid beetles inhabiting cereal fields: a field tracing study. *Oecologia (Berl.)* 77: 39-43.

Wallin, H. & Ekbom, B. 1994. The influence of hunger level and prey densities on movement patterns in three species of *Pterostichus* beetles. *Environ. Entomol.* 23: 1171-1181.

species of agricultural importance. But to what extent is it possible to extrapolate the results from single species to the relevant taxonomic group or even the ecological guild of species?

To look at as many species as possible within the whole complex of predatory arthropods in an agricultural landscape, we applied the trap transect method over a distance of 5 km in an intensely farmed landscape. By comparing the distribution patterns in space and time of both surface activities and flight, we can assign specific dispersal types to the more abundant species. The type of the distribution pattern and, accordingly, the presumed dispersal activity allows an assessment of the regional habitat requirements for the various types of predatory species in agricultural areas. The main purpose of this analysis is to quantify the contribution of the different dispersal and hibernation types in order to set priorities for landscape planning in agricultural extensification programs.

Materials and methods

The transect in the Limpach valley northwest of Bern, Switzerland, consisted of twenty trap stations. It started with two trap stations in an isolated piece of wetland (the nearest other wetland being at a distance of 7 km) and ended with one trap station in another isolated habitat, a dry meadow (Mesobrometum) along the edge of a mixed forest. Between these two regionally unique natural or seminatural habitats, there was a flat plain with a mosaic of small (maximum 1.5 ha) cultivated fields and fertilized grassland. The trap stations in the cultivated land were located at distances of 200 - 300 m in the center of either wheat or maize fields, or in fertilized grassland. To obtain data from an equally long period of trapping in all habitat types, the traps in the wheat fields were all removed after wheat harvest in July and reinstalled in the nearest maize fields. Each trap station consisted of a set of one window (interception) trap (Fürst & Duelli 1988, Stöckli & Duelli 1989), one sticky grid trap (Huber & Duelli 1987), one yellow water pan trap and three pitfall traps. Two of the pitfall traps were funnel traps with a diameter of 15 cm, one was a standard plastic cup of 7 cm diameter. Sweep net samples every two weeks consisted of 100 sweeps per trap station.

The traps were operated for one full year in 1987 and emptied at weekly intervals. The most important groups of predatory arthropods were identified to the species level. The weather conditions in 1987 were somewhat unusual, with snow and ice until the end of March, a rainy spring, and an extremely warm autumn with temperatures up to 18°C around Christmas.

All species with five or more individuals collected were displayed in a two-

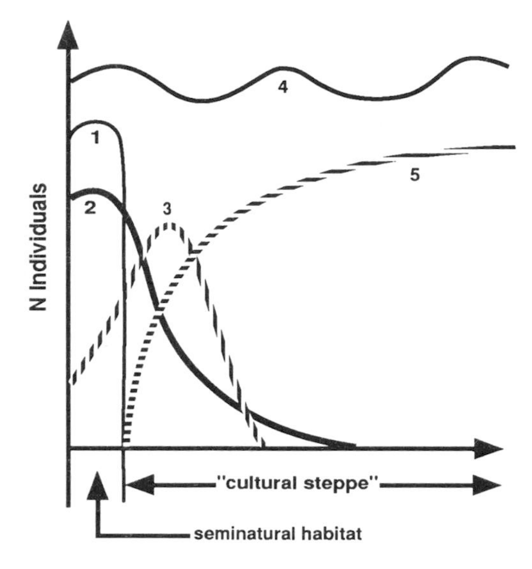

Fig. 1. The five categories to characterise transect distribution types. 1 - captures only in seminatural habitats; 2 - maximum captures in seminatural habitats, with gradual decrease into agricultural areas; 3 - maximum captures very close to seminatural habitats; 4 - species distributed in all habitats; 5 - captures almost exclusively in cultivated areas.

dimensional transect distribution graph (Duelli et al. 1992). Species with more than ten specimens were displayed in a three-dimensional graph, with axes for number of individuals, space and time.

Results

The role of natural habitats as "ecological compensation areas"

The primary goal of this transect study in 1987 was to quantify the contribution of natural or seminatural areas to the overall biodiversity of an agricultural landscape. So we first focussed on species with a distribution pattern in the transect which indicated that the natural areas are a vital source or an important hibernation habitat for those species. For this purpose each species, including all the non-predatory species, was assigned to one of five distribution types (Fig. 1). Species of the types 1 - 3 (stenotopic in natural areas, diffusion from a maximum in the natural areas into the cultivated fields, maximum in cultivated areas bordering the natural areas) were considered to depend, at least at one point in time during their lives, on the presence of natural areas. An average of 60% of all species in seven arthropod groups belonged to the distribution types 1, 2 or 3 (Duelli et al. 1992).

We can assume that a lot of the species considered to be "ubiquists" or "cultural species" (types 4 and 5) also depend on natural habitats such as hedgerows or forest edges for overwintering. Thus the contribution of natural or seminatural areas to the overall biodiversity of an agricultural landscape was certainly underestimated with our average of 60%, based solely on the two-dimensional transect distribution, which does not show seasonal differences.

Estimating the potential for genetic exchange between isolated natural habitats

A second goal of the transect distribution graphs was to develop a dispersal model for species with a stronghold in natural habitats, and a measurable diffusion into the cultivated areas (type 2 in Fig.1). Distribution patterns for all type 2 species were fitted to various published dispersal functions (Taylor 1978). The equation with the best fit was used to calculate the distances from a natural habitat, where 1%, 0.1% or 0.01% of the total number of individuals captured in the trap station in the natural area were expected to be collected (Duelli et al. 1992). In a model we can thus estimate the probability for a given population to reach another potentially suitable habitat. Furthermore, we can roughly estimate how many of the species from a taxonomic group will be able to reach a habitat in a given distance with at least one propagule per year etc. So far, only preliminary calculations for spiders and aculeate Hymenoptera have been performed.

Table 1. Numbers of species collected either in flight (window traps, sticky grids, yellow pans) or on the surface and in the vegetation layer (pitfall traps, sweepnet samples, suction samples). Only species with five or more individuals (left side in each column) were analysed with the help of transect distribution graphs.

Group	Total No. sp.		Flying		Crawling		Only in flight		Only Crawling	
	≥5ind.	<5ind.	≥5ind.	<5ind.	≥5ind.	<5ind.	≥5ind.	<5ind.	≥5ind.	<5ind.
Araneae	105	71	22	26	105	52	0	19	83	45
Carabidae	82	15	40	0	71	15	11	0	42	15
Staphylinidae	124	90	84	53	104	46	20	44	40	37
Heteroptera[1]	13	29	13	27	5	2	8	27	0	2
Syrphidae[2]	11	20	11	20	7	0	4	20	0	0
Coccinellidae	10	6	10	5	3	2	7	4	0	1
Neuroptera	8	15	8	15	2	0	6	15	0	0
Total Predators	599		334		414		185		265	
	353	246	188	146	297	117	56	129	165	100
% of total in class	59%	41%	56%	44%	71%	29%	34%	66%	63%	38%

[1]predatory; [2]aphidophagous

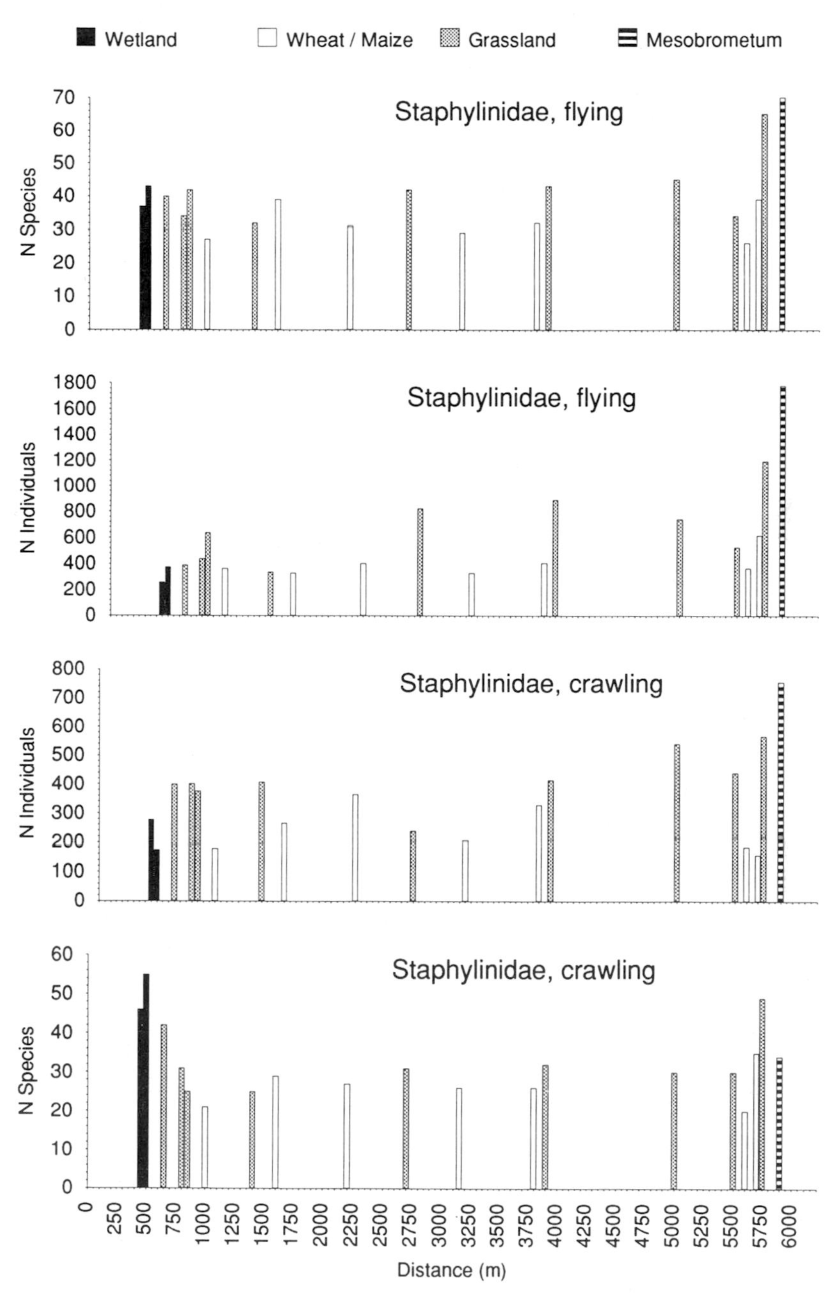

Fig. 2. Two-dimensional transect distribution graphs of staphylinid beetles. Habitat types are indicated with different filling patterns, including the wetland (left), the dry meadow (Mesobrometum, right), fertilized grassland and wheat, where the traps were moved into the nearest maize fields after wheat harvest in July.

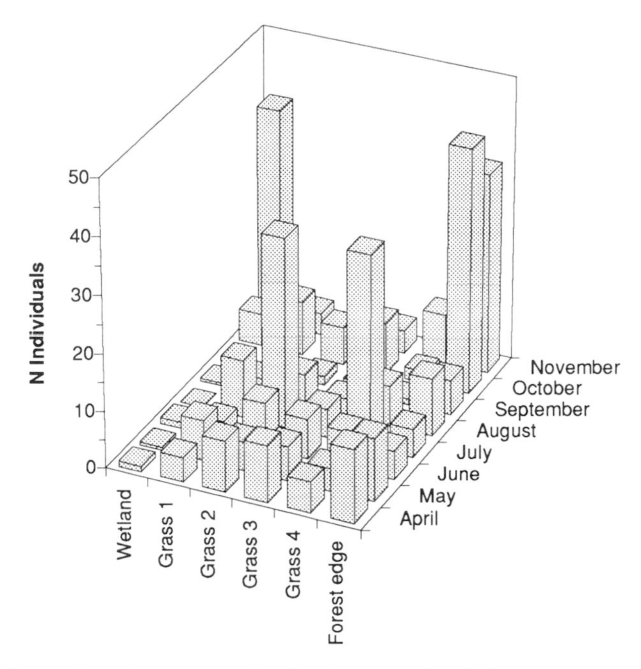

Fig. 3. Three-dimensional transect distribution graph of the common green lacewing, *Chrysoperla carnea* (Neuroptera), as an example of a nomadic species (dispersal type 5).

Dispersal types of predatory arthropods

In a third approach to the transect data, we address the following questions:

- What proportion of the predatory arthropod fauna (regarding either species or individuals) moves between natural habitats and cultivated areas, in other words: how many of them depend on natural or seminatural habitats outside the cultivated areas at certain times of their lives?
- What part of the predatory fauna develops within the cultivated fields and hibernates there or in the immediate surroundings in field margins?

On the one hand, we analysed the highly mobile groups of more or less specialized aphidophagous insects (Syrphidae, Coccinellidae, Neuroptera), on the other hand, we identified as many species as possible of the more polyphagous arthropods, where the spectrum ranges from flightless species to highly mobile fliers. The most important groups of polyphagous predators in cultivated areas are spiders, Carabidae, Staphylinidae and parts of the Heteroptera. The percentage of species collected in flight is highest in the Heteroptera and staphylinids, and lowest in the spiders.

We compared the number of species collected in flight traps like window

1. **Stenotopic species, spending their entire life cycle within one type of habitat.** (Identification: restricted to one habitat type, distribution type 1 in Fig.1)
2. **Species with a stronghold in or around natural or seminatural habitats, showing various degrees of diffusion into cultivated areas.** They are most likely not of agricultural importance, since they never reach high population levels in cultivated areas. They depend on natural or seminatural habitats for reproduction, development and hibernation. (Identification: maximum numbers in or close to natural areas, decreasing away from maxima; more than 33% in or around natural areas, distribution types 2 and 3 in Fig.1)
3. **Poor fliers or flightless species, hibernating at field margins or within fields.** Predators of high agricultural importance, independent of the presence of natural habitats. They require hibernation places in or close to cultivated areas. (Flight index below 10%, less than 33% of individuals in natural areas, distribution type 5 in Fig.1)
4. **Active fliers, present in all trap types at certain times of the year.** They hibernate in natural habitats outside of the cultivated areas, but have their main reproductive efforts in the fields. (Flight index above 10%, abundances in flight traps in spring and fall in most cases highest in or close to natural areas)
5. **Nomadic species, which depend on hibernation places outside the cultivated areas,** often quite far away, but mainly reproduce in the fields. The difference to dispersal type 4 is that their dispersal activity goes on during the reproductive period: they are true nomads and may change not only the fields regularly, but also may cover various regions. (Flight index above 50%, evenly spread over the entire transect. Abundances in spring and fall highest in or close to natural areas)
6. **Migrating species with no affinities to the habitat types in the transect area.** (Present only in flight traps, spread randomly over all habitat types)

Fig. 4. Dispersal types defined according to species parameters such as the percentage of specimens collected in flight traps, the percentage of specimens collected in or close to natural areas, and the transect distribution graphs.

traps, sticky grid traps and yellow water pan traps, with capture results from surface methods such as pitfall traps, sweepnet samples, and a suction device (Table 1). Of the total of almost 600 predatory arthropod species identified, 59% were collected in sufficient numbers to be displayed in a two-dimensional transect. Rare species were dominant in the Heteroptera, Syrphidae and Neuroptera (first column in Table 1). More than half of the 599 species were collected in flight (second column in Table 1), and even more (414) were caught crawling on the surface (third column in Table 1). 185 species were only collected in flight (fourth column) and 265 only in surface traps (fifth column).

For every taxonomic group, transect distributions for the number of species and the number of individuals per trap station were calculated. An example is given in Fig. 2 for the Staphylinidae. The species distribution of staphylinids collected in flight shows a maximum in the dry meadow along the forest edge (top transect, right), which is even more evident in the distributions of numbers of flying and crawling individuals. Species numbers of staphylinids caught crawling, on the other hand, show their maximum in the wetland (bottom transect, left), which indicates that a large number of less frequent species are living in the wet area. The high numbers of species and individuals flying along the forest edge can be interpreted as migration flights to and from the hibernation places.

Further evidence for seasonal migration flights is given by three-dimensional transect distributions for species with more than a total of ten individuals collected. Many dispersive species show their maximum in the natural areas in early spring and/or autumn, while the maxima in the cultivated areas occur in late spring or summer. An example is given in Fig. 3, where the nomadic common green lacewing, *Chrysoperla carnea*, is active in the cultivated areas during the vegetation period, and after diapause induction in August accumulates in natural habitats for hibernation.

In addition to the distribution patterns in space and time for all species with more than ten individuals, a list with species specific information was created with the aim to form groups for distinct dispersal types for all the predatory spe-

Table 2. Example of a species list (staphylinid beetles) with the information leading to the assignment of a dispersal type (see Fig. 4) to every species.

Species List (Staphylinidae)	N	% in flight	Naturalness index	Distribution type 1-5	Dispersal type 1-6
Arpedium quadrum	2630	55.02%	36%	5	5
Anotylus tetracarinatus	2388	95.77%	11%	3	5
Carpelimus corticinus	2099	93.85%	61%	5	5
Paederus fuscipes	1509	28.56%	100%	3	4
Anotylus rugosus	1339	86.33%	65%	5	5
Philonthus carbonarius	946	57.51%	100%	4	5
Tachyporus hypnorum	912	29.50%	100%	5	4
Philonthus cognatus	905	25.41%	100%	5	4
Tachyporus nitidulus	603	40.30%	0%	5	4
Carpelimus gracilis	592	98.99%	48%	5	5
Platystethus arenarius	378	89.95%	100%	3	5
Xantholinus longiventris	378	8.73%	0%	3	2
Scopaeus laevigatus	364	81.59%	83%	3	4
Staphylinus caesarius	362	0.00%	71%	3	3
Tachyporus chrysomelinus	285	44.91%	100%	4	4
Lathrobium fulvipenne	262	4.96%	100%	4	2
Stenus biguttatus	226	4.87%	100%	5	2
Gabrius pennatus	220	59.09%	61%	4	5
Staphylinus dimidiaticornis	200	0.00%	79%	5	2
Carpelimus rivularis	192	100.00%	50%	2	3
Paederus litoralis	183	0.00%	100%	2	3
...	.	.	.	.	.
...	.	.	.	.	.

cies. The lists for the various taxonomic groups contain information on the total number of individuals collected per species, the percentage of individuals collected in flight, the percentage of individuals collected in natural or seminatural habitats, and the specific transect distribution type. With regard to all the available information, each species was assigned to one of six dispersal types (Fig. 4), and this information was also entered into the list, as shown in an example for the Staphylinidae in Table 2.

The six dispersal types shown in Fig. 4 were defined here for the purpose of grouping predatory arthropods according to their population movements between natural or seminatural habitats and cultivated areas. To avoid an interpretative bias from published observations in other parts of Europe, where certain species may show different habitat requirements, only informations emanating from the transect data were used for the grouping process.

The dispersal types 4, 5 and 6, if only looked at in the two-dimensional transect distribution, all look similar and would be regarded as "ubiquists" (distribution type 4 in Fig. 1).

Table 3. Contribution of the different taxonomic groups to the six dispersal types defined in Fig. 4 (only species with ≥5 indiv. classified). The species numbers in the classified total do not all correspond to the figures in Table 1, because some of the rare species could not be assigned to a specific dispersal type.

	Dispersal type							
	1	2	3	4	5	6	Sum	TOTAL
Araneae								
No. species	16	55	20	5	0	0	96	176
No. individuals	274	5202	44145	724			50345	50601
Carabidae								
No. species	9	30	24	5	10	2	80	97
No. individuals	376	5633	24343	6388	8949	84	45773	45830
Staphylinidae								
No. species	3	38	16	29	10	5	101	214
No. individuals	40	1865	1562	5272	10827	45	19611	20086
Heteroptera								
No. species	1	6	0	4	2	0	13	42
No. individuals	21	135		48	188		392	446
Coccinellidae								
No. species	3	2	0	0	5	0	10	16
No. individuals	254	718			2794		3766	3775
Syrphidae								
No. species	1	2	0	0	8	0	11	31
No. individuals	7	14			2157		2178	2218
Neuroptera								
No. species	1	3	0	0	3	1	8	23
No. individuals	6	30			671	7	714	743
All predators								
No. species	34	136	60	43	38	8	319	599
No. individuals	978	13597	70050	12432	25612	136	122805	123699

The various columns in the lists for each taxonomic group (example shown in Table 2) can be sorted according to different criteria. The species column can be sorted alphabetically or systematically. Sorting the second column for numbers of individuals allows us to group the dominant species and to visualize the correlation between abundance (number of catches, in fact) and flight activity, "naturalness", or dispersal type. Sorting the column dispersal type in turn allows an assessment of the correlation between a given dispersal type and the other parameters.

For each taxonomic group, the number of species and individuals per dispersal type was calculated (Table 3). Considering all predatory species, the dominant dispersal type is 2, i.e. species with a stronghold in or close to the natural habitats, but with dispersal capacities into the cultivated areas. The dominant dispersal type with regard to the number of individuals, however, is clearly type 3, with species staying predominantly in or close to cultivated areas.

Discussion

In the contribution presented here, the focus is on methodological aspects rather than on results and their interpretation.The 5 km long Limpach-transect of 1987 was, and still is, an extremely time consuming and therefore costly study, which involved numerous students, specialized taxonomists and computer experts. But the trade-off is obvious: there has hardly been any project in an agricultural landscape, where (1) several different habitat types have been investigated at the same time, with the same standardized methods, and for one full year, and (2) a wide spectrum of collecting methods has allowed a comparison within and between different groups of predators.

The methods are standardized in the sense that they were operated in the same manner in all transect stations. For a particular species the chance to get caught should be as similar as possible with regard to local abundance. We are aware of the fact that the same method may be of variable efficacy in different habitat types. This is particularly true for pitfall traps, and less so for the flight traps. The efficacy varies considerably for different species, but this does not affect the transect distribution at the species level. Another problem is, that in spite of the big efforts taken, not all predatory arthropod groups were adequately sampled (e.g. small spiders with nets close to the surface), or have not been identified so far (many Diptera and Coleoptera). Furthermore, only about half of the species were collected in sufficient numbers to interpret with some confidence their distribution in space and time. But in spite of these shortcomings, the large number of standardized collecting methods and the large number of taxa considered will make this data set a useful empirical base for modelling spatial and temporal aspects of dispersal and metapopulation dynamics.

The proposed grouping of predatory species into dispersal types will help to quantify and functionally interpret population movements on a regional scale. To avoid oversimplification or even misinterpretation in the process of modelling, additional information from detailed studies on exemplary species have to be taken into consideration.

Of the six dispersal types defined in this paper, not all have the same importance for agriculture or nature protection, and not all have the same habitat requirements for reproduction, development and hibernation.

If we want to increase the biodiversity of predators, we should try to augment:

- type 1 and 2 dispersers by increasing natural habitats and creating a mosaic landscape,
- type 4 and 5 dispersers by increasing hedgerows, ecotone structures along forest edges and other arboreal habitats.

If we want to increase the overall abundance of predators (numbers of individuals, biomass, impact on pest insects etc.), we should try to augment:

- polyphagous type 3 disperser with grassy field margins, headlands and herbaceous strips along hedges,
- aphidophaga of type 4 and 5 with suitable hibernation sites in arboreal habitats and attract the nomadic species into fields with flowers or aphids on non-crop hosts.

References

Duelli, P., Blank, E. & Frech, M. 1992. The contribution of seminatural habitats to arthropod diversity in agricultural areas. In: *Proceedings of the Fourth European Congress of Entomology, Gödöllö,* 1991. Hungarian Natural History Museum, Budapest 1992: pp. 29-38.

Fürst, R. & Duelli, P. 1988. Fensterfallen und Klebgitterfallen im Vergleich: Die flugaktive Insektenfauna einer Kiesgrube. *Mitt. dtsch. Ges. allg. angew. Ent.* 6: 194-199.

Huber, M. & Duelli, P. 1987. Vergleich der flugaktiven Coleopterenfauna über naturnahen Biotopen und Intensivkulturen. *Rev. Suisse de Zoologie* 94 (3): 525-532.

Stöckli, E. & Duelli, P. 1989. Habitatbindung und Ausbreitung von flugfähigen Wanzenarten in naturnahen Biotopen und Kulturlandfächen. *Mitt. deutsch. Ges. allg. angew. Entomol.* 7: 221-224.

Taylor, R.A.J. 1978. The relationship between density and distance of dispersing insects. *Ecol. Entomol.* 3: 63-70.

How do immigration rates affect predator/prey interactions in field crops? Predictions from simple models and an example involving the spread of aphid-borne viruses in sugar beet

Wopke van der Werf

Wageningen Agricultural University, Department of Theoretical Production Ecology
P.O. Box 430, NL-6700 AK Wageningen, The Netherlands

Abstract

Sugar beet viruses are spread by aphid populations which are preyed upon by natural enemies. This paper discusses the hypothesis that aphid vector population buildup and the spread of viruses can be decisively affected by early activity of natural enemies, e.g. immigrant adult ladybirds. The hypothesis is analysed with two models of the interaction between aphids and coccinellids. The models are based on exponential growth of aphid populations and on a predator density that depends on the balance between immigration and emigration. In the first model, predator feeding rate is assumed to have a maximum level, β. Under this assumption, there will be a vector outbreak (defined as exponential growth in the long term) if

$$\beta\left(y_0 + \frac{\delta}{\alpha}\right) < (\alpha + \varepsilon)\left(x_0 + \frac{\gamma}{\alpha}\right)$$

In this threshold rule, x_0 is initial aphid density, y_0 is initial predator density, α is the relative growth rate of the pest, β is the feeding rate of the natural enemy, γ is the immigration rate of the pest, δ is the immigration rate of the natural enemy and ε is the relative emigration rate of natural enemy. In the second model, mortality by predation is proportional to prey density. In this case there will be an outbreak if

$$\alpha > \kappa\, y^*$$

Here κ is the relative mortality rate of aphids per unit of predator density and y^* the equilibrium density of the predator. In this threshold rule, the occurrence of aphid outbreaks is determined by aphid population growth rate, predator searching efficacy, predator immigration and emigration rate, but not by initial densities.

Both models point out a sensitivity to the timing of immigration. In the first model, a sufficiently early immigration of the prey can give it a decisive advance on the predator. In the second model, predators can always catch up with prey

***Arthropod natural enemies in arable land* · I** *Density, spatial heterogeneity and dispersal*
S. Toft & W. Riedel (eds.). *Acta Jutlandica* vol. 70:2 1995, pp. 295-312.
 ISBN 87 7288 492

dynamics, but the parameter quantifying searching efficacy (κ) may decrease in time as leaf area increases.

Results of field experiments on the spread of viruses in sugar beet are interpreted in the light of the model results. The field observations tend to confirm the concept of a critical predator/prey ratio (Model 1) rather than that of a critical predator density (Model 2), but this conclusion remains tentative because immigration rates are unknown.

Key words: Aphids, Coccinellidae, natural enemies, predation, biological control, immigration, emigration, spatial, virus spread, model

Introduction

Many above ground pests and diseases in annual crops overwinter outside the field, e.g. in other fields or in more or less natural habitats. The timing and intensity of immigration into crops, relative to crop phenology, affect the following spread and damage. Several groups of natural enemies or antagonists of pest and diseases also overwinter outside the field. The timing and intensity of their immigration, in relation to the timing and intensity of pest/pathogen immigration, may determine whether a pest or disease outbreak or biological control will occur. It would be desirable to have quantitative criteria to assess the effects of immigration rates and timing on pest and disease dynamics because such criteria would make it possible to relate trends in pest and disease occurrence in crops to spatial dynamics in the larger context of the entire agroecological landscape (Fig. 1; Galecka 1966). Two questions should be asked:

1. Which criteria (e.g. a predator/prey ratio; Janssen & Sabelis 1992, van der Werf et al. 1994) distinguish situations with and without biological control?
2. How are critical values of such criteria affected by times and rates of immigration and other factors such as crop development stage and weather?

In this paper I use two simple models to explore possible answers to these two questions. Results of these models are expressed as threshold rules that mark the transition from natural control to pest outbreak. The model predictions are compared to field data on the spread of sugar beet viruses by the green peach aphid, *Myzus persicae*. This aphid is preyed upon by a complex of predators and parasitoids. This paper focuses on coccinellids, a predator group known to be capable of reducing aphid population buildup (Hodek et al. 1965, Frazer & Gilbert 1976) and thereby the spread of aphid-transmitted viruses (Ribbands 1963, Kershaw 1965, van der Werf et al. 1992).

In Europe, two yellowing viruses occur in sugar beet. The most prevalent

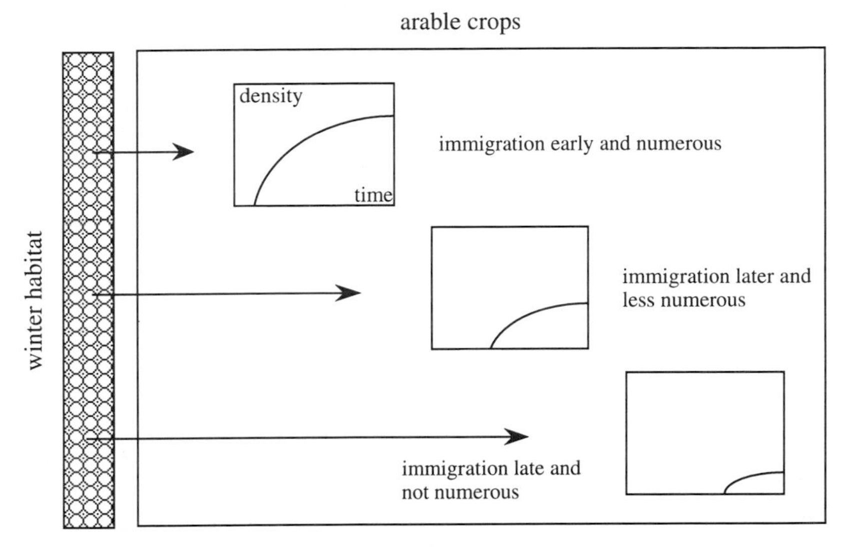

Fig. 1. Time and rate of immigration of pests and natural enemies into crops may depend upon distance from overwintering habitats. If migration is a random process, the rate of immigration will rise earlier and to higher levels at short distances from the source than at long distances. The magnitude of these differences and the scale at which they occur depends on circumstances and on species. For instance, *Propylaea quatuordecimpunctata* in an arable field would originate predominantly from nearby overwintering sites (Basedow 1990) whereas in a field experiment with syrphids (Groeger 1992) no differences were observed in time and rate of immigration between sites at different distances, up to 700 m, from an assumed overwintering site.

one is beet mild yellowing virus, BMYV, the other is beet yellows virus, BYV. BMYV belongs to the luteovirus group and is transmitted in the persistent manner. BYV belongs to the closterovirus group and is transmitted in the semi-persistent manner. Upon early and complete infection, sugar beet crops infected with BMYV incur about 30% yield reduction while crops infected with BYV incur 50% yield reduction (Smith & Hallsworth 1990). The two viruses are taxonomically unrelated but their ecologies are similar. Practice considers the two pathogens as a single disease complex (Harrington et al. 1989). The epidemiology of both viruses is characterized by a distinct year cycle. During the crop season, the viruses occur in sugar beet, where they cause leaf yellowing. Their main vector within sugar beet is *M. persicae*. A vector of secondary importance for BYV (but not for BMYV) is the black bean aphid, *Aphis fabae*. During winter, when in Western Europe no beet plants are available in the field, BYV and BMYV occur in a wide range of weedy hosts (Peters 1988). *M. persicae* can overwinter as viviparous females (i.e. anholocyclic) on those weedy hosts, but suffers high mortality. *M. persicae* also overwinters in the form of eggs, produced by sexual males and females (i.e. holocyclic) and then uses woody host plants, such as peach tree, *Prunus persica*, which are not hosts for yellowing viruses. In spring, *M. persicae* populations build up on both types of winter host. In spring and early summer, winged aphids are produced that migrate to (other) weedy hosts. It is during this phase that beet crops become infected. This is called *primary* infection. Infections that are made by aphids dispersing within the crop are called *secondary* infections. Other aphid species than *M. persicae,* e.g. the potato aphid *Macrosiphum euphorbiae*, may play a role in causing primary infections.

In Dutch field crops, three species of coccinellids can rise to abundance levels of significance in aphid control: seven-spot ladybird, *Coccinella septempunctata,* two-spot ladybird, *Adalia bipunctata* and 14-spot ladybird, *Propylea quatuordecimpunctata*, in order of importance. The life-cycle of these species is predominantly univoltine (Hodek 1973, Majerus & Kearns 1989). Adults overwinter in sheltered places, e.g. under leaf litter, in grass tussocks or in bark crevices. They come out of shelter on sunny days in late winter and early spring to forage. Aerial dispersal of adults is frequent throughout spring and summer, so that fresh vegetation, including newly emerged crops with associated aphid populations are colonized at an early stage. Mating and egg laying results in larval populations being abundant in early summer which is the period of highest aphid densities. New adults emerge later in summer. They disperse to hibernation sites in autumn.

Structure of the paper

First two simple models are introduced that broadly represent predator/prey dynamics of aphid/ladybird systems in early spring, before ladybird larvae start to make a significant contribution to overall mortality due to predation. Threshold criteria for biological control (defined as the non-occurrence of exponential growth) are derived for both models. These criteria are interpreted in biological terms and the likely importance of timing in both models is shown. Relationships between model parameters and cumulative aphid density, as a measure for pest and vector pressure, are also analysed. Model results are then compared to experiments.

Models

Model 1. A prey-predator model with a fixed predator feeding rate and immigration

The model is based on the following set of assumptions:

1. stage differentiation in prey and enemy can be neglected
2. life cycle parameters are constants
3. prey population growth is density-independent
4. the enemy has a fixed feeding rate, independent of density (the plateau of the functional response is taken; this overestimates predation at low prey density)
5. there is constant immigration of both prey and predator
6. prey and predator immigration can start at different times
7. predator emigration is proportional to predator density
8. predators do not reproduce

$$\begin{cases} \frac{dx}{dt} = \alpha x - \beta y + \gamma \\ \frac{dy}{dt} = \delta - \varepsilon y \end{cases} \qquad (1)$$

In these equations, x is pest density, y is predator density, t is time, α is the relative growth rate of the pest ($[x]\,[x]^{-1}\,d^{-1}$), β is the feeding rate of the natural enemy ($[x]\,[y]^{-1}\,d^{-1}$), γ is immigration rate of pest ($[x]\,d^{-1}$), δ is the immigration rate of the natural enemy ($[y]\,d^{-1}$) and ε is the relative emigration rate of natural enemy ($[y]\,[y]^{-1}\,d^{-1}$). $[x]$ and $[y]$ denote dimensions of pest and predator density, e.g. numbers of individuals per m^2. The model can be analytically integrated (cf. Edelstein-Keshet 1988, Janssen & Sabelis 1992):

$$\begin{cases} x_t = A e^{\alpha t} - B e^{-\varepsilon t} - C \\ y_t = y^* - (y^* - y_0) e^{-\varepsilon t} \end{cases} \qquad (2)$$

where:

$$A = x_0 + B + C$$

$$B = \frac{\beta}{\alpha + \varepsilon}(y^* - y_0)$$

$$C = \frac{1}{\alpha}(\gamma - \beta y^*) \qquad (3)$$

$$y^* = \frac{\delta}{\varepsilon}$$

The predator equation (2) describes a negative exponential convergence from the initial density y_0 to an equilibrium density $y^* = \delta/\varepsilon$. The prey equation consists of three terms, of which only the first one increases in magnitude in time. The second term extinguishes exponentially to zero, while the third

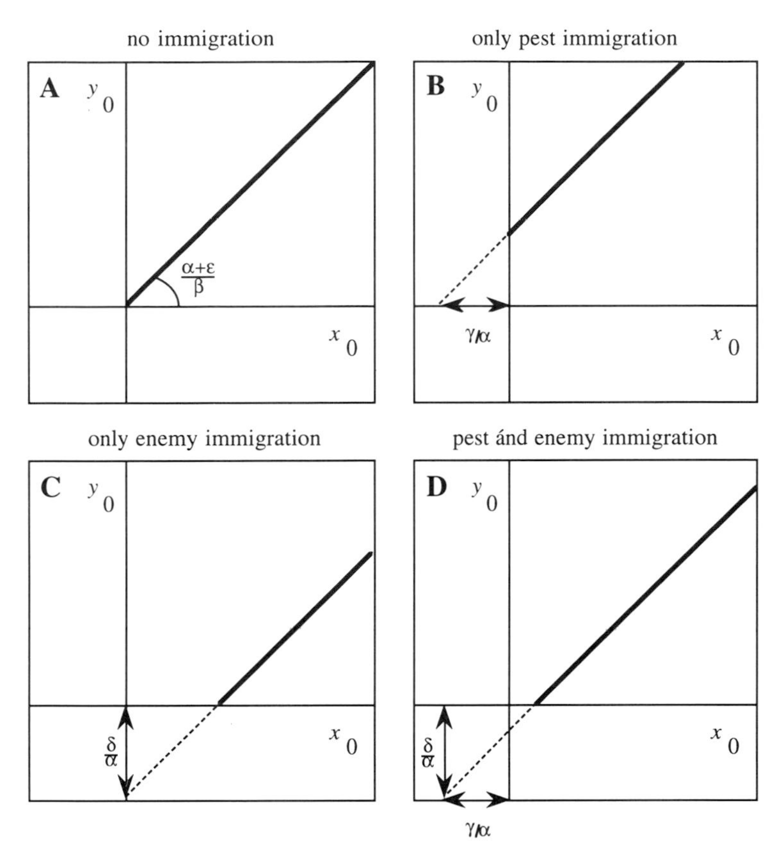

Fig. 2. The influence of immigration rates of a pest and its natural enemy on enemy/pest ratios required for preventing pest outbreak in Model 1. Meaning of symbols: x_0: initial pest density ([x]), y_0 : initial natural enemy density ([y]) α: relative growth rate of pest ([x] [x]$^{-1}$ d^{-1}), β: feeding rate of natural enemy ([x] [y]$^{-1}$ d^{-1}), γ: immigration rate of pest ([x] d^{-1}), δ: immigration rate of natural enemy ([y] d^{-1}), ε: relative emigration rate of natural enemy ([y] [y]$^{-1}$ d^{-1}). Bold drawn lines indicate initial densities of the enemy that are just sufficient for preventing pest outbreak at a given pest density. Figures A-D illustrate how this "critical" line is shifted by changing pest and enemy immigration rates. In figure A, immigration rates are 0. The bold line goes through the origin and has a slope of $(\alpha+\varepsilon)/\beta$. In figure B, the pest has non-zero immigration rate γ; this shifts the line to the left such that higher initial enemy densities are required. In figure C, the natural enemy has positive immigration rate δ; this shifts the critical line downwards. Lower initial enemy densities give pest control. In figure D, both the pest and the enemy have positive immigration rates. The position of the critical line depends on the relative sizes of γ and δ, as well as on the slope of the critical line, $(\alpha+\varepsilon)/\beta$. The intercept of the critical line is defined by $\frac{\alpha+\varepsilon}{\beta}\ \frac{\gamma}{\alpha} - \frac{\delta}{\alpha}$.

term is a constant. The prey will be eradicated if the coefficient A is negative. This leads to the following condition for biological control:

$$y_0 + \frac{\delta}{\alpha} > \frac{\alpha + \varepsilon}{\beta}\left(x_0 + \frac{\gamma}{\alpha}\right) \tag{4a}$$

Prey density will approach an equilibrium $x^* = -C$ if

$$y_0 + \frac{\delta}{\alpha} = \frac{\alpha + \varepsilon}{\beta}\left(x_0 + \frac{\gamma}{\alpha}\right) \tag{4b}$$

There will be a prey outbreak if

$$y_0 + \frac{\delta}{\alpha} < \frac{\alpha + \varepsilon}{\beta}\left(x_0 + \frac{\gamma}{\alpha}\right) \tag{4c}$$

Condition (4a) states that the predator/prey ratio should be greater than $(\alpha+\varepsilon)/\beta$ to prevent a pest outbreak when the pest and predator immigration rates (γ and δ) are 0 (Fig. 2A). The existence of this critical ratio was pointed out by Janssen & Sabelis (1992). When immigration rates are non-zero, the "critical" line, indicating for a given pest density the lowest predator density that will prevent an outbreak, is shifted along the axes (Fig. 2B-D). The slope of the line remains the same.

When $y_0 = y^*$, the time at which pest extinction takes place is:

$$\tau = \frac{1}{\alpha} \ln\left(\frac{C}{A}\right) \tag{5}$$

When $y_0 < y^*$, this is an underestimate of the time until extinction, else it is an overestimate. A precise estimate of τ can be obtained by simulation or iteration.

Fig. 3 explores the range of dynamics possible in Model 1. Fig. 3A shows how trajectories starting at initial densities (x_0, y_0) fulfilling (4a) converge to the equilibrium (x^*, y^*). Small deviations from the rule lead either to prey extinction or to a prey outbreak (Figs. 3B,C). Because of the sensitivity to initial conditions, the system is sensitive to timing. In Fig. 3D, simulations are shown in which the prey is given an advance on the predator of $\Delta t = 7, 8, 9$ or 10 days. During the predator-free period, the prey grows exponentially. When $\Delta t = 7$ days, the prey is eradicated after 8 days. When $\Delta t = 8$ days, eradication takes 23 days. When $\Delta t = 9$ or 10 days, there is a prey outbreak.

The integral of prey density over time provides a useful measure of the direct and indirect effects of aphids on crop growth and of vector pressure.

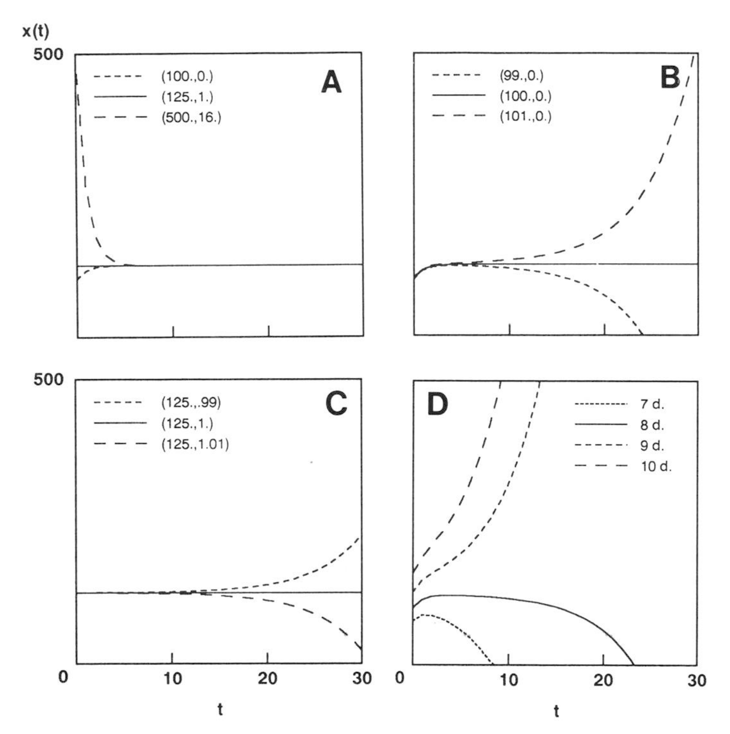

Fig. 3. Dynamics shown by Model 1 for parameter values $\alpha = 0.2\ d^{-1}$, $\beta = 30$ aphids predator^{-1} d^{-1}; $\gamma = 5$ aphids $m^{-2}\ d^{-1}$; $\delta = 1$ ladybird $m^{-2}\ d^{-1}$; $\varepsilon = 1.0\ d^{-1}$. A: convergence to equilibrium $x^* = -C$ when condition 4b is fulfilled. B: divergence from equilibrium when prey density is 1% higher or lower than required by condition 4b. C: divergence from equilibrium when predator density is 1% higher or lower than required by condition 4b. D: Sensitive dependence of long term dynamics on the timing of predator immigration relative to that of the prey.

This integral (Area Under Curve) is:

$$\text{AUC} = \frac{A}{\alpha}\left(e^{\alpha t} - 1\right) + \frac{B}{\varepsilon}\left(e^{-\varepsilon t} - 1\right) - C t \tag{6}$$

As in the threshold rule for biological control, all parameters and initial conditions have influence in this equation. Fig. 4 shows contour plots of log(AUC+1) over 30 days time in graphs of predator immigration rate, δ,

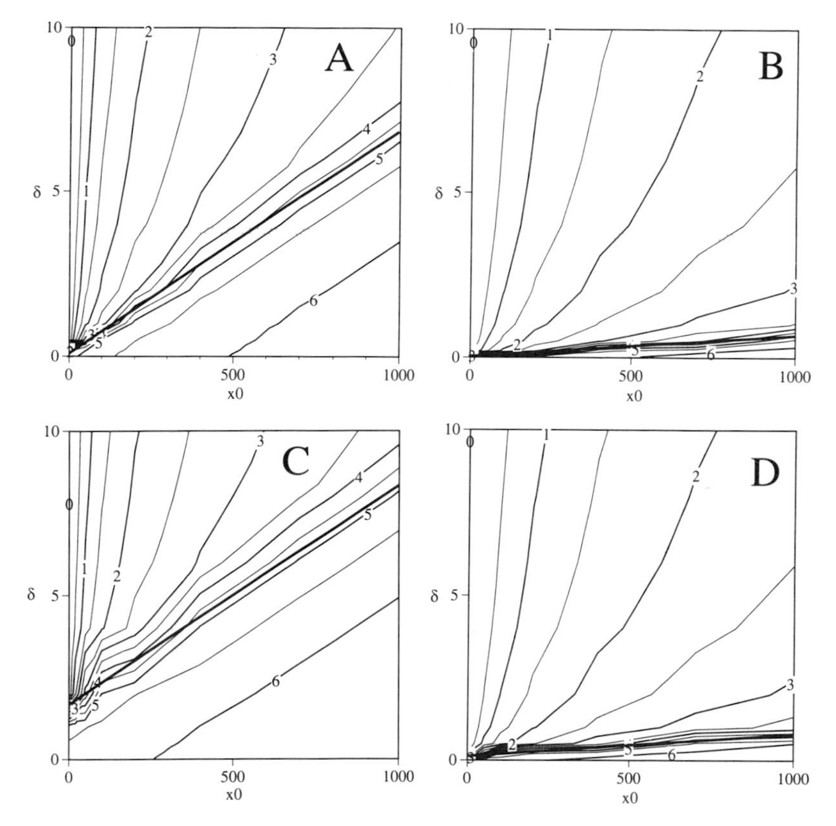

Fig. 4. Contour plots of log(AUC + 1) for four parameter settings in Model 1. The top and bottom two figures are characterized by *low* and *high* aphid immigration rates, respectively, while the left and right figures are characterized by *rapid* and *slow* predator departure. The parameter values are: (A) γ = 5 [x] d^{-1}; ε = 1 [y] $[y]^{-1}$ d^{-1}; (B) γ = 5 [x] d^{-1}; ε = 0.1 [y] $[y]^{-1}$ d^{-1}; (C) γ = 50 [x] d^{-1}; ε = 1 [y] $[y]^{-1}$ d^{-1}; (D) γ = 50 [x] d^{-1}; ε = 0.1 [y] $[y]^{-1}$ d^{-1} . The values of the other parameters are: α = 0.2 d^{-1}; β = 30 [x] $[y]^{-1}$ d^{-1} and $y_0 = y^*$. The thick drawn line in each of the figures is the critical line for eradication .

against x_0. Parameters are set to α = 0.2 [x] $[x]^{-1}$ d^{-1}, β = 30 [x] $[y]^{-1}$ d^{-1}, γ = 5 or 50 [x] d^{-1}, ε = 1 or 0.1 [y] $[y]^{-1}$ d^{-1} and $y_0 = y^*$. In addition to the contour lines for log(AUC+1), a "critical" line is drawn in these figures that forms the distinction between combinations of x_0 and δ that do and do not result in ultimate aphid extinction. Below these "critical" lines, there is aphid outbreak, above them there is extinction. The critical lines are derived from Eqn 4. The intercept of these lines is $\gamma\epsilon/\beta$ while the slope is $\alpha\epsilon/\beta$. Comparison of the top and bottom figures illustrates the effect of *low* and *high* aphid immigration rates (γ) on the predator immigration rate required for biocontrol (defined in terms of extinction or AUC), as a function of initial aphid density. Comparison of the left and right figures illustrates the effect of *rapid* and *slow* predator departure (ϵ). Predator departure is quite influential at all initial aphid densities; aphid immigration rate is only of substantial influence at low initial aphid densities. In all four figures, the critical line for eradication lies in the neighbourhood of the AUC contour for 10^4 aphid days during 30 days. These results indicate that, by and large, predator immigration rates required for either definition of biological control (eradication or acceptable AUC) may be affected in a similar manner by model parameters. If tactical management is the objective, a criterion for biological control must be chosen. The criterion of final prey eradication has the advantage of being more simple to analyse in a model, but AUC as a criterion will often be more relevant from a practical viewpoint.

Model 2. A prey-predator model with the predator feeding rate proportional to prey density

In the first model, prey mortality due to predation is limited by predator feeding capacity, β. This allows the pest to escape eradication if the initial density is high enough. Such limited predation capacity may not be a fair description of reality. First, at low densities, prey finding is more limiting to the overall predation rate than is feeding capacity. Second, at high densities, coccinellids may kill more specimens than is required for meeting their food demand. The more prey are captured the less of each specimen is actually consumed (e.g. Kareiva & Odell 1987). Due to this "wasteful killing", the killing rate at high densities may not reach a plateau. In both situations, the killing rate is determined by the frequency of encounters between predator and prey. The relative death rate of prey is then proportional to prey and predator density:

$$\begin{cases} \frac{dx}{dt} = (\alpha - \kappa y)x + \gamma \\ \frac{dy}{dt} = \delta - \varepsilon y \end{cases} \tag{7}$$

In the above equation, the effective relative growth rate of prey, α-κy, decreases linearly with y. The solution for prey equation (7) is:

$$x_t = \exp\left(-\frac{K}{\varepsilon}e^{-\varepsilon t} + r^* t\right)\left[x_0 \exp\left(\frac{K}{\varepsilon}\right) + \gamma F(t)\right] \tag{8}$$

with

$$K = \kappa\left(\frac{\delta}{\varepsilon} - y_0\right) = \kappa(y^* - y_0)$$

$$r^* = \alpha - \kappa\frac{\delta}{\varepsilon} = \alpha - \kappa y^*$$

$$F(t) = \int_0^t \exp\left(\frac{K}{\varepsilon}e^{-\varepsilon t} - r^* t\right) dt$$

$$y^* = \frac{\delta}{\varepsilon} \tag{9}$$

The predator equation is the same as before (cf. Eqn 2). The derived parameter r^* is the relative growth rate of the prey in the long run. Equation (8) can be simplified making additional assumptions. For instance, in the absence of immigration of prey ($\gamma=0$), equation (8) simplifies to:

$$x_t = x_0 \exp\left(r^* t - \frac{K}{\varepsilon}(e^{-\varepsilon t} - 1)\right) \tag{10}$$

When t is several times greater than $1/\varepsilon$, equation (10) becomes approximately

$$x_t = x_0 \exp\left(\frac{K}{\varepsilon}\right) e^{-r^* t} \tag{11}$$

Equation (11) denotes an exponential increase or decrease, depending on the sign of r^*.

For the assumptions $y_0 = y^*$ and $r^* \neq 0$, equation (8) simplifies to:

$$x_t = x_0 e^{r^* t} + \frac{\gamma}{r^*}(e^{r^* t} - 1) = \left(x_0 + \frac{\gamma}{r^*}\right)e^{r^* t} - \frac{\gamma}{r^*} \tag{12}$$

When r^* is greater than 0, this equation describes an exponential growth with relative rate of increase r^*. When r^* is negative, Eqn 12 describes an exponential decline of prey density to a plateau level of $-\gamma/r^*$. When r^* is equal to 0, prey population growth ultimately becomes linear with a rate equal to the rate of immigration, γ. Prey eradication proceeds as an exponential decline if there is no immigration. Fig. 5 gives examples of these types of dynamics.

For $y_0 = y^*$ and $r^* \neq 0$, cumulative prey density is given by

$$\mathrm{AUC} = \frac{1}{r^*}\left(x_0 + \frac{\gamma}{r^*}\right)(e^{r^* t} - 1) - \frac{\gamma}{r^*}t \tag{13}$$

Of the four unknowns in this equation, r^* is by far the most influential one in the long run, when $r^* t >> 1$. Hence, in the second model, the components

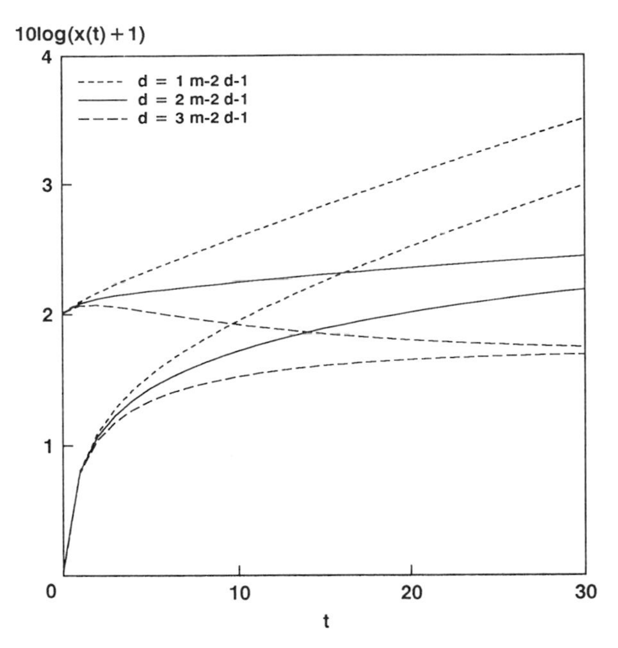

Fig. 5. Dynamics of Model 2 for parameter values α = 0.2 d^{-1}, κ = 0.1 d^{-1}; γ = 5 aphids m^{-2} d^{-1}; ε = 1.0 d^{-1} and δ = 1, 2 or 3 ladybirds m^{-2} d^{-1} (i.e. y^* = 0.1, 0 or -0.1 ladybirds m^{-2} and r^* = 0.1, 0 or -0.1 d^{-1}. The starting conditions are (x_0, y_0) = (0, 0) and (100, 0).

of r^* (α, κ, δ and ε) have the biggest effect on AUC. Initial densities and prey immigration are of less importance.

Experiments

Three experiments were done in the Netherlands in 1985 and 1986 to study the relationship between the date of primary virus infection with BYV and BMYV in sugar beet and the rate of virus spread by *M. persicae*. Details of these experiments are given by van der Werf et al. (1992). These experiments demonstrated substantial differences between fields in vector establishment and virus spread. In Experiments 1 and 2, aphid densities in the centre of artificially started virus foci stayed below 5 per plant while the final number of infected plants in a focus was c. 50. In Experiment 3, aphid densities reached a peak of c. 70 in early-inoculated and 35 in late-inoculated plots while the corresponding numbers of infected plants were c. 2000 and 100. Differences in virus spread between the three experiments were obviously related to the number of vectors. Differences in vector establishment were attributed to different impact of natural enemies, such as coccinellids, because reproduction of clipcaged aphids was

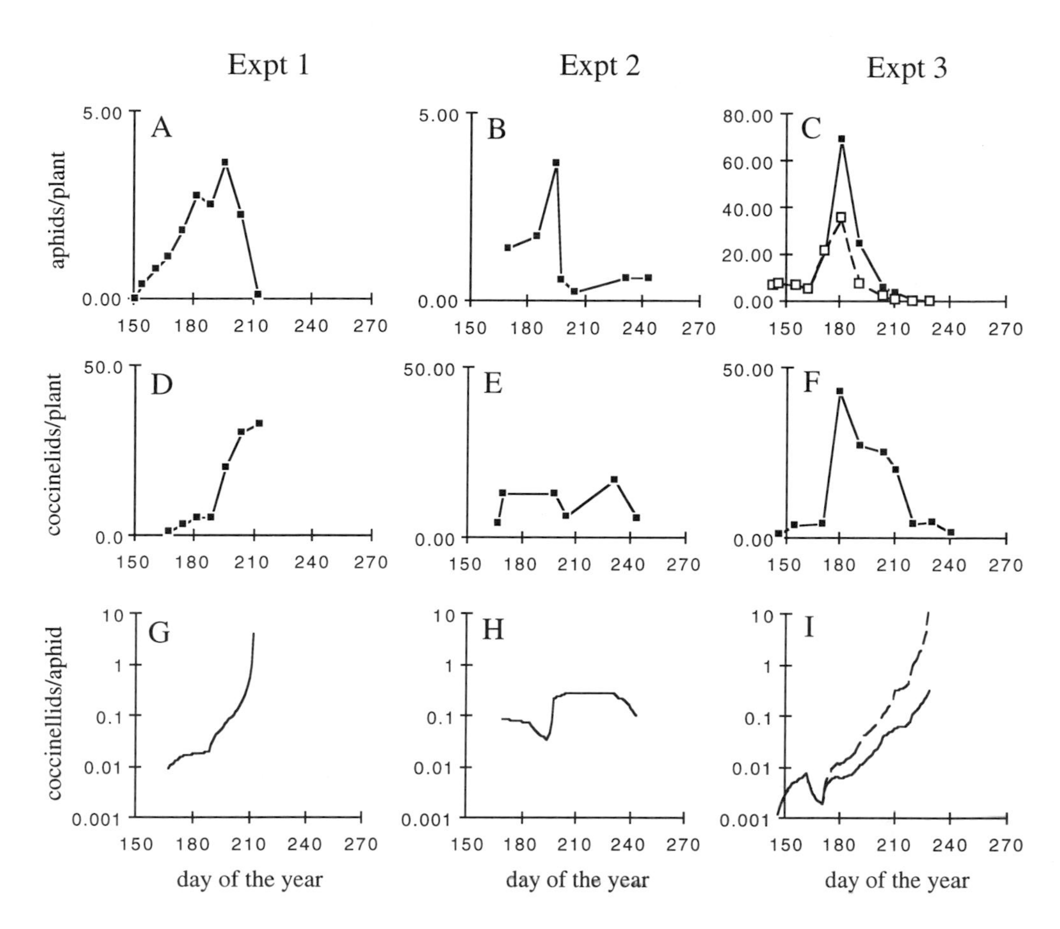

Fig. 6. Overview of aphid and coccinellid densities in three experiments on the spread of sugar beet yellowing viruses by the green peach aphid, *Myzus persicae*. The top row of graphs (A,B,C) indicates the time course of aphid density (*M. persicae* per plant; all stages lumped) in the three experiments. In Expt. 3, distinction is made between early-inoculated plots (inoculations until 10 June; —■—) and late-inoculated plots (inoculations after 20 June;- -□- -). The middle row of graphs (D,E,F) indicates coccinellid density (all species and stages lumped: an egg cluster counted as one). In Expt. 1, coccinellid density (individuals/plant) is derived from observed incidence (fraction occupied plants), assuming that the two are equal. The bottom row of graphs (G,H,I) indicates the ration of coccinellid to aphid density in the course of time. For Expt. 3, distinction is made between early inoculated plots (full line) and late inoculated plots (hatched line). Time is expressed in day of the year. The episodes marked by days 150, 180, 210, 240 and 270 broadly correspond to the months of June, July, August and September.

found to be similar in the three fields. Observations on aphid density, coccinellid incidence and coccinellid/aphid ratio are summarized in Fig. 6.

If natural enemies were the main cause of the field to field differences in vector establishment and if coccinellids were — early on — (one of) the main predator group(s), then the information in Fig. 6 may give clues as to what criteria distinguish the initial predator-prey interaction in the three experiments, e.g. predator density, predator timing or a predator/prey ratio. The experimental results will first be interpreted in terms of Model 1, which led to a threshold rule that involved a predator/prey ratio with immigration terms $\left(\frac{\gamma}{\alpha} \text{ and } \frac{\delta}{\alpha}\right)$. The highest initial predator/prey ratio was observed in experiment 2. This ratio was well above a critical level for biological control of, say:

$$\frac{\alpha + \varepsilon}{\beta} = \frac{0.2 + 1.0}{30} = \frac{1}{25} \tag{14}$$

In estimating this critical ratio, prey and predator immigration are set to 0. In Experiment 1, the initial ratio was only 0.01, yet biological control occurred. Possibly, this discrepancy occurs because the assumption of zero immigration is unrealistic. If we take $y_0 = 0.01$, $x_0 = 1$ and $\gamma = 0$, then an immigration rate of only 0.006 coccinellid m^{-2} d^{-1} would suffice to let the ratio rule (4) predict biocontrol. This low rate is easily achieved in practice (cf. Kareiva & Odell 1987). For lower emigration rates than 1 d^{-1}, lower rates of immigration would suffice for natural control. The importance of the predator immigration rate in Model 1 can be elucidated by assuming that the observed initial predator density is equal to the equilibrium density y^*. The threshold rule (4b) can then be written as:

$$y_0 + \frac{\varepsilon y_0}{\alpha} > \frac{\alpha + \varepsilon}{\beta}\left(x_0 + \frac{\gamma}{\alpha}\right) \tag{15}$$

If we assume an emigration rate of coccinellids $\varepsilon = 1$ d^{-1}, then the immigration-dependent term $\varepsilon y_0/\alpha$ is five times greater than the density term y_0. For a lower emigration rate $\varepsilon = 0.1$ d^{-1}, the immigration dependent term is smaller than y_0, but it is still of importance. Thus, if this model represents reality, immigration rate substantially modifies the predator-prey ratios predicting natural control. The lowest predator prey ratios were observed in Experiment 3. In the light of the outcomes of Model 1, it is not surprising that aphid population increase reached the highest peak densities in this experiment. The data in a broad sense support Model 1.

When Model 2 (with the feeding rate proportional to prey density) is considered applicable to the experiments, r^* becomes the parameter of most interest. According to equation (9):

$$r^* = \alpha - \kappa y^* \quad (16)$$

If the initial density of coccinellids is assumed to be y^*, then, to explain the differences in vector establishment, this density should be higher in Experiments 1 and 2 than in Experiment 3, the other parameters, α and κ, being equal. Experiment 2 has indeed the highest initial coccinellid density, but Experiments 1 and 3 show different aphid population trends in spite of similar coccinellid densities. The disaccordance can be used to reject the model or to expand the hypothesis. In Experiment 3 (with BYV) *M. persicae* reproduction may have been enhanced by the effects of this virus on host plant quality (Baker 1960), while in Experiment 1 (with BMYV for which such an effect has not been described) this might not have occurred. Another explanation for the lower *M. persicae* densities in Experiment 1 may be that in this experiment there was a substantial population of *A. fabae*. The availability of alternative prey may have decreased the emigration rate of coccinellids.

Discussion

The models provide a framework to interpret the field experiments, but too many parameters are unknown in the experiments to reject any of the models. The models are also too abstract to warrant formal rejection because they can only represent main features of the system and not the smaller details (as virus-enhanced reproduction) that are important in reality but that would hamper a lucid theoretical analysis.

The theoretical analysis of the (admittedly too) simple models yields some interesting results. In both models a critical criterion can be derived that distinguishes situations that will result in an exponential prey outbreak from cases in which the prey will be eradicated (Model 1) or kept at a constant level (Model 2). In Model 1, where predator feeding rate is a constant, this threshold rule weights predator density, immigration and emigration against prey density, population growth and immigration. Initial densities are important here, as are immigration rates, relative growth or emigration rates and predator feeding rate. In the second model, where predator feeding is proportional to prey density, initial densities are not important. Here prey relative growth rate compared to predator equilibrium density and searching capacity determine the occurrence of outbreaks. The first model shows a threshold behaviour in which the prey can escape from predation and cause an outbreak if it establishes populations early enough. Timing is crucial in this model. In Model 2, predators can always catch up with a prey advance because their per capita feeding rate is unlimited. Model 1 seems to be more in accordance with reality than Model 2. The threshold rule (Eqn 4) may prove useful for estimating the beneficial effect of natural enemies immigrating from nearby reservoirs.

It is important to relate immigration rates to landscape because landscape may be managed such that biological control is favoured and dependence on pesticides reduced. The fields where *Myzus* population buildup was prevented and hence virus spread limited, were situated south of Wageningen, in an environment with hedgerows and overwintering hosts of black bean aphid, *Aphis fabae*. The vicinity of hedgerows and the early infestation of the sugar beet crop with *Aphis fabae* could cause a greater and/or earlier settlement and impact of natural enemies in the experiments done here. The experiment with the greater amount of spread was laid out in the open landscape of the polder Oostelijk Flevoland, which presumably provides fewer overwintering sites for coccinellids.

To substantiate the beneficial effect of natural enemy winter reservoirs in agroecosystems, quantitative insight will be needed in immigration rates of natural enemies in relation to distance between source and destination. The emphasis should be on quantifying process parameters, not on registering anecdotal information such as maximum possible dispersal distances. More important are the processes underlying spatio-temporal (re)distribution of natural enemies in the early growing season when pests and diseases are about to take either an outbreak or an extinction course. Studies on local dynamics of pests and diseases need to be supplemented with measurements of immigration. To be useful for biological control analysis, studies on spatial dynamics of natural enemies need to be carried out with a fine time resolution in the early growing season because the early system dynamics may be quite sensitive to timing. This is currently not always done. For instance, Basedow (1990) reported differences in natural enemy catches during two month periods in arable fields in different landscapes. To estimate the importance of these differences for pest dynamics it is necessary to know when these differences occurred. It is to be expected that a small difference in the density of natural mortality agents in May is much more important for the dynamics of pest and diseases than a much greater difference in July. Another point that deserves continued and wider attention is the scaling up of functional responses from the level of the petri dish to the level of the crop. One way to do this is to make explicitly spatial models (Kareiva & Odell 1987). This is, however, no practical way ahead for models to be used in tactical management. For such purposes, simulation and experimentation approaches towards deriving descriptive equations for functional responses at greater spatial scales (van der Werf et al. 1989, Mols 1993) are more promising.

Acknowledgements

I am grateful to Rudy Rabbinge, Kees Eveleens and Peter Mols for useful comments on a draft of this paper.

References

Baker, P.F. 1960. Aphid behaviour on healthy and on yellows-virus-infected sugar beet. *Ann. Appl. Biol.* 48: 384-391.

Basedow, Th. 1990. Zum Einfluss von Feldreinen und Hecken auf Blattlausraüber, Blattlausbefall und die Notwendigkeit von Insektizideneinsätzen im Zuckerrübenanbau. *Gesunde Pflanzen* 42: 241-245.

Edelstein-Keshet, L. 1988. *Mathematical models in biology.* Random House, New York, NY, 586 pp.

Frazer, B.D. & Gilbert, N. 1976. Coccinellids and aphids; a quantitative study of the impact of adult ladybirds (Coleoptera: Coccinellidae) preying on field populations of pea aphids (Homoptera: Aphididae). *Journal of the Entomological Society of British Columbia* 73: 33-56.

Galecka, B. 1966. The role of predators in the reduction of two species of potato aphids, *Aphis nasturtii* Kalt. and *Aphid frangulae* Kalt. *Ekol. Polska* (A) 16: 245-274.

Groeger, U. 1992. Undersuchungen zur regulation von Getreideblattlaus-populationen unter dem Einfluss der Landschaftsstruktur. *Agraökologie* Bd 6, Verlag Haupt, 169 pp.

Harrington, R., Dewar, A.M. & George, B. 1989. Forecasting the incidence of virus yellows in sugar beet in England. *Ann. Appl. Biol.* 114: 459-469.

Hodek, I. 1973. *Biology of Coccinellidae.* Dr. W. Junk, Den Haag, Netherlands.

Hodek, I., Novak, K., Skuhravy, V. & Holman, J. 1965. The predation of *Coccinella septempunctata* L. on *Aphis fabae* Scop. on sugar beet. *Acta Entomologica Bohemoslovaka* 62: 241-253.

Janssen, A. & Sabelis, M.W. 1992. Phytoseiid life-histories, local predator-prey dynamics, and strategies for control of tetranychid mites. *Experimental & Applied Acarology* 14: 233-250.

Kareiva, P. and Odell, G. 1987. Swarms of predators exhibit "preytaxis" if individual predators use area-restricted search. *Amer. Natur.* 130: 233-270.

Kershaw, W.J.S. 1965. The spread of yellows viruses in sugar beet. *Ann. Appl. Biol.* 56: 231-241.

Majerus, M. & Kearns, P. 1989. *Ladybirds.* Naturalists' Handbooks 10, Richmond Publ. Co. Ltd, Slough, UK, 103 pp.

Mols, P.J.M. 1993. Walking to survive; searching, feeding and egg production of the carabid beetle *Pterostichus coerulescens* L. (= *Poecilus versicolor* Sturm). PhD thesis, Wageningen Agricultural University, 201 pp.

Peters, D. 1988. A conspectus of plant species as host for viruses causing beet yellows disease. In: *Virus Yellows Monograph*. Institut International de Recherches Betteravieres, p. 87-117.

Ribbands, C.R. 1963. The spread of apterae of *Myzus persicae* (Sulz.) and of yellows viruses within a sugar-beet crop. *Bull. Ent. Res.* 54: 267-283.

Smith, H.G. & Hallsworth, P.B. 1990. The effects of yellowing viruses on yield of sugar beet in field trials, 1985 and 1987. *Ann. Appl. Biol.* 116: 503-511.

Van der Werf, W., Rossing, W.A.H., Rabbinge, R., de Jong, M.D. & Mols, P.J.M. 1989. Approaches to modelling the spatial dynamics of pests and diseases. In: Cavalloro, R. & Delucchi, V. (eds): *Parasitis 88*; proceedings of a Scientific Congress, Barcelona, 25-28 October 1988. *Boletin de Sanidad Vegetal*, Fuera de Serie no 17, 549 pp.

Van der Werf, W., Westerman, P.R., Verweij, R. & Peters, D. 1992. The influence of primary infection date and establishment of vector populations on the spread of yellowing viruses in sugar beet. *Ann. Appl. Biol.* 121: 57-74.

Van der Werf, W., Nyrop, J.P. & Hardman, J.M. 1994. Sampling predator/prey ratios to predict cumulative pest density in the mite - predatory mite system *Panonychus ulmi - Typhlodromus pyri* in apples. In: *Sampling to make decisions. Asp. Appl. Biol.* 37: 41-51.